Springer Climate

Series Editor

John Dodson, Institute of Earth Environment, Chinese Academy of Sciences, Xian, Shaanxi, China

Springer Climate is an interdisciplinary book series dedicated to climate research. This includes climatology, climate change impacts, climate change management, climate change policy, regional climate studies, climate monitoring and modeling, palaeoclimatology etc. The series publishes high quality research for scientists, researchers, students and policy makers. An author/editor questionnaire, instructions for authors and a book proposal form can be obtained from the Publishing Editor.

Now indexed in Scopus® !

More information about this series at http://www.springer.com/series/11741

Riyanti Djalante · Joni Jupesta · Edvin Aldrian
Editors

Climate Change Research, Policy and Actions in Indonesia

Science, Adaptation and Mitigation

Springer

Editors
Riyanti Djalante
United Nations University
Institute for the Advanced Study
of Sustainability (UNU-IAS)
Tokyo, Japan

Joni Jupesta
SMART Research Institute
Sinarmas Agribusiness and Food
Pekanbaru, Indonesia

Edvin Aldrian
Agency for Assessment and Application
of Technology (BPPT)
Jakarta, Indonesia

ISSN 2352-0698 ISSN 2352-0701 (electronic)
Springer Climate
ISBN 978-3-030-55535-1 ISBN 978-3-030-55536-8 (eBook)
https://doi.org/10.1007/978-3-030-55536-8

This Springer imprint is published by the registered company Springer Nature Switzerland AG
The registered company address is: Gewerbestrasse 11, 6330 Cham, Switzerland

Preface

The latest Intergovernmental Panel on Climate Change (IPCC) Special Report on 1.5 Degree in 2018 highlights that impacts on natural and human systems from global warming have already been observed. Many land and ocean ecosystems and some of the services they provide have already changed due to global warming. Adaptation and mitigation actions are already in progress. Sea level rise will continue beyond 2100 even if global warming is limited to 1.5 °C in the twenty-first century. Future climate-related risks would be reduced by the upscaling and acceleration of far-reaching, multilevel, and cross-sectoral climate mitigation and by both incremental and transformational adaptation. Climate-related risks to health, livelihoods, food security, water supply, human security, and economic growth are projected to increase with global warming of 1.5 °C, and increase further with 2 °C.

Indonesia is one of the most vulnerable countries to climate change. Its geographical situation in the Equatorial line, in between the Indian and Pacific oceans, leads the country to experience strong and severe El-Nino in the past. Its physical conditions of being low-lying coastal islands put the country to be extremely vulnerable to floods and sea level rise like other small island states. The country also hosts some of the most protected biodiversity of the Coral Triangle and tropical rain forest that are under threat from climate change due to ocean acidification and forest fires. At the same time, the country is also one of the largest emitters of GHGs. Among various challenges are the rapid deforestation and forest fires for land clearing in the past that had emitted an enormous ton of Greenhouse gasses (GHGs).

There are many initiatives to understand and deal with the impacts of climate change in the country. The national government, led by the Ministry of Environment and Forestry, has issued key documents that include National Action Plan for Climate Change. International agencies together with local stakeholders are working on strengthening the capacity in the policy formulations and implementing actions to build community resilience. Indonesian universities are conducting researches on climate change at different scales. Cities and local governments are implementing innovations in adapting to the impacts of climate change and transitioning toward a green economy.

This is by far the first collection of edited volumes that try to collect and review the latest research, including advances in policies and actions in understanding the science, impacts, and management of climate change in Indonesia. This edited book which is the result of the collaboration between Indonesian authors with international authors could strengthen capacity and increase further collaboration in the future. I do hope this book could be useful for the policy-makers, scientists, and practitioners of climate change actions worldwide.

Jakarta, Indonesia

Dr. Edvin Aldrian
IPCC Vice Chair Working Group I

Contents

Chapter 1
Introduction

Riyanti Djalante, Joni Jupesta, and Edvin Aldrian

Abstract Climate change is an important issues in indonesia. It is one of the largest emitters of greenhouse gasses (GHGs) while experiencing the negative impacts of climate change. This book aims to provide the latest research and advances in policies and actions in understanding the science, impacts, and management of climate change in Indonesia. The editors encourage involvement of and collaboration between Indonesian and international authors to strengthen capacity and increase access for international publications. The primary readership of the book includes policymakers, scientists, and practitioners of climate change actions in Indonesia. This book is divided into two main parts. Part one (Chapter 2–9) highlights the climate change impact and adaptation. Part 2 (Chapter 10–15) highlights the climate change mitigation and cross-cutting issues (legal, finance, and communication).

Keywords Climate change · Impact · Adaptation · Mitigation · Indonesia

Climate Change in Indonesia

Indonesia is located in the equatorial between 7° 44′ 35.11″ North latitude and 13° 55′ 59.99″ South latitude, and stretches from 91° 38′ 25.55″ West longitude and 144° 24′ 00″ East latitude. It lies between the Pacific and the Indian Oceans and between Asia and Australia continents. This country has a total area of 820 Mha in which around 200 Mha is terrestrial. As one of the largest archipelagic countries in the world, there are 13,466 islands of which only 6,000 islands are inhabited,

R. Djalante (✉)
United Nations University—Institute for the Advanced Studies of Sustainability (UNU-IAS), Tokyo, Japan
e-mail: djalante@unu.edu

J. Jupesta
SMART Research Institute (SMARTRI), Riau, Indonesia

Sinarmas Agrobusiness and Food, Pekan Baru, Indonesia

E. Aldrian
Agency for the Assessment and Application of Technology (BPPT), Jakarta, Indonesia

R. Djalante et al. (eds.), *Climate Change Research, Policy and Actions in Indonesia*, Springer Climate, https://doi.org/10.1007/978-3-030-55536-8_1

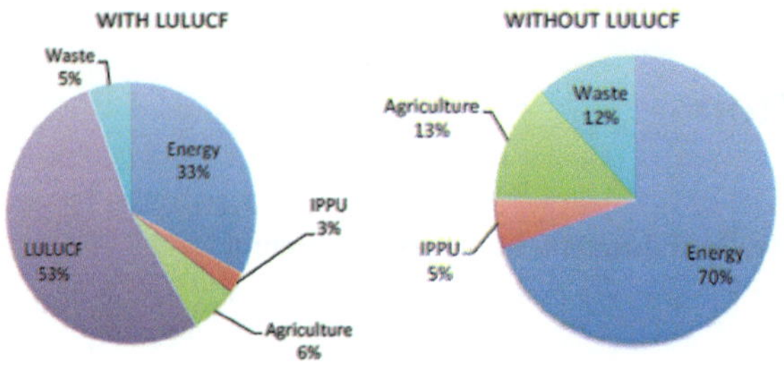

Fig. 1.1 National Greenhouse Gases (GHG) emissions by sectors in 2014 (MOEF 2017)

including the five main islands of Sumatera, Java, Kalimantan, Sulawesi, and Papua. Of the 200 Mha, the four largest land cover areas are Lowland Forest, Upland Forest, Shrub, and Seasonal Crop on Dry Land. Indonesia is a democratic country lead by President Joko Widodo, with decentralized, multi-level governance arrangement of 34 provinces and more than 500 districts/cities. Indonesia's population has been increasing from 119.2 million people in 1971 to 252.16 million people in 2014. The population is projected to increase and reach an estimation of 305 million people in 2035. The life expectancy in Indonesia has improved significantly in the past four decades, from only 47.9 years old in 1970 to 70.6 years old in 2014 (REF).

Monsoon dominates Indonesia's climate (more than 50%) which gives a degree of homogeneity across the region. Overall, the wet season varies between 10 to 110 days (short) and 280 to 300 days (long period) with rainfall varying from 4,115 mm to as low as 640 mm. Besides, strong temporal variation combining with climatic changes create severe and disturbing variability in the region. According to National Agency for Disaster Management, more than 75% of disasters in the country are hydrometeorological disasters. Rainfall variability in Indonesia is influenced by many large-scale climate phenomena, one of them is El Nińo Southern Oscillation (ENSO). El Nińo is considered as one of the causes of forest and land fire in the region. Outbreaks of crop pests and diseases are often connected with this phenomenon. In 2016, the economic loss due to floods and forest/land fire has reached 2.5 billion US$. The rehabilitation cost for infrastructure damages due to floods occurred in 2016, and reached 275 million US$. There are also increasing risks on rising sea level that will threaten small islands and communities who lives extremely close to the coasts (Hoegh-Guldberg et al 2009). Fisheries and agriculture are the two sectors that are expected to be highly impacted by climate change, threatening the food security of millions of Indonesian populations (Naylor et al 2007; Glaeser and Glaser 2010). Studies also examine the impacts of climate change along with rapid deforestation on terrestrial biodiversity (see Paoli et al 2010; Murray et al 2015), or the impacts of ocean warming on coral bleaching in the Indonesian coral triangle (e.g., Keller et al 2009, McClanahan et al 2012). Disasters are undermining the progress of development gain in the country and the impacts of climate change are expected to worsen in the future (Djalante et al 2012, 2017).

Figure 1.1 shows the impacts of disasters, comparing geophysical and climatic disasters between 1900 and 2020 (Source: modified from EMDAT). Climate-related disasters occurred more often, affected more people, and caused 60% economic damages. There were 142 disasters, causing 243,704 deaths, affecting 32.1 million people, and costing 33.7 Billion USD between 1900 and 2019 in Indonesia (EMDAT 2020).

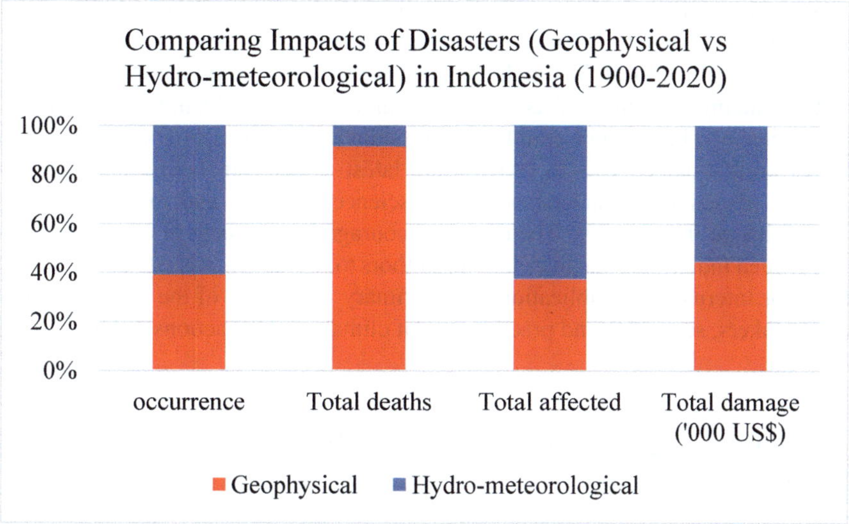

Some of the latest policy developments in the last years to deal with climate change include the adoption of Low Carbon Development Plan (Ministry of National Development Planning (Bappenas) 2019). Climate Action Tracker reports that as of 2019, the Indonesian government's commitment toward climate action is highly insufficient and if all government targets were in this range, warming would reach between 3 and 4 °C (CAT 2019). It is the world's fifth-largest emitter of greenhouse gases and largest contributors of forest-based emissions (WRI 2017). The country pledged to reduce its GHG emissions by 29% by 2030 against a business-as-usual baseline scenario, and up to 41% subject to international assistance and financial support (Ministry of National Development Planning (Bappenas) 2019). From now, it is important that the country notes and implements the findings of the IPCC Special Report on Impacts of 1.5° Warming, which calls for a rapid reduction in GHGs to start in 2030 and toward net-zero emission in 2050 (IPCC 2018). It is not clear how the country aims to implement its strategy in a rapid and transformative manner.

The Indonesia National Greenhouse Gases Inventory for the period 2000–2014 was estimated by methodologies that comply with the IPCC Guideline for National GHG Inventories and IPCC GPG for Land Use, Land Use Changes and Forestry (LULUCF) (Ministry of Environment and Forestry (MOEF) 2007). In 2014 Indonesia's total GHG emission for the three main GHG emissions (CO_2, CH_4, and N_2O) including LULUCF and peat fire sector was estimated at approximately

1,844,370 Gg CO_2eq. Figure 1.1 shows that LULUCF contributes 53% of Indonesia GHG emissions. The next contributors are coming from energy (33%), agriculture (6%), waste (5%), and industrial processes and product use (IPPU) (3%) sectors. Without LULUCF, energy sector contributes the highest in terms of GHG emissions; 70% followed by the agriculture sector (13%), waste sector (12%), and IPPU sector (5%).

Further, the low carbon growth path can deliver an average Gross Domestic Products (GDP) growth rate of 6% annually until 2045 (Ministry of National Development Planning (Bappenas) 2019). The benefit of low carbon development pathway is global as well as local. Through sustainable utilization of its natural resources, and by reducing utilization of its natural resources, and by reducing its carbon and energy intensity, Indonesia's total GHG emissions can fall by nearly 43% by 2030.

This book's objective is to provide the latest research including advances in policies and actions in understanding the science, impacts, and management of climate change in Indonesia. The editors encourage involvement of and collaboration between Indonesian and international authors to strengthen capacity and increase access for international publications. The primary readership of the book includes policy-makers, scientists, and practitioners of climate change actions in Indonesia.

The Methodological Approaches to Addressing Climate Change in Indonesia

Due to its geographical location, topography, and socio-economic aspects, Indonesia is suffering the impacts of climate variability and climate change in food security and economic livelihood. This book was divided into two main parts: part one (Chaps. 2–9) highlights the climate change impact and adaptation.

Part I: Climate Change Impact and Adaptation

Chapter 2 by Triyanti et al. review the current status of socio-economic scenario on climate change and disaster and risk reduction. This chapter had four objectives: First, to highlight the current status of socio-economic scenario on climate change, both in existing theory and empirical case studies, and the impact on different disasters and risk reduction efforts in scholarly literature. Second to assess the challenges and opportunities for upscaling existing shared climate and socio-economic projection models for policy design and implementation. Third is to identify gaps and opportunities of the socio-economic scenarios for better policy-making at the national level. And finally, fourth, to reflect on the broader implication toward CCA theories and future research needs. The methodological approach is a semi-structure method for the literature review. Scopus scientific databases are used in addition to Google scholars to list the past research on this topic with two sets of keywords. It is found that In the context of Indonesia, there is a need of a stronger emphasis on climate mitigation and adaptation and disaster risk reduction and adjustment of socio-economic variables that goes beyond the economic indicators (i.e., GDP), such as of welfare, health, education, social capital human development, participation, and the role of disruptive technology. It is therefore important to promote interdisciplinary research collaboration for a new generation of The Representative Concentration Pathways (RCPs) and Shared Socioeconomic Pathways (SSPs) through available methods, including the Integrated Assessment Modelling (IAM), and leveraging the role of social science and governance studies in future climate change research.

Nurhidayah in Chap. 3 investigates the impact of sea-level rise on human security and human rights in Indonesia using legal, institutional, and policy analysis. This chapter aims at several objectives: first is to investigate SLR and its impacts on human security and human rights in Indonesia. In addition, this chapter will also investigate challenges in responding to the impact of climate change using human security and human rights approach. The research uses a qualitative research method. Literature review and empirical study are both employed as methodologies for this research. This study found that despite global recognition of the implication of climate change on human security and human rights, at the national level the protection and empowerment of local communities against sea-level rise are hindered by overlapping laws, gaps, and laggard behind of legislation. Appropriate measures to improve the involvement of local people in climate change adaptation decision-making is needed. Policy options for adaptation are always consulted with the people affected by the decisions. This involvement will give a legitimate adaptation policy that can be accepted by local communities and improve social justice.

Turner Walker et al. in Chap. 4 outline the local experiences of farmers in two key adaptation projects implemented at the village level as part of a broader nation-wide program. The chapter uses a conceptual framework approach to describe adaptation activities (under agency, resilience, resourcefulness, and adaptive capacity) through which international climate frameworks are translated into national development planning for integration into local activities. Two cases are drawn upon as comparative case studies for investigation and as examples of "putting into practice" adaptation in the development sector through national programming. The study shows that the approaches used in programming, particularly in engaging communities, largely determine the success of programming outcomes, particularly in the longer term.

In Chap. 5, Partiwi et al. examine the effects of company–community partnership upon the climate change adaptation and mitigation practices of the smallholders' coffee producers in Lampung, Indonesia. The methodological applied is statistical methods Propensity Score Matching (PSM) and Inverse Probability Weighting Regression (IPWR). This chapter shows that well-connected farmers with better access for mobility have a higher propensity to join the company–community partnership initiatives, as determinants of membership in the partnership include strong network intensity with fellow farmers group members and possession of motorized transportation. The methodology applied is statistical methods Propensity Score Matching (PSM) and Inverse Probability Weighting Regression (IPWR).

Chapter 6 by Neise et al. examine the enablers and barriers to convene SMEs around collective flood adaptation measures in order to alleviate flood risk more substantially in Jakarta and Semarang. This study uses a mixed-method approach and takes Jakarta and Semarang as examples for empirical research. Interviews with representatives of firms, business associations, industrial parks management, and public authorities give an understanding of Small and Medium Enterprise (SMEs)' adaptation challenges. The analysis shows that SMEs rely on incremental flood adaptation measures (e.g., elevating, pumps) and barely engages in more substantive or transformative risk reduction measures. The SMEs hesitate to engage in collective flood adaptation strategies due to a lack of joint governance mechanisms and sparse support from public authorities and managers of industrial parks.

Chapter 7 by Warsilah elaborates on the government's effort to make the coastal area such as Benoa Bay in Bali as a better social resilience in coping with climate change. The objective of this paper is to analyze the social resilience of coastal areas in the face of global climate change impacts to enrich regional climate studies and climate change policy in Indonesia. This research employs qualitative research approaches and using the concept of Bengen, we found that the key to coastal areas of governance are four aspects as follows: (1) ecological integration; (2). integration of sectors; (3) the integration of scientific disciplines; and (4) the integration of stakeholders and using ICM Concept by Adrianto and Warsilah (2017–2018). Data collection is obtained from tiered focus group discussions. The study of climate change impacts is important to enrich regional climate studies and climate change policy, particularly to Indonesia where its sea is a two-third of its mainland. Social resilience will be achieved if communities' adaptation capacity is growing stronger in coping climate change challenges.

Kasim in Chap. 8 explores the level of vulnerability of several coastal ecosystems such as coral reefs, mangroves, and seagrasses. He is also identifying the best choice of livelihoods in order to adapt to climate change in Eastern Indonesia's coastal areas. This chapter aims to explore the level of vulnerability of several coastal ecosystems such as coral reefs, mangroves, and seagrasses and to identify the best choice of livelihoods in order to adapt to climate change in Eastern Indonesia's coastal areas. Methods to explore some impacts of climate change on the coastal community were explored by Focus Group Discussion (FGD). This chapter found that the vulnerability of coral reefs occupies a significant level of decline compared to mangrove forests and seagrass habitats. In the end, the impact of climate change is resilience and the spread of seagrasses.

Further, Sinapoy and Djalante in Chap. 9 develop adaptation strategies that can be applied to the fishing communities of the Bajo Tribe as an effort to tackle ecological changes as well as adaptation efforts in the form of responsive actions to minimize and anticipate impacts by way of technology adoption, infrastructure development, altering arrestment area, network and mobilization of family members, and diversification of livelihoods.

Part II: Climate Change Mitigation and Cross-Cutting issues (finance, communication, law)

Part two of this book (Chap. 10 till Chap. 15) highlights the climate change mitigation and cross-cutting issues.

In Chap. 10, the land use change was examined in the case of peat forest conversion into oil palm plantation by Novita et al. Over past decades, large areas of tropical peat forests have been converted to oil palm plantation in Indonesia resulting in significant carbon emissions into the atmosphere. This chapter aims to quantify the ecosystem's carbon stocks in a total of six sites: two primary peat swamp forest sites, one secondary peat swamp forest site, and three young oil palm plantation sites in Tanjung Puting, Central Kalimantan, Indonesia. Using a stock difference approach, we estimated potential carbon emissions from vegetation change resulting from the conversion of primary peat swamp forest to oil palm plantation of 640 ± 114 Mg CO_2e/ha. Finally, while restoring peatlands is important, avoiding peat conversion is imperative for Indonesia's climate change mitigation effort.

Later, Goh et al. in Chap. 11 provide advices on transforming exploitative land-based activities in order to reduce terrestrial carbon stock loss. This chapter using proposal approaches on transforming land use based activities draw upon previous studies. The conventional exploitative land-based economies are facing a predicament: how to maintain economic growth not only without causing further environmental impacts but also repairing the damage done in the past. This overview identified and discussed four key transformative strategies with the elaboration of the experience in Kalimantan as well as the associated opportunities and challenges. The first two strategies with wealth creation as the center of policy-making may prevent further degradation but are inadequate to repair the previous environmental damage. Similarly, the last two strategies that emphasize restoration have limited contribution to economic growth.

In Chap. 12, Sari et al. proposed strategic and innovative funding sources in order to fund the peat restoration that was aimed by the Government of Indonesia. The amount estimation is US$ 1.70 billion for the whole 2.49 million ha target. This chapter aims to present business cases for peatland restoration and to propose a potential innovative financial instrument to finance the restoration activities in Indonesia. The creation of investment instrument is a necessity by means of structuring a peat bond where a combination of cash and carbon asset returns can be gained by investors. The second author of this chapter, Alue Dohong PhD, is now the Vice Minister at the Ministry of Environment and Forestry, the Government of Indonesia.

Mutiara et al. in Chap. 13 also explore climate finance as a tool to assess public expenditure in East Kalimantan province, i.e., climate budget tagging as defined by UNDP. Using climate budget tagging as defined by UNDP as a public expenditure assessment tool, insights from East Kalimantan Province and West Kutai District were attained. The results of climate budget tagging in the province and one of its districts consecutively depict that 7 and 47% of budget allocation in 2015 were potentially in support of low carbon development; while in 2016 both allocations in

East Kalimantan (24%) and West Kutai (16%) encouraged further discussions on the dynamics of climate mitigation in the sub-national level of governance. Nonetheless, such seemingly high allocations of the regional government budget that were earmarked for achieving climate mitigation targets were not equipped with clear outputs and outcomes that were sensitive to climate mitigation.

Rosemary and Evensen in Chap. 14 examines the types of information, members of the public received on environmental and climate challenges in Aceh from the key local mass media outlet in Aceh. The objective of this chapter is first to examine and analyze coverage from "Serambi Indonesia" about the AGV—this is the most read newspaper by the locals in Aceh; second, we present our methodology: the background of the local newspaper—Serambi Indonesia, the study objectives, the data collection process, and data analysis procedures. This study indicates the importance of climate change communication, and potentially for societal action on climate change, to understand the role of media in addressing climate change.

Further, Wibisana and Setyorini in Chap. 15 aim to assess the recent legal development in the protection and management of peatlands in Indonesia and analyzes how regulatory measures addresses the issue of peatland rehabilitation. This study found that peatland degradation and peatland fires in Indonesia have triggered various responses from the Indonesian government in the last five years. During these years, the government has enacted various regulations related to peatland protection, ratified the Paris Agreement, and formed an institution specifically aimed at accelerating peatland restoration in seven provinces seriously affected by fires called Peat Restoration Agency, and launched severe law enforcement against companies contributing to forest fires since 2013. However, there are two critiques related to the response on the peatland protection, namely the use of strict liability for plantation along with the "abnormally dangerous activities" criteria and the requirement of water-table level of a minimum of 0.4 m from the surface which constitutes a scientific uncertainty.

Conclusion

This edited book presented the latest on the climate change related studies on Indonesia. There are different methodological approaches stray with different expertise such as disaster risk, energy, coastal, small and medium enterprises (SME), smallholder farmers, and peat rehabilitation. The scopes of studies are covering all parts of climate change; climate adaptation, climate, climate law, and climate finance.

Indonesia is a developing country in adapting and mitigating climate change. There are still needs to further work on governing the action. Improvement is needed in multi-sectoral coordination and in legal construction of climate action. Adaptation and mitigation will mean a sustainable development processes and doing so will in longer run eradicate the poverty, increase equality, and preserve the environment. There are still hopes of further improvement of the processes; however, the lesson learnt to further incorporate needs to be streamlined further in more development issues.

There is importance to the realization of low carbon economy, yet Indonesia must balance the tradeoff between social, economic, and environmental benefits in order to achieve low carbon pathway. Engaging multiple stakeholders, removing the fossil fuel subsidies, and simplifying the bureaucracy could accelerate the transition toward low carbon development.

References

Djalante R, Thomalla F, Sinapoy MS, Carnegie M (2012) Building resilience to natural hazards in Indonesia: progress and challenges in implementing the Hyogo Framework for Action. Natural Hazards 62(3):779–803

Djalante R, Garschagen M, Thomalla F, Shaw R (eds) (2017) Disaster risk reduction in Indonesia: progress, challenges, and issues, Springer

EMDAT (2020) https://public.emdat.be/

Glaeser B, Glaser M (2010) Global change and coastal threats: the Indonesian case. An attempt in multi-level social-ecological research. Human Ecol Rev 1:135–147

Hoegh-Guldberg O, Hoegh-Guldberg H, Veron JEN, Green A, Gomez ED, Ambariyanto A, Hansen L (2009) The coral triangle and climate change: ecosystems, people and societies at risk

IPCC (2018) Global warming of 1.5 °C. An IPCC special report on the impacts of global warming of 1.5 °C above pre-industrial levels and related global greenhouse gas emission pathways, in the context of strengthening the global response to the threat of climate change, sustainable development, and efforts to eradicate poverty. In: Masson-Delmotte V, Zhai P, Pörtner HO, Roberts D, Skea J, Shukla PR, Pirani A, Moufouma-Okia W, Péan C, Pidcock R, Connors S, Matthews JBR, Chen Y, Zhou X, Gomis MI, Lonnoy E, Maycock T, Tignor M, Waterfield T (eds), In Press

Keller BD, Gleason DF, McLeod E, Woodley CM, Airamé S, Causey BD, Friedlander AM et al. (2009) Climate change, coral reef ecosystems, and management options for marine protected areas. Environ Manage 44(6):1069–1088

McClanahan TR, Donner SD, Maynard JA, MacNeil MA, Graham NA, Maina J, Baker AC et al. (2012) Prioritizing key resilience indicators to support coral reef management in a changing climate. PloS one 7(8):e42884

Ministry of Environment and Forestry (MOEF) (2007) Third National Communication under the United Nations Framework Convention on Climate Change. Republic of Indonesia, Ministry of Environment and Forestry

Ministry of Environment and Forestry (MOEF) (2017) Third National Communication under the United Nations Framework Convention on Climate Change. Republic of Indonesia, Ministry of Environment and Forestry

Ministry of National Development Planning (Bappenas) (2019) Low carbon development: a paradigm shift towards a green economy in Indonesia. Republic of Indonesia, Ministry of National Development Planning

Murray JP, Grenyer R, Wunder S, Raes N, Jones JP (2015) Spatial patterns of carbon, biodiversity, deforestation threat, and REDD+ projects in Indonesia. Conserv Biol 29(5):1434–1445

Naylor RL, Battisti DS, Vimont DJ, Falcon WP, Burke MB (2007) Assessing risks of climate variability and climate change for Indonesian rice agriculture. Proc Natl Acad Sci 104(19):7752–7757

Paoli GD, Wells PL, Meijaard E, Struebig MJ, Marshall AJ, Obidzinski K, Tan A, Rafiastanto A, Yaap B, Slik JF, Morel A (2010) Biodiversity conservation in the REDD. Carbon Balance Manage 5(1):7

Wijaya A, Chrysolite H, Ge M, Wibowo C, Pradana A, Utami A, Austin K (2017) How can Indonesia achieve its climate change mitigation goal? An analysis of potential emissions reductions from energy and land-use policies. Working Paper. Jakarta, Indonesia: World Resources Institute

Part I
Climate Change Impacts and Adaptation

Chapter 2
Review of Socio-Economic Development Pathway Scenarios for Climate Change Adaptation in Indonesia: Disaster Risk Reduction Perspective

Annisa Triyanti, Muh Aris Marfai, Estuning Tyas Wulan Mei, and Irina Rafliana

Abstract The work of Intergovernmental Panel on Climate Change (IPCC) on a Special Report on Emission Scenarios has pioneered the methods for greenhouse gas emission scenario associated with socio-economic development pathways in the coming century, followed by other models such as the Shared Socio-economic Pathways (SSPs) in climate change and disaster risk. This scenario is useful to understand how human society develops the future assessment of climate change and to provide possible mitigation and response strategies. This chapter is aimed to review the current status of socio-economic scenario on climate change and disaster and risk reduction effort in scholarly literatures and to identify gaps and opportunities for future research and decision-making based on the reflection of existing Climate Change Adaptation (CCA) and Disaster Risk Reduction (DRR) theories and emblematic case studies. We have conducted a semi-structured literature review and content analysis. The result of our analysis revealed that there is still a dearth of study on the application of different models of socio-economic forecasting scenarios to understand how would each pathway affect the vulnerability of certain type of disaster and its potential as a decision-making tool in Indonesia. However, there are opportunities to expand the methods and define socio-economic variables that go beyond the economic indicators (i.e. GDP), such as of welfare, health, education, social

A. Triyanti (✉)
Copernicus Institute of Sustainable Development, Utrecht University, Princetonlaan 8a, 3584 CB Utrecht, The Netherlands
e-mail: a.triyanti@uu.nl

M. A. Marfai · E. T. W. Mei
Faculty of Geography, Universitas Gadjah Mada, Bulaksumur, Yogyakarta, Indonesia
e-mail: arismarfai@ugm.ac.id

E. T. W. Mei
e-mail: estu.mei@gmail.com

I. Rafliana
Indonesian Institute of Sciences, LIPI, Jakarta, Indonesia
e-mail: irina_rafliana@hotmail.com

R. Djalante et al. (eds.), *Climate Change Research, Policy and Actions in Indonesia*, Springer Climate, https://doi.org/10.1007/978-3-030-55536-8_2

capital human development, participation and technology. Challenges are also identified, including the limitation of methodology, availability of data, lack of synergy between CCA and DRR, lack of interdisciplinarity, space for science–policy interface and political support. Future research on SSPs should pay attention to the aspect of multi-hazard approaches to climate change impact, emerging technology and its adverse impacts. We argued that projection is a highly important tool; however, largely reliable at the global scale rather than regional or national scale. To understand that climate variability and change is high, it is important to raise self-awareness on adaptation to future disasters.

Keywords Disaster risk reduction · Climate change impact · Socio-economic scenario · Indonesia

Introduction

Real-Life Problem and Existing Global Policies

Climate change is responsible for the increasing number and intensity of disaster, especially of hydro-meteorological causes (CRED and UNISDR 2016). There are growing evidence that certain typology of disasters such as floods, storms, fires and droughts are increasing due to climate variability and change. The latest WMO report on the State of Global Climate stated that in 2018, nearly 63 million people were affected by extreme weather associated with climate change leaving 2 million people displaced (WMO 2018). This shows that climate variability and change affect both fast- and slow-onset disaster and are responsible for the loss of lives, livelihoods, culture and heritage.

One of the latest deadly disaster events is the Cyclone Idai that ripped into Southern Africa in March 2019. With totalities of more than 1000 human loss and more than 1$ billion US economic loss, the Cyclone Idai event has laid an evidence of how climate change affects the intensity of tropical storms. It demonstrates the vulnerability and exposure of low-lying cities and also the disruption of normal weather patterns due to climate change impact (UNDRR 2019).

In terms of climate change mitigation, the issue of clean energy and emerging innovation are the core focus of the effort to decarbonize development. There is an issue of tradeoffs between mitigation and adaptation, and between reducing disaster risk in the short term and adapting to a long-term climate change impact. For example, in terms of land use prioritization and policy for conserving resources and ecosystem, improving health, well-being of communities, while at the same time also promote the use of clean energy and promoting innovation. These are hard choices to be made and would require thinking to anticipate the impact of anthropological activities and disruptive technology, which could increase the risk of adverse climate change impact and disaster.

In the global policy arena, the year of 2015 marked a very important year for disaster risk reduction, climate change actions and sustainable development. The Sendai Framework for Disaster Risk Reduction adopted in March 2015 (UNISDR 2015) and Paris Agreement in December 2015 (UNFCCC 2015) and the SDGs were agreed by 193 countries (UNGA 2015). The interlinkages among these three frameworks are crucial to ensure orchestrated efforts to achieve sustainable development.

The Sendai Framework for Disaster Risk Reduction focuses on 4 priorities for actions, namely understanding of risk, governance, investment and the building back better for resilience. The latest UNFCCC Conference of the Parties 24 (COP24) in Katowice, Poland, has produced a rulebook which clearly mentioned that parties should advance their effort and consider future climate risk in implementing their plans and strategies to reduce disaster risk:

> 10(b) To take into consideration future climate risks when developing and implementing their relevant national plans and strategies that seek to avert, minimize and address loss and damage and reduce disaster risks, as appropriate. (UNFCCC, 2018a, Draft decision -/CP.24 para 10.b)

The Katowice rulebook is also further calling out for more transparency of countries in terms of reporting mechanism and accountability of their climate actions.

The SDG 13 focuses on taking urgent action to combat climate change and its impacts. This particular SDG will be reviewed in the High-Level Political Forum 2019 on climate change at later this year. More efforts to ensure synergy is anticipated.

These global policy processes have strongly encouraged countries to prepare, adapt and mitigate the unknowable adverse impact of climate change in the future, which could lead to increasing number and intensity of disasters, jeopardizing the efforts to achieve the sustainable development goals. It is, therefore, crucial for science to contribute to understanding disaster risk and to unpack the connection between climate change adaptation and disaster risk reduction. This includes the understanding of socio-economic factors and variables affecting the future pathways in climate change adaptation and mitigation efforts.

Theoretical Reflections

We selected four main theories or concepts relevant to the disaster risk reduction and climate change research, including the resilience and vulnerability theories, theory of societal change and transformation and adaptive governance.

The resilience and vulnerability theories are most often referred in disaster risk reduction and climate change research. The resilience concept has been around for about 5 decades since Holling's groundbreaking paper (1973). The definitions of resilience, however, is not easy to grasp and often times are confusing (see Walker et al. 2004). The general definition of resilience has been laid out by many scholars, indicating the capacity of a system to retain to original function within critical threshold after experiencing shocks or changes (see Walker et al. 2004; Folke et al. 2010; Folke 2006; Olsson et al. 2006; Voss 2008).

According to Folke et al. (2010), the component of adaptability and transformability are considered as crucial to build resilience. While adaptability ensures stability throughout development trajectory when experiencing changing external drivers such as climate change impacts and disaster risk, transformability ensures crossing threshold to a new, better development trajectory (Folke et al. 2010). While transformation is crucial, the preparation for changes is at most important to guide transformability. Walker et al. (2004) has also argued that transformability as the last, crucial phase of resilience, where a stability of a new system is created. Olsson et al. (2006) argued that the component of building knowledge, networking and leaderships are the key factors for preparing changes in the socio-ecological systems. Using the case study of Everglades, a new novel system configuration and windows of opportunity are found through building knowledge and network, where actors manage to suspend extant beliefs, question perceptions and contrast possible futures (Olsson et al. 2006: 18).

The vulnerability theory on the other hand focuses on both environmental and social condition. It emphasizes the complex mixture and dynamics of variables which affects the vulnerability of such systems. Environmental system consists of physical and biophysical factors such as land, climate, ecosystem and its functions. The factor of social condition includes diverse aspects such as capital distribution, institution, economic and political system, as well as technology and the impact of emerging technology to society (see Voss 2008: 49). Voss (2008) also argued that vulnerability could not be only seen from the perspective of physical aspect, and that of social aspect is often overlooked. Using the case study of Indonesia, they highlighted an example of the importance of social aspect in defining the level of vulnerability. The research discovered that the existing customary rule such as "adat" could shape society and affect the level of vulnerability of certain community groups. It constructs the way people understand reality and the value of their lives in the local environment, including on how they perceive certain event as a disaster as a threat or not.

In the context of environmental and climate change policy, the theory of societal change ABC (Attitude-Behaviour—and choice) is believed to be a suitable entry point for social scientists to contribute to the discourse (see Shove 2010: 1273). The ABC theory argued that the responsibility for responding to climate change lies on individuals whose behavioural choices will make the difference (Shove 2010: 1283; Wilson and Chatterton 2011).

Park et al. (2012), extends the transformative concept into decision-making tools by designing adaptation action cycles. The cycle consists of four phases including (1) problem structuring and establishing the adaptation arena; (2) developing the adaptation agenda, vision and pathway; (3) implementing adaptation actions; and (4) evaluation, monitoring and learning. These phases are tested through a case study of the wine industry in Australia and their incremental adaptation and transformative adaptation. The incremental adaptation actions are defined as short term by nature, meanwhile the research revealed that transformative adaptation helped to ensure long-term changes and enable learning. The research also concluded that the contemporary challenges in transformative adaptation are the lack of understanding of socio-economic environmental conditions in the present time and the future.

Lastly, the concept of adaptive capacity and adaptive governance further highlights the need of new thinking in good governance when it comes to finding the best mechanism on how to respond to global environmental change, including climate change and disaster risk (see Smit and Wandel 2006; Pahl-Wostl 2007; Tompkins and Adger 2004; Adger et al. 2005a, b; Djalante et al. 2011). Adaptive governance emphasized factors such as co-learning (Pahl-Wostl 2007; Djalante et al. 2011), co-management, participation, and collective action (Tompkins and Adger 2004; Djalante et al. 2011), polycentric and multilayered institutions, and self-organization and network (Djalante et al. 2011). Tompkins and Adger (2004) further argued that the social elements are strongly influencing the present and future vulnerabilities as both are a function of adaptive capacity, which is in turn dependent on social capital, institutions, and resources distributions. In addition, an adaptive governance also emphasizes the improvement of the capacity of institutions to better coordinate relief operations, public awareness and risk reduction policy in case of disasters, by encouraging learning from experience (Bakkour et al. 2015).

The review on existing theories revealed that resilience bounds mitigation and adaptation in regards to climate change and disaster risk reduction. There is an increasing focus on societal change and transformation to ensure resilience. And at the same time also serves as perquisite variables to increase adaptive capacity and adaptive governance for climate change adaptation and disaster risk reduction. We discovered that learning and building knowledge is a catalytic element to enable resilience and transformation for successful adaptation and mitigation of climate change impact and disaster risk in the future. The knowledge of socio-economics and environmental conditions is identified as detrimental to the process of adaptive transformation and resilience.

The Objective

Against these backgrounds (i.e. real-life problem, global policy status and theoretical reflections), there are four objectives of this paper. *First*, to highlight the current status of socio-economic scenario on climate change, both in existing theory and empirical case studies, and the impact to different disasters and risk reduction effort in scholarly literatures. *Second* to assess the challenges and opportunity for upscaling existing shared climate and socio-economic projection models for policy design and implementation. *Third* is to identify gaps and opportunities of the socio-economic scenarios for better policy-making at the national level. And finally, *fourth*, to reflect on the broader implication towards CCA theories and future research needs.

Structure of This Chapter

This chapter has introduced the real-life problem, existing global policies related to climate change and disaster risk reduction, theoretical reflections and objectives. Section "Methods" describes the methods used in this chapter. Section "Literature Review" presents the results from the literature review on SSPs and socio-economic impacts and future pathways and scenario of CC and DRR. Section "Results: Indonesian Case Studies" elaborates the context of Indonesia based on the review of existing literature and policy documents. Section "Discussion" discusses the opportunities and challenges on using SSPs and other models of socio-economic scenarios to help as decision-making tools in DRR. Section "Conclusion" finally provides a general conclusion of this chapter.

Methods

We have selected a semi-structure method for the literature review. Scopus scientific databases are used to list the past research on this topic with two sets of keywords. In addition, the Google scholars are used to allow authors to find prominent literatures and sources which are not listed in International journals or databases.

There are five sets of keywords used to generate an inventory of resources. Four sets are operated in Scopus database:

(1) For worldwide-general case, keywords selected are (TITLE-ABS-KEY ((climate PRE/0 change AND disaster AND projection OR scenario)) AND (socioeconomic OR socio-economic), depicts the general scholarly literatures in climate change and disaster with 136 results;
(2) For Worldwide-Shared Socioeconomic Pathways (SSPs), keywords selected are TITLE-ABS-KEY ((climate PRE/0 change AND disaster AND projection OR scenario)) AND (socioeconomic OR socio-economic) AND SSPs;
(3) For Indonesia-general case, keywords selected are TITLE-ABS-KEY ((climate PRE/0 change AND disaster AND projection OR scenario))) AND ((socioeconomic OR socio-economic)) AND (Indonesia), resulted into 16 scholarly literatures specifically on Indonesian case studies;
(4) For Indonesia-SSPs, keywords selected are TITLE-ABS-KEY ((climate PRE/0 change AND disaster AND projection OR scenario))) AND ((socioeconomic OR socio-economic)) AND (Indonesia) AND SSPs.

In addition, one set of keywords is operated with Google search (SRES scenario, disaster, Indonesia) which are later filtered and selected based on relevancy to the topic and intended purpose of this research.

In addition to the literature review, we also conducted analysis on policy documents. We have selected the policy documents relevant to climate change policy especially at the national level in the form of regulations, laws, and policy paper. For

Indonesia, four documents are analyzed including (1) Country National Communication document submitted to the UNFCCC (1st, 2nd and 3rd); (2) Nationally Determined Contribution (NDC) Country Report of Indonesia (Republic of Indonesia 2016); and (3) Other relevant documents (e.g. disaster, land use and forest management policy).

Literature Review

Introduction to SSPs

The Special Report on Emission Scenario (Nakicenovic et al. 2000) provides four storylines of development with regard to the way world population, economies and political structure may evolve over the next few decades (see also Adger et al. 2005a, b). There are four storylines included in the SRES, including A1, A2, BI and B2. Each is characterized by different conditions in regards to economic growth, population, the use of technology and scale focus in development. There is also a distinction in regards to the technological emphasis on the use of energy sources such as fossil, non-fossil and a more balanced source (For detail storylines, see Nakicenovic et al. 2000; pp. 4–5).

The pathways have been evolved since first published in 2000. The Representative Concentration Pathways (RCPs) and Shared Socioeconomic Pathways (SSPs) are established as an expansion of the earlier pathways (see Fig. 2.1). The pathways include differentiations and consideration of the factors of socio-economic challenges for mitigation as well as adaptation. SSPs served as an methodological approach for reference, elaborated by narratives and quantitative model and pathways, and when combined quantitatively with Climate RCPs model, it became a scenario (see O'Neill et al. 2014 and Van Vuuren et al. 2013). However, scholars argued that there is still a problem of the combinatorics between climate and socio-economic models in the new RCPs and SSPs model (see Van Vuuren et al. 2011).

The SSPs' pathways combined nine socio-economic categories, including demographics, economic development, welfare, environmental and ecological factors, resources, institutions and governance, technological development, broader societal factors and policies (see O'neill et al. 2014; pp. 396).

The SRES scenario outlines the four pathways (i.e. A1, A2, B1 and B2) (Nakicenovic et al. 2000). Some of them are still relevant in the new SSPs illustrations. SSP1, identified as low challenges for mitigation and adaptation happened when sustainable development was proceeding at a high pace with low inequalities and more environmentally friendly technology which would increase the productivity of land. Meanwhile, in the other extreme, high challenges for mitigation and adaptation will be experienced when the rate of emissions are high and unmitigated due to moderate economic growth and growing population, which led to high inequality and slow technological change in the energy sectors, with low adaptation. In between

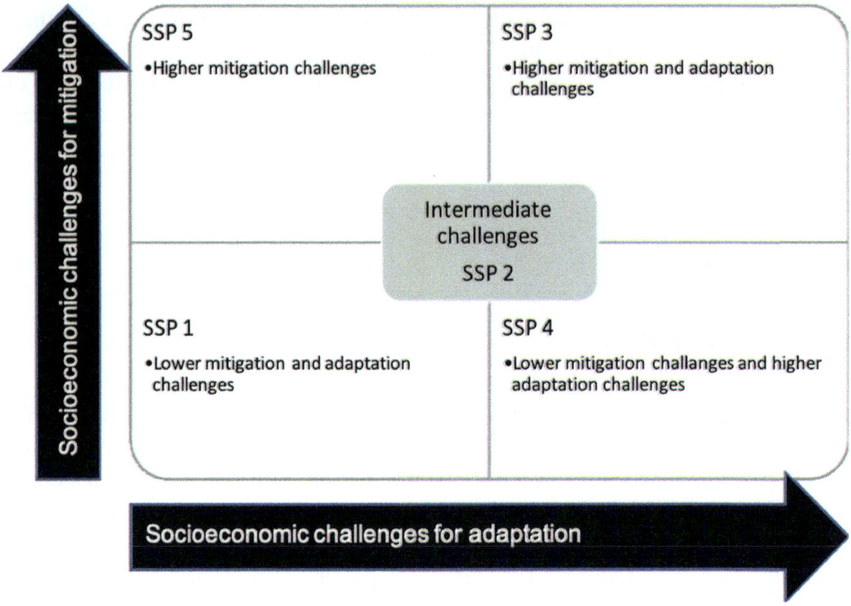

Fig. 2.1 SSPs model. Adopted from O'neill et al. (2014)

these extremes there are SSP4 and SSP 5 with an analogue that high challenges on adaptation happened when there is inequality, but rapid technology on clean technology. Whereas the high challenges on mitigation are experienced when there is a lack of alternative energy technology, but a more improved human capital with less vulnerability. Based on the two models, it is clear that there is an existing storyline and illustration on narratives of how socio-economic trajectory develop and would help to understand better action according to each challenge faced.

Review from the Worldwide Case Studies on Socio-Economic Impacts and Future Pathways and Scenario of CC and DRR

The result from the database query shows that there is an increasing number of scholarly research on the use of climate change scenarios for measuring impact in terms of disaster (see Fig. 2.2). However, in the case of Indonesia, there is still a dearth of study published in the international publication on the socio-economic scenario to climate change impact in Indonesia and its policy usage and/or implementation.

This section will review the query results on socio-economic impacts/variables of CC and Disaster. However, it will be mainly focussing and referring to the models of socio-economic scenarios pathways and implications on disaster risk and vulnerabilities (Nakicenovic et al. 2000; Birkmann et al. 2013). Based on the review, we

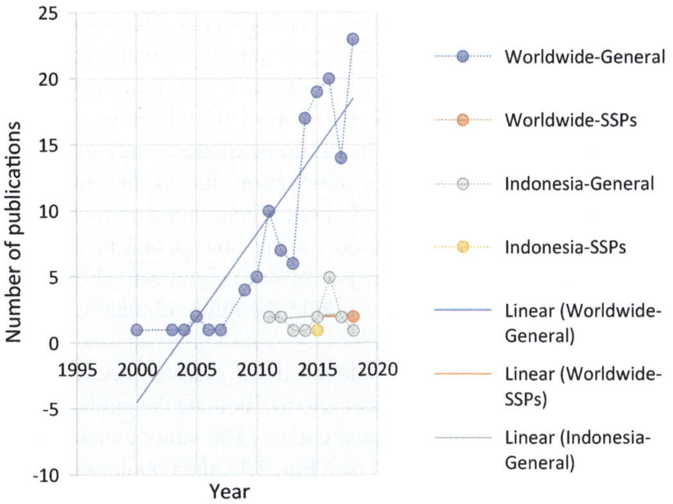

Fig. 2.2 Result of literature review analysis. *Source* Scopus database analysis

discovered that there is still strong domination of physical scenarios on climate change projections and impact rather than socio-economic-related scenario. Out of the 136 results from worldwide case, only four results showed the direct and specific usage of SSPs (Mochizuki et al. 2018; Chen et al. 2018; Birkmann et al. 2015; Cuaresma and Lutz 2016).

In the case of Indonesia, only one result shows the direct usage of SSPs out of the 16 results, (case of flooding in Jakarta) (Birkmann et al. 2015), meanwhile other publications mentioned or referred to another type of future climate scenario. The usage of SSP is still limited to the impact of climate change on a very limited type of disaster (in this case flooding in urban and coastal areas). Review on the existing emblematic case studies around the world using the models of socio-economic scenarios to understand the implication of each pathway to disaster vulnerabilities from the scholarly literature are listed as below.

Chen et al. (2018) conducted studies on the population exposure to droughts in China under the 1.5 °C global warming target proposed by the Paris Agreement (UNFCCC 2015). It uses the Standardized Precipitation Evapotranspiration Index (SPEI) to calculate drought frequencies in the period of 1986–2005 and 1.5 °C global warming scenario (2020–2039 in RCP 2.6). The research revealed a specific outcome that population are more exposed to drought in the East China rather than the west. However, generally the total drought frequency will be decreasing if the 1.5 °C global mean temperature target is achieved. Finally, it suggested that reaching the 1.5 °C target is a potential way for mitigating the impact of climate change on both drought hazard and population exposure (Chen et al. 2018).

Melkonyan (2014) also presented the case of drought in Armenia. Armenia has experienced different types of water-related disasters, such as droughts, floods, and storms, which have a direct influence on economy and are expected to occur more frequently in terms of climate change, raising the need to estimate economic vulnerability especially in agricultural sector. One of the most important sectors in Armenia, which possess 21% share in the GDP is the agricultural sector. The agricultural resources are vulnerable to the current and future climate change impact. The study assesses the economic loss of the crop production due to the climate variability and elements, including temperature, radiation, precipitation and wind field. Future climate projections in the region are used for the time period of 2011–2040. The research revealed that temperature increase causes a significant shift, and predicted that the economic loss in the drier condition in the future climate within the period 2011–2040 will be more than doubled.

Another study conducted in the field of flood disaster insurance in Austria (Mochizuki et al. 2018) uses the SSPs scenario to calculate the public cost of demography and expenditures in climate-related events. The study concluded that in the SSP2 scenario or intermediate scenario (see Fig. 2.1), the population in Austria will be increasing from 8.6 million in 2015 to 9.2 million in 2050 with increasing flood risk due to socio-economic development and climate change. It revealed that this will cause increasing public debt and insufficient current budgetary arrangement to deal with the rising risk of extreme floods in the future.

The analysis in Patt et al. (2010) suggests that climatic disaster risk (as measured by the number of persons affected by a climate disaster) is highest in the countries with Human Development Index (HDI) levels of around 0.5, after controlling for other determinants of vulnerability. The number of affected individuals starts to decline only after countries reach this level of development (Cuaresma and Lutz 2016). Through a set of 50-year scenarios of human losses due to climate change in Mozambique, which later extended to a sample of 23 least-developed countries, Patt et al. (2010) suggested the relationship between an increase in income and adaptive capacity. The results of their research suggested that in the second quarter of the century, the effects of socio-economic development trends may begin to offset rising climate exposure. This implies that vulnerability will be increasing rapidly in the period between now and then, which will be desperately in need of international assistance to finance adaptation.

In terms of the interlinkages between mitigation and adaptation, the concept and case studies of Ecosystem function have served as co-benefits. McVittie et al. (2018) in their paper on EbA for DRR and CCA, particularly in Europe emphasized the need for lessons learnt from implementing EbA across a range of land uses. The research evidence indicates that adaptation and DRR are achievable, cost-effective and would attract acceptability and funding when these co-benefits are demonstrated clearly. In Asia, a study regarding the application of EbA for DRR and CCA has been conducted in Indonesia (Triyanti et al. 2017). They conclude that the key to the success of EbA is the involvement and participation of stakeholders. However, the gaps in the EbA for DRR and adaptation are still existing, especially in terms of knowledge of the biophysical and economic benefits, or negative impacts of EbA.

The review revealed that the approach which considers both potential changes in countries' exposure to climatic extreme events and socio-economic development trends that influence countries' own adaptive capacities is lagging. This is potentially caused largely by the absence of data and reliability, especially due to extreme dynamics and complexity of socio-economic development (Freire et al. 2016; Hoeppe 2016).

Results: Indonesian Case Studies

Country Status on Climate Change

Climate variability and change has impacted different sectors in Indonesia, including agriculture, water, and increasing risk of disaster. In terms of agriculture, climate change alters precipitation, evaporation, run-off water and soil moisture. This will severely affect the production of main crops and threaten food security (see World Bank 2011). In the water sector, climate change has affected the water quantity and quality. The shortage of water during the dry season is estimated in 2009, particularly in the urban area, which will cause 2% total economic loss annually, due to limited access to water and sanitation. Another water problem is including poor water quality affecting the spread of diseases and raising vulnerability of people. Furthermore, industries continue to use the groundwater causing land subsidence and increasing the risk of seawater intrusion and flooding (World Bank nd). This is shown in the case of big low-lying coastal cities in Indonesia including Jakarta and Semarang (see Marfai and King 2008; Marfai et al. 2015).

In terms of risk of disaster, a study conducted by IPCC (2018) revealed that by the end of the twenty-first century, it is very likely that sea level will rise in more than about 95% of the ocean area, and about 70% of the coastlines worldwide are projected to experience a sea-level change within ±20% of the global mean. This condition will increase the vulnerability of the community to climate-related natural hazards and disaster including, flood, drought, storm and hurricane (IPCC 2018).

In terms of contribution to mitigate and adapt to climate change, based on the NDCs submitted to the UNFCCC, Indonesia has voluntarily committed to reduce unconditionally 29% of its greenhouse gasses emissions against the business as usual scenario by the year of 2030 (Climate Watch Data 2019). Indonesia could increase its contribution up to 41% reduction of emissions by 2030, subject to availability of international support for finance, technology transfer and development and capacity building. Meanwhile based on the target found in the Decree 62/2013 regarding a Managing Agency for the Reduction of Emission (sic) from Deforestation and Degradation of Forest and Peat lands, there will be 26% cut in GHG emissions, 41% cut in GHG emissions with international assistance by 2020 (Climate Watch Data 2019).

Indonesian current effort is highly insufficient, with the projection that warming could reach 3 to 4 degrees, far from the 2 and 1.5 degrees (Climate Action Tracker 2018). Indonesia has been increasing emission at a faster rate and might be doubled in 2030 compared to 2014 values if the policy does not change due to extensive use of coal (Climate Action Tracker 2018).

In terms of adaptation, Indonesia identified eight sectors including Agriculture, Coastal Zone, Cross-Cutting Area, Disaster Risk Management (DRM), Education, Energy, Environment and Health.

Policy progress and country commitment in the latest COP24 in Katowice submitted by the Indonesian government could be a good start for transforming policy to support more ambitious climate action. As an outcome of this conference, Indonesia supported the National Adaptation Communication and also transparency and compliance, with consideration of the case of force majeure in the face of challenges to natural hazard prone countries to implement the NDCs (UNFCCC 2018a, b).

Review on the Contemporary Use of SSPs in Climate Change and Disaster Risk in Indonesia

Review of the Scholarly Literatures

Our literature review shows that the emission scenario and projection are abundant in terms of methodology; however, there is still a lack of understanding on how these physical projections affect the socio-economic condition, now and in the future (Wilby et al. 2009).

The number of populations is a crucial factor. General population growth projection in the coastal area (Neumann et al. 2015), Indonesia and other four Asian countries, including China, India, Bangladesh and Vietnam accounted for more than half of the global low elevation coastal zone population in 2000 and will continue growing vastly under the future scenarios.

In terms of the relation to disaster, several cases such as the case of Jakarta (Birkmann et al. 2015) shows the exposure and susceptibility due to increasing population due to floods, cyclones, droughts and sea-level rise.

The sea-level rise is identified as a major threat and increases risk of flooding in Asian Countries, including the Maldives, Vietnam, Bangladesh, India, China, Thailand, Indonesia, Philippines and Myanmar.

Through quantitative analysis using the World Risk Index and participatory analysis in Jakarta, Birkmann et al. (2015) suggests that adaptation under B1 SRES scenario will reduce the number of people affected by flooding by 136 million people in 2100. Main socio-economic parameters in the case of disaster vulnerability are migration, poverty, social security, labour, energy use and governance (Birkmann et al. 2015).

Review of the Policy Document

The NCs submitted by the government of Indonesia to some extent has utilized the scenario projection. The second NC utilized the SRES scenario. It suggests that under the scenario of SRES A2, where GHG emission is increasing, most of the models are suggesting that in 2025, there will be increasing wet seasonal rainfall in several regions in Indonesia (i.e. Java, Bali, NTB, NTT and Papua), and other regions will decrease. Moreover, it is predicted that by 2050 and 2080, Indonesian regions will mostly experience higher rainfall, with exceptions in the northern parts of Sumatra and Kalimantan. Meanwhile, for dry season rainfall, especially in West Java and South Sumatra, it might be decreasing in 2025, increasing in 2050, and decreasing by 2080. If it achieves the low emission scenario (SRESB1), the dynamics will be lower (see Republic of Indonesia 2010; pp. IV-4). Following the first NC (Republic of Indonesia nd), the third NC (Republic of Indonesia 2017) uses the RCPs (i.e. temperatures, sea-level rise, seasonal rainfall among others), while the socio-economic as drivers and impact is still centred around the GDP (Republic of Indonesia 2017). Despite the apparent use of the scenario, however, there is no detail elaboration in the third NC on the SSPs method and its values for future policy design. Based on our analysis, although there is an increasing effort to mainstream climate change into different sectors, it is still perceived in singularity. There is an opportunity for future research agenda to test implementation across governing actors and jurisdictions.

Discussion

Opportunities and Challenges on Using SSPs and Other Models of Socio-Economic Scenarios to Help as Decision-Making Tools in DRR

Opportunities

The advanced list of elements derived from worldwide case study reviews on socio-economic impacts of climate change and disaster (see Table 2.1) shows that there is a huge opportunity to use the SSP for future projection. Socio-economic variables relevant for determining pathways and scenario for CC in disaster perspective are more or less similar with the SSPs determined variables, with a strong focus on risk component (e.g. socio-economic vulnerability of certain category of population, the susceptibility, coping capacity and adaptive institutions determinants such as preparedness, social capital, institutional settings and policies). Furthermore, the current focus on socio-economic drivers and impact for future pathways and scenarios on CCA-DRR in Indonesia is still too general and mainly focusing on economic variables (i.e. GDP) instead of other important variables such as loss and damage factors (i.e. both economic and social assets), environment (i.e. ecosystem approach for DRR, CCA

Table 2.1 Potential socio-economic parameters in DRR and CCA case

Socio-economic categories	Elements of the scenario (O'neill et al. 2014)	Interpreted elements in DRR and CCA case
Demographics	• Population including total and age structure	• Population by age (Cuaresma and Lutz 2016) • Vulnerable population (Birkmann et al. 2015) • Migration (Birkmann et al. 2015)
	• Urban versus rural populations and forms	• Ibid (Preston 2013)
	• Other specified geographical locations (e.g. coastal versus inland)	• Vulnerability of coastal areas (Neumann et al. 2015)
Economic development	• Global and regional GDP or productivity trends	• GDP (Birkmann et al. 2015; Cuaresma and Lutz 2016)
		• Economic losses (Preston 2013)
	• Regional, national, and sub-national distribution of GDP (i.e. economic development of developing countries)	• Insurance for CC and DRR
	• Specific sectoral structure of national economies (i.e. the share of agriculture, and land productivity)	• International aid
	• Share of population in extreme poverty	
	• Nature of international trade	
Welfare	• Human development	• Ibid (Cuaresma and Lutz 2016)
	• Educational attainment	• Ibid (see Cuaresma and Lutz 2016)
	• Health including access to public health and health care infrastructure	
Environmental and ecological factors	• Air, water, soil quality	
	• The function of ecosystem	• Ecosystem function to reduce disaster risk (McVittie et al. 2018)
Resources	• Fossil fuel resources and renewable energy potentials	• Energy and potential disruption (Fan et al. 2014)
	• Other key resources, such as phosphates, freshwater, etc	

(continued)

Table 2.1 (continued)

Socio-economic categories	Elements of the scenario (O'neill et al. 2014)	Interpreted elements in DRR and CCA case
Institutions and governance	• Existence, type and effectiveness of national/regional/global institutions	• Ibid
	• Degree of participation	• Community engagement (Opitz-Stapleton and MacClune 2012)
	• Rule of law	Ibid
		• Corruptions (Birkmann et al. 2015)
Technological development	• Type (e.g. slow, rapid, transformational) and direction (e.g. environmental, efficiency, productivity improving) of technological progress	• Early Warning System (EWS)
	• Diffusion of innovation in particular sectors, e.g. energy supply, distribution and demand, industry, transport, agriculture	
Broader societal factors	• Attitudes and world views to environment/sustainability/equity	• Preparedness
	• Lifestyles (i.e. diets)	• Social capital
	• Societal tension and conflict levels	
Policies	• Non-climate policies (i.e. policies on development, technology policies, urban planning, transportation, energy security, environmental policies to protect air, soil and water quality) and policy objectives (e.g. welfare improvement)	• Relevant policies regarding CC and DRR

and CCM), socio-cultural aspects (i.e. welfare, health, education, social capital and human development), technology (i.e. technology in early warning systems) and political complexity and dynamics (i.e. participation, inclusiveness in DRR governance) among others (see Table 2.1). Finally, there is a need for disaggregation and thorough socio-economic studies which would be helpful for future assessment such as through integrated assessment modelling (IAM).

Challenges

Based on the review, there are several challenges identified: (1) Methodological limitations on the extension of SRES, SSPs and combination with climate RCPs.

Although there are existing references and study which elaborate methods and socio-economic variables, there is difficulties in terms of mismatch of scales between natural and social systems. Socio-economic aspect of human development is very dynamic and complex in nature, therefore oftentimes intangible; (2) Availability, inconsistencies and reliability of the data; (3) Lack of synergy of CC and DRR due to sectoral and policy segregation, as well as gaps on terminology and understanding of the scope of disaster risk, climate change adaptation and mitigation; (4) Lack of interdisciplinarity and continuous and monitored space for advancing science–policy interface; and (5) Lack of political support. There are still challenges in communication between scientists and policy-makers, which affect the implementation of CCA and DRR research. Furthermore, lack of coordination among governmental sector and complex bureaucracy oftentimes also hindered the effectiveness of CCA and DRR efforts.

Conclusion

This section highlights the contribution of the finding towards CCA and DRR theories, gaps and opportunities and future research direction, both apply to global and national level using the case of Indonesia. We concluded that vulnerability and resilience are largely controlled by human dynamics. Transformation and adaptive capacity to governed interlinked issues such as climate change and disaster risk reduction are dependent on socio-economic conditions. Therefore, there is an urgent need of building knowledge and increasing understanding of the future based on the socio-economic factors that increase vulnerability of disaster, exacerbated by climate change. Interdisciplinary and holistic research is needed for the short and long term. There is a new development in terms of future climate change and socio-economic scenarios (RCPs and SSPs) which is based on the SRES scenario. However, the new combination of RCPs and SSPs is still daunting (Nakicenovic et al. 2014; Van Vuuren et al. 2011), therefore, for future research, it would be useful to unpack the SRES SSPs to develop new insights through robust empirical research.

In the context of Indonesia, there is a need of a stronger emphasis on climate mitigation and adaptation and disaster risk reduction and adjustment of socio-economic variables that goes beyond the economic indicators (i.e. GDP) such as of welfare, health, education, social capital human development, participation and the role of disruptive technology. It is also crucial to include the aspect of multi-hazard approach to climate change impact, emerging technology and its adverse impact, and research related to process and mechanism of inclusive development and sustainable impact of research to community-based programme. It is therefore important to promote interdisciplinary research collaboration for a new generation of RCPs and SSPs through available methods, including the integrated assessment modelling (IAM), and leveraging the role of social science and governance studies in future climate change research.

A more advance strategy is needed in science communication and dissemination of future climate modelling and uptake of valuable finding to inform policy and community. This will be particularly useful to serve as a basis for monitoring and evaluation of Paris Agreement transparency mechanism and SDGs targets and indicators implementation.

We argued that projection is one of the highly important tools; however, largely reliable at the global scale rather than regional or national scale. To leverage understanding of high climate variability and change, it is important to raise self-awareness on adaptation to its impact, including future disasters.

References

Adger WN, Hughes TP, Folke C, Carpenter SR, Rockström J (2005) Social-ecological resilience to coastal disasters. Science 309(5737):1036–1039

Adger WN, Arnell NW, Tompkins EL (2005) Successful adaptation to climate change across scales. Global Environ Change 15(2):77–86

Bakkour D, Enjolras G, Thouret JC, Kast R, Mei ETW, Prihatminingtyas B (2015) The adaptive governance of natural disaster systems: Insights from the 2010 mount Merapi eruption in Indonesia. International journal of disaster risk reduction 13:167–188

Birkmann J, Cutter S, Rothman DS, Welle T, Garschagen M, van Ruijven B, O'Neill BC, Preston B, Kienberger S, Cardona OD, Siagian T, Hidayati D, Setiadi N, Binder C, Hughes B, Pulwarty R (2013) Scenarios for vulnerability: opportunities and constraints in the context of climate change and disaster risk. Climate Change, 1–16

Birkmann J, Cutter SL, Rothman DS, Welle T, Garschagen M, Van Ruijven B, Siagian T (2015) Scenarios for vulnerability: opportunities and constraints in the context of climate change and disaster risk. Climatic Change 133(1):53–68

Chen J, Liu Y, Pan T, Liu Y, Sun F, Ge Q (2018) Population exposure to droughts in China under the 1.5 °C global warming target. Earth System Dyn 9(3):1097–1106

Climate Watch Data (2019) https://www.climatewatchdata.org/contained/countries/IDN. Accessed 18 Mar 2019

Climate Action Tracker (2018) https://climateactiontracker.org/countries/indonesia/. Accessed 7 Mar 2019

CRED, UNISDR (2016) The Human Cost of Weather Related Disasters 1995–2015. https://www.unisdr.org/2015/docs/climatechange/COP21_WeatherDisastersReport_2015_FINAL.pdf

Cuaresma JC, Lutz W (2016) The demography of human development and climate change vulnerability: a projection exercise. Vienna Yearbook of Population Research 1:241–261

Djalante R, Holley C, Thomalla F (2011) Adaptive governance and managing resilience to natural hazards. Int J Disaster Risk Sci 2(4):1–14

Fan JL, Liang QM, Liang XJ, Tatano H, Kajitani Y, Wei Y. M. (2014). National vulnerability to extreme climatic events: the cases of electricity disruption in China and Japan. Natural hazards, 71(3), 1937–1956. Folke, C. (2006). Resilience: The emergence of a perspective for social–ecological systems analyses. Global environmental change, 16(3), 253–267.

Folke C, Carpenter S, Walker B, Scheffer M, Chapin T, Rockström J (2010) Resilience thinking: integrating resilience, adaptability and transformability. Ecology Soc 15(4)

Freire P, Tavares AO, Sá L, Oliveira A, Fortunato AB, Dos Santos PP, Pinto PJ (2016) A local-scale approach to estuarine flood risk management. Nat Hazards 84(3):1705–1739

Hoeppe P (2016) Trends in weather related disasters–Consequences for insurers and society. Weather and climate extremes 11:70–79

Holling CS (1973) Resilience and stability of ecological systems. Annu Rev Ecol Syst 4(1):1–23

IPCC (2018) Global Warming of 1.5°C. An IPCC Special Report on the impacts of global warming of 1.5°C above pre-industrial levels and related global greenhouse gas emission pathways, in the context of strengthening the global response to the threat of climate change, sustainable development, and efforts to eradicate poverty. In: Masson-Delmotte V, Zhai P, Pörtner HO, Roberts D, Skea J, Shukla PR, Pirani A, Moufouma-Okia W, Péan C, Pidcock R, Connors S, Matthews JBR, Chen Y, Zhou X, Gomis MI, Lonnoy E, Maycock T, Tignor M, Waterfield T (eds)

Marfai MA, King L (2008) Tidal inundation mapping under enhanced land subsidence in Semarang. Central Java Indonesia. Natural hazards 44(1):93–109

Marfai MA, Sekaranom AB, Ward P (2015) Community responses and adaptation strategies toward flood hazard in Jakarta. Indonesia. Natural hazards 75(2):1127–1144

McVittie A, Cole L, Wreford A, Sgobbi A, Yordi B (2018) Ecosystem-based solutions for disaster risk reduction: Lessons from European applications of ecosystem-based adaptation measures. International journal of disaster risk reduction 32:42–54

Melkonyan A (2014) Environmental and socio-economic vulnerability of agricultural sector in Armenia. Sci Total Environ 488:333–342

Mochizuki J, Schinko T, Hochrainer-Stigler S (2018) Mainstreaming of climate extreme risk into fiscal and budgetary planning: application of stochastic debt and disaster fund analysis in Austria. Regional Environ Change 18(7):2161–2172

Nakicenovic N, Alcamo J, Grubler A, Riahi K, Roehrl RA, Rogner HH, Victor N (2000) Special report on emissions scenarios (SRES), a special report of Working Group III of the intergovernmental panel on climate change. Cambridge University Press

Nakicenovic N, Lempert RJ, Janetos AC (2014) A framework for the development of new socio-economic scenarios for climate change research: introductory essay. Climatic Change 122(3):351–361

Neumann B, Vafeidis AT, Zimmermann J, Nicholls RJ (2015) Future coastal population growth and exposure to sea-level rise and coastal flooding-a global assessment. PLoS ONE 10(3):e0118571

Olsson P, Gunderson L, Carpenter S, Ryan P, Lebel L, Folke C, Holling CS (2006) Shooting the rapids: navigating transitions to adaptive governance of social-ecological systems. Ecol Soc 11(1)

O'Neill BC, Kriegler E, Riahi K, Ebi KL, Hallegatte S, Carter TR, van Vuuren DP (2014) A new scenario framework for climate change research: the concept of shared socioeconomic pathways. Climatic change 122(3):387–400

Opitz-Stapleton S, MacClune K (2012) Chapter 11 Scientific and Social Uncertainties in Climate Change: The Hindu Kush-Himalaya in Regional Perspective. In Climate Change Modeling for Local Adaptation in The Hindu Kush-Himalayan Region. Emerald Group Publishing Limited, pp 207–237

Pahl-Wostl C (2007) Transitions towards adaptive management of water facing climate and global change. Water Resour Manage 21(1):49–62

Park SE, Marshall NA, Jakku E, Dowd AM, Howden SM, Mendham E, Fleming A (2012) Informing adaptation responses to climate change through theories of transformation. Global Environmental Change 22(1):115–126

Patt AG, Tadross M, Nussbaumer P, Asante K, Metzger M, Rafael J, Brundrit G (2010) Estimating least-developed countries' vulnerability to climate-related extreme events over the next 50 years. Proc Natl Acad Sci 107(4):1333–1337

Preston BL (2013) Local path dependence of U.S. socioeconomic exposure to climate extremes and the vulnerability commitment. Global Environmental Change 23 (4):719–732

Republic of Indonesia. (nd). The First National Communication: under the UNFCCC. https://unfccc.int/sites/default/files/resource/Indonesia%20INC.pdf. Accessed 1 May 2019

Republic of Indonesia (2010) Second National Communication: under the UNFCCC. November, 2010. Jakarta. https://unfccc.int/files/national_reports/non-annex_i_natcom/submitted_natcom/application/pdf/indonesia_snc.pdf. Accessed 1 May 2019

Republic of Indonesia (2016) First Nationally Determined Contribution. https://www4.unfccc.int/sites/ndcstaging/PublishedDocuments/Indonesia%20First/First%20NDC%20Indonesia_submitted%20to%20UNFCCC%20Set_November%20%202016.pdf. Accessed 1 May 2019

Republic of Indonesia (2017) Third National Communication: under the UNFCCC. Jakarta. https://unfccc.int/sites/default/files/resource/8360571_Indonesia-NC3-2-Third%20National%20Communication%20-%20Indonesia%20-%20editorial%20refinement%2013022018.pdf. Accessed 1 May 2019
Shove E (2010) Beyond the ABC: climate change policy and theories of social change. Environ Plan A 42(6):1273–1285
Smit B, Wandel J (2006) Adaptation, adaptive capacity and vulnerability. Global Environ Change 16(3):282–292
Tompkins E, Adger WN (2004) Does adaptive management of natural resources enhance resilience to climate change? Ecology and society 9(2)
Triyanti A, Walz Y, Marfai MA, Renaud F, Djalante R (2017) Ecosystem-based disaster risk reduction in Indonesia: unfolding challenges and opportunities. In: Disaster Risk Reduction in Indonesia, Springer, Cham, pp 445–467
UNDRR (2019) Cyclone Idai highlights urgent need to ensure resilience of infrastructure to extreme weather. https://www.undrr.org/news/cyclone-idai-highlights-urgent-need-ensure-resilience-infrastructure-extreme-weather. Accessed 6 Mar 2019
UNFCCC (2018a) Proposal by the president. https://unfccc.int/sites/default/files/resource/Informal%20Compilation_proposal%20by%20the%20President_rev.pdf. Accessed 7 January 2019.
UNFCCC (2018b) Statement by Indonesia at the joint closing plenary of cop24, cmp14, cma1 2018. https://www4.unfccc.int/sites/SubmissionsStaging/Documents/201812190916---INDONESIA%20Fin%202_Closing%20Joint%20Plenary%2014Dec.docx_rev3.pdf. Accessed 10 Apr 2019.
UNFCCC (2015) Adoption of the Paris agreement. Draft decision. Retrieved from https://unfccc.int/resource/docs/2015/cop21/eng/l09r01.pdf. Accessed 8 Feb 2019
UNGA (2015) Resolution adopted by the General Assembly on 25 September 2015: 70/1. Transforming our world: the 2030 Agenda for Sustainable Development. https://www.un.org/ga/search/view_doc.asp?symbol=A/RES/70/1&Lang=E. Accessed 5 Mar 2019
UNISDR (2015) Sendai Framework for Disaster Risk Reduction 2015–2030. UNISDR, Geneva
Van Vuuren DP, Edmonds J, Kainuma M, Riahi K, Thomson A, Hibbard K, Masui T (2011) The representative concentration pathways: an overview. Climatic change 109(1–2):5
van Vuuren DP, Kriegler E, O'Neill BC, Ebi KL, Riahi K, Carter TR, Edmonds J, Hallegatte S, Kram T, Mathur R, Winkler H (2013) A new scenario framework for Climate Change Research: scenario matrix architecture. Climatic Change, Special Issue, Nakicenovic N, Lempert R, Janetos A (eds) A Framework for the Development of New Socioeconomic Scenarios for Climate Change Research
Voss M (2008) The vulnerable can't speak. An integrative vulnerability approach to disaster and climate change research. Behemoth-A J Civilisation 1(3), 39–56
Walker B, Holling CS, Carpenter, S, Kinzig A (2004) Resilience, adaptability and transformability in social–ecological systems. Ecol Soc 9(2)
Wilby RL, Troni J, Biot Y, Tedd L, Hewitson BC, Smith DM, Sutton RT (2009) A review of climate risk information for adaptation and development planning. Int J Climatology: J Royal Meteorological Soc 29(9):1193–1215
Wilson C, Chatterton T (2011) Multiple models to inform climate change policy: a pragmatic response to the 'beyond the ABC'debate. Environment and Planning A 43(12):2781–2787
World Bank (2011) Vulnerability, risk reduction, and adaptation to climate change: Indonesia. April 2011. https://climateknowledgeportal.worldbank.org/sites/default/files/2018-10/wb_gfdrr_climate_change_country_profile_for_IDN.pdf. Accessed 10 Mar 2019
World Bank (nd) Country Profile Indonesia. Climate Change Knowledge Portal. https://climateknowledgeportal.worldbank.org/country/indonesia. Accessed 12 Mar 2019
World Meteorological Organization (2018) WMO Statement on the State of the Global Climate in 2018. World Meteorological Organization (WMO)

Chapter 3
Sea-Level Rise (SLR) and Its Implication on Human Security and Human Rights in Indonesia: A Legal Analysis

Laely Nurhidayah

Abstract Sea-level rise impacts have threatened Indonesia's environment, economy, social and cultural aspects. As an archipelagic state, Indonesia is extremely vulnerable to climate change impact especially in coastal areas. Highly populated and low-lying delta areas such as Jakarta, Semarang and Demak are extremely vulnerable to sea-level rise (SLR). Climate change has exacerbated the existing vulnerability such as poverty, threatens the food security of people living in coastal areas. Local poor coastal communities are the most vulnerable and suffer the most from the effect of flooding because of coastal inundation. Environmental migration is also an inevitable consequences. Climate change has put both human security and human rights such as the rights to healthy environment at risk. This chapter aims to investigate the impact of SLR on human security and human rights in Indonesia. Legal, institutional, and policy response and strategies of central and local government in addressing SLR will be discussed. The discussion to linkage between climate change impact on human security and human rights in Indonesia is lacking. Therefore, this chapter provides new insight on how SLR as a threat to human security and human rights perspective. Human security is an approach in identifying and addressing widespread and cross-cutting challenges to survival, livelihood, and dignity of the people. Human security and human rights theory will be used to examine the policy response of government in addressing SLR. The article finds that despite it is not explicitly mentioned in the legislation about the implication of SLR as a threat to human security and human rights, in the implementation, Indonesian government put some efforts to reduce the impact of CCA. However, the government needs to improve legal framework to support climate change adaptation as a means of protecting human security and human rights including the adaptive capacity of the poor and improving ecosystem-based adaptation approach. The current legislation is laggard behind in its approach to build adaptive capacity for vulnerable groups. Appropriate measures to improve involvement of local people in climate change adaptation decision-making are needed.

L. Nurhidayah (✉)
Indonesian Institute of Sciences (LIPI), Jakarta, Indonesia
e-mail: lae_ly@yahoo.com

R. Djalante et al. (eds.), *Climate Change Research, Policy and Actions in Indonesia*, Springer Climate, https://doi.org/10.1007/978-3-030-55536-8_3

Keywords Sea-level rise · Human security · Human rights · Climate change · Semarang · Demak

Introduction

Context

Global sea-level rise (SLR) is accelerating around the world and the annual rate could increase more than triple every year by 2100 (UN Climate Change News 2018). This phenomenon is likely to accelerate in the future (Schofield and Freeston 2013). According to the research that was conducted by the University of Colorado, melting ice sheet is responsible for half of the 7 cm of increase observed since 1993 (UN Climate Change News 2018). The Intergovernmental Panel on Climate Change (IPCC) Fifth Assessment Report (AR6) suggested that "By 2100 global mean sea-level rise is projected to be around 0.1 m lower with the global warming of 1.5 °C compared to 2 °C (IPCC 2018). Sea-level rise will continue to rise well beyond 2100, the magnitude and rate of this rise depends on the future emission pathways" (IPCC 2018).

Indonesia as an archipelagic country is very vulnerable to the impact of SLR. Based on analysis, the sea-level rise rate in Indonesia is ranging from 0.2 cm/year to 1 cm/year with an average of rate 0.6 cm/year (ICCSR 2010). In addition, land subsidence will exacerbate the impact of sea-level rise. These impacts have been observed in several coastal cities in Indonesia such as Jakarta, Semarang, Pekalongan, and Demak. These impacts include loss of land and property because their land has been permanently inundated, increased cost of rehabilitation of damaged house, loss of their economy and livelihood, increased disaster illness, reduced the quality of life people affected by SLR, loss and damage of infrastructure, loss of social and cultural significance and forced environmental migration.

Sea-level rise is slow and has less dramatic impacts on affected communities not like earthquake and tsunami. For example, it takes 20 years for local communities to lose their land due to SLR in Semarang and Demak. Despite this slow impact, SLR might have deeper impact on peace and security. Prof Walter Kaelin a humanitarian stated that SLR will likely weaken or even jeopardize the statehood of affected countries and thus undermine the very foundation of the present international order (Security Implications of Climate Change: Sea Level Rise—Presentation by Prof. Walter Kaelin to the Security Council 2016). The IPCC suggested that "without adaptation a hundred million people will be affected by coastal flooding and will be displaced due to land loss by year 2100" (Wong et al. 2014).

Rationale

The rationale of this chapter is that there is a growing concern of sea-level rise that affected the human security and human rights of Indonesia people. The impacts are undermining people's rights to life, security, food water, health, shelter, and culture. This chapter will investigate whether the emerging concept of human security and human rights can provide a useful framework for identifying and analyzing the risks and crafting response to those risks. However, there is some limitation to use human security as an approach as this situation is shaped by political choices of the States. In addition, there is a gap by using human rights approach, such human rights violation could never be truly remedied in courts of law.

Gaps

While there is growing concern on the impact of climate change on human security and human rights at current literature and studies. There is limitation of existing international framework in responding to the needs of those displaced by climate change. Similarly at national level, little protection has been given by domestic law to the people affected and displaced by climate change such as the impact of SLR.

It is argued that there should be more protection given by domestic law to the vulnerable people affected by the impact of climate change. These vulnerable groups include poor people, people living in disaster-prone areas.

Aim and Objectives of the Chapter

This chapter aims several objectives: first is to investigate SLR and its impacts on human security and human rights in Indonesia. In addition, this chapter will also investigate challenges in responding to the impact of climate change using human security and human rights approach. Environmental changes including climate change present a new threat to human security (United Nations University—Institute for Environment and Human Security 2008). Many coastal cities are located in areas prone to SLR.

Structure of the Chapter

The chapter is divided into several sections: after the introduction, the second section examines the general literature review on adaptation in coastal areas from human security and human rights perspectives and its implications. The third section analyzes Indonesia's coastal adaptation legal framework from human security and human rights perspective. The fourth section is a case study of SLR impact on human security and human rights in Semarang and Demak, and the fifth and sixth on CCA policy options and human security and human rights implications and the way forward.

This chapter aims at several objectives: first is to investigate SLR and its impacts on human security and human rights in Indonesia. In addition, this chapter will also investigate challenges in responding to the impact of climate change using human security and human rights approach. Environmental changes including climate change present a new threat to human security (United Nations University—Institute for Environment and Human Security 2008). Many coastal cities are located in areas prone to SLR.

Methodology

The research uses qualitative research method. Literature review and empirical study are both employed as methodologies for this research. Empirical research through site observation, in-depth interview with government officials, and local communities were conducted. Two villages in Demak were chosen. Bedono and Sri Wulan were chosen as case studies as these villages are the worst affected villages by SLR and land subsidence. Many of the villagers have migrated and as a result their land has been permanently inundated.

Adaptation in Coastal Areas: Human Security and Human Rights Implications (General Literature Review)

Human Security

Security is a multifaceted concept. Human security is now acknowledged as a legitimate governance issue within the international climate regime (Mason 2015). However, this has not yet led to any formal policy decision by the UNFCCC to protect and enhance human security in the context of climate change (Mason 2015). The concept of human security was popularized in 1994 in the UNDP Human Development Report (UNDP 2018). UNDP has put forward the concept of human security to assist in the framing of development and equity issue. The 1994 HDR highlighted two major components of human security: Freedom from fear and freedom from want. The 1994 HDR further listing seven essential dimensions of human security include economic, health, personal, political, food, environmental, and community. Environmental change has been increasingly understood as a security issue (Barnett 2001) Environmental change nowadays has been associated with the impact of climate change. Security is accentuated with a discourse on vulnerability. To some nations sea-level rise has been considered as security issue such as in small islands developing states (SIDS). The small island will be threatened of its existence and sovereignty in the long term by SLR. The fear of whole islands drowned by SLR has raised the concern of national security.

Freedom from Fear, Freedom from Want, and Freedom of Dignity

The UN Commission on Human Security (CHS) stated that human security is to protect the vital core of human lives in ways that enhance human freedoms and human fulfillment. Human security integrates three freedoms: freedom from fear, freedom from want, and freedom of dignity. Freedom from fear refers to protecting individuals from threats directed at their security and physical integrity and includes various forms. Freedom from wants refers to the protection of individuals so they might satisfy their basic needs and their economic, social, and environmental aspects of life and livelihood. And freedom of dignity refers to the promotion of improved quality of life and enhancement of human welfare that permit people to make choices and seek opportunities that empower them. Sea-level rise has threatened the fulfillment of freedom of fear, freedom from want, and freedom of dignity as everyday people fear that their house is flooded and they suffer the loss of productive land because of permanent inundation (Table 3.1).

Indeed, human security has evolved and is being transformed from traditional one which focuses more on state and the obligation of states to protect its citizen against war or violence to non-traditional threats including environmental, social and, economic threats. In the broadest sense, human security encompasses human rights, good governance, access to education, health, freedom from fear, reducing poverty, and achieving economic growth.

IPCC AR 5 has a special report on human security in Chap. 12. IPCC has suggested in high agreement that human security will be progressively threatened by climate change. Climate change will have significant impacts on forms of migration. Furthermore, IPCC stated that mobility is a widely used strategy to maintain livelihood in response to social and environmental changes.

Table 3.1 Types of security. Adapted from UNDP' Human Development Report (1994)

Economic security	Persistent poverty and unemployment
Food security	Hunger, famine
Health security	Infectious diseases, unsafe food, and water
Environmental security	Environmental degradation, resources depletion, lack of access to water, natural disaster, pollution

Threat to Health and Food Security

Coastal inundation, storm surge as a result of SLR followed by land subsidence has increased the risk of health including the increasing of outbreak of dengue fever, tuberculosis as well as physiological stresses due to extreme climate events. Climate change will influence yields and sea-level rise can inundate producing land. According to a study by Chen et al., combination of crops yield and sea-level effects causes significant reduction in production and an increase in rice prices which may have important policy implications for food security (Chen et al. 2012). In addition, the study suggested that rice production is sensitive to sea-level rise resulting in annual welfare losses up to US$10.59 billion. The more significant negative impacts fall in Bangladesh, Japan, Taiwan, Myanmar, Egypt, and Vietnam (Chen et al. 2012).

Threats to Livelihoods and Culture

Sea-level rise (SLR) has negative impacts on local communities' livelihoods and cultures. People are losing their agricultural lands because of SLR and are now shifting their livelihoods from being farmers to become fishermen or as working laborers in industries as they do not have any productive lands anymore. In addition, SLR is also a threat to transmission of people's culture. There are irreversible and devastating impacts on climate change on cultures, land, and livelihoods. SLR is washing out the significance of people's homes and cultural significance such as graveyards.

Human Rights

State Duties to Protect and Fulfill Facilitate the Enjoyment of Basic Human Rights

Universal Declaration of Human Rights and the International Covenant on Economic, Social and Cultural Rights have recognized basic human rights. States have the responsibility and duty to protect, promote, and implement all human rights and fundamental basic human rights by adopting steps which are necessary to create all conditions in the social, economic, as well as legal guarantee required to ensure that all persons under its jurisdiction are able to enjoy all those rights and freedom.

Climate change threatens human ability to achieve sustainable development and human survival. States have an obligation to respect, protect, fulfill, and promote all human rights for all persons without discrimination. Failure to take affirmative measures to prevent human rights from harms caused by climate change, including foreseeable long-term harms, breaches this obligation.

In the Resolution of UN No 10/4 2009, the council noted that climate change impacts have a range of implications both direct and indirect for the effective enjoyment of human rights. And that such an effect will be felt most acutely by those segments of population who are already in vulnerable situations. Flavia Pansieri United Nations Deputy High Commissioner for Human Rights, stated that "climate change, human induced climate change is obviously an assault of the ecosystem that we all share has the added feature of undercutting rights, important rights like the right to health, the right to food, to water and sanitation, to adequate housing, and, in a number of small island States and coastal communities, the very right to self-determination and existence".

The Linkage of Human Security and Human Right and CCA

Human security put people and communities at the center of analysis and response strategies. They are both agents and beneficiaries of interventions. Human security has two strategies of actions: protection and empowerment. Protection which is defined by the Commission on Human Security as "Strategies are set up by States, international agencies, NGOs and private sectors to shield people from menace". It implies establishing top-down measures recognizing that people face threat that are beyond their control (natural disasters, financial crises, and conflicts). Human security requires systematic, comprehensive, and preventive protection. States have the primary responsibility to implement such protection while other actors such as international bodies, civil society, and NGOs play a pivotal role. Empowerment is the strategy that enables people to develop their resilience to a difficult situation and implies bottom-up measures that aim to develop the capacity of individuals and communities. The implementation of human security approach calls for the protection and empowerment of population implying the need for both top-down and bottom-up measures. Security providers range from self-protection (individual, family, community), horizontal protection by non-state actors (lifeline, humanitarian organization), and vertical protection by state, foreign state, and intergovernmental actors.

There is a linkage between human security and the Sendai Framework of DRR/CCA. Sendai Framework for action is in line with the protection of human security and human rights, namely preventing new and reducing existing disaster risk through the implementation of integrated and inclusive economy; structural, legal, social, health, cultural, educational, environmental, technological, political and institutional measures that prevent and reduce hazard exposure and vulnerability to disaster; increase preparedness for response and recovery, and thus strengthen resilience.

Appropriate legal and policy frameworks and operational measures are needed to help people remain in situ if possible and desirable or move elsewhere in anticipation of the harm (McAdam et al. 2016). In addition, comprehensive approaches for disaster risk reduction that simultaneously responds to economic, food, health,

Fig. 3.1 Relation between human rights, climate change adaptation, and human security

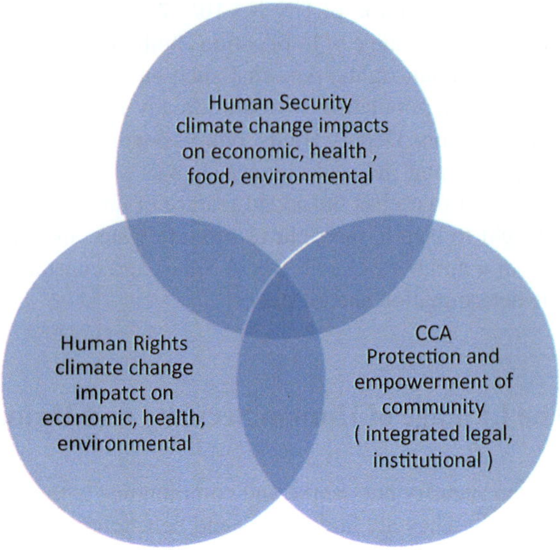

and environmental insecurity of individuals and communities in multiple contexts are needed.

Integrating human rights in climate action will necessitate higher levels of ambitions and improve mitigation and adaptation strategies by making them more effective and inclusive. A human rights based approach should be integrated into any climate change adaptation or mitigation measures, such as the promotion of alternative energy sources, diversification of income-generating activities in agriculture and fishing system, forest conservation or tree-planting projects, resettlement schemes, and others (Fig. 3.1). Despite the clear human rights implications of failure to act to prevent climate change, the international community has not taken adequate preventive action.

An Analyze of Indonesia's Coastal Adaptation Legal Framework from Human Security and Human Rights Perspective

Indonesia does not have a definition of national security which is similar to the concept of human security. The concept of national security is more on territorial integrity. While the issue of the impact of climate change on national security has been aware by academics and policy-makers in Indonesia. It has not been reflected in the national legislation. Despite the laggard behind legislation to respond to the impact of climate change, the government focusses on mainstreaming the climate change adaptation into sectoral development planning. Bappenas, for example, has established Indonesia's national action plan on climate change adaptation (RAN-MAPI) with the strategy for improving national security on food security, energy, health, shelter, infrastructure, ecosystem services, urban, and small islands. Several legislations which are related to coastal adaptation from human security and human rights perspective include.

Law no 39/1999 on Human Rights

Climate change impact affects and sets to undermine many basic human rights undercutting the rights to health, food, shelter, culture, develop, security, safety, and dignity for freedom to pursue their development. This adverse effect of climate change is most acutely felt by those segments of population that are already in vulnerable situation owing to factors such as geography, poverty, gender, age, and disability (McAdam et al. 2016). The basic recognition of human rights has been recognized in Law No 39/1999. Under this legislation several basic human rights are recognized such as article 9 Law No 39/1999 stated that everyone has the right to life, well-being, and adequate and healthy environment. The right to life according to International Covenant on Civil and Political Rights (ICCPR) is described as a supreme right which cannot be derogated from, even in time of public emergency. Protection of the right to life is closely linked to other rights such as the right to adequate food, adequate health, adequate water, and adequate housing. Article 28–35 stated on the right to security. Article 30 stated everyone has the right to security and protection against the threat of fear from any act or omission.

Despite this recognition, little has been linked between human rights and climate change impact in Indonesia's legislation particularly the impact of SLR on human rights in Indonesia. Human rights issue in Indonesia is currently most associated with the case of human rights against violence.

Law on 24/2007 Disaster Management

To respond to the disaster, Indonesia has enacted disaster management law. Law No 24/2007 is the forefront of the legal basis for disaster management in Indonesia. It covers all disasters including tsunamis, earthquakes, volcanic eruptions, flooding, drought, storms, landslides, and other natural and man-made disasters. The legislation uses a proactive approach to disaster risk reduction (DRR) and recognizes that people have a basic right to protection from disaster, particularly groups who are vulnerable to disaster (Djalante 2012). Even though the legislation does not explicitly make any reference to climate change impacts and adaptation measures, It is an important legislation which deals with disaster risk reduction in Indonesia and covers the full disaster cycle of mitigation, emergency response, compensation, and rebuilding. Likewise, in terms of disaster risk reduction, other legislation only provides protection and accommodation options such as spatial planning implementation and building coastal dyke or sea wall; Law No 24/2007 provides retreat options in DRR. Article 32 states that the government can decide that disaster-prone areas are restricted for settlement and/or remove the land title or the ownership of the land and pay compensation to the people that land title being removed from.

Despite Law No 24/2007 recognized flood as disaster, the drawback of this legislation is coastal flooding due to SLRwhich is not categorized as disaster. The Law only recognizes flooding as a result of heavy rain from the land. According to Government Regulation No. 64/2010 on Disaster Mitigation in Coastal Areas and Small Island, disaster needs to be sudden and creating shock and damage at certain times. While coastal flooding is occurring daily. Therefore, there is no effective response from BNPB or BPBD to this kind of disaster as resulting of SLR.

In addition, there is no effective response to the forced migration as a result of SLR. As the legislation only covers rehabilitation and reconstruction in terms of post-disaster in the area of disaster and does not cover the permanent mobility of the people from the disaster areas to other safe places. Therefore, there should be amendments to national disaster management Law No 24/2007 to include a reference to climate change and outline possible responses to climate change impacts.

Law of the Republic of Indonesia, No 32/2009: Environment Protection and Management

Law No 32/2009 is an important legislation related to ecosystem-based approach to adaptation for climate change in coastal areas. Recognition of the concern of climate change impact has been stated in Law No 32/2009 on Environmental Management. Article 21 (4) stated that the criteria of environmental damage due to climate change impact is based on increasing of temperature, sea-level rise, storm, and drought. Law No 32/2009 is adopting protection of soft structures within mangrove forests. Mangrove forests could play an important role in protecting coastal areas from sea-level rise. According to a study conducted by the University of Southampton, mangroves have the ability to create a buffer zone between the sea and the land (Southampton 2015). The protection of mangroves has been regulated in several other legislations including Law No 5/1990 on Conservation of Living Resources and their Ecosystem and Law No 41/1999 on Forestry Law. However, degradation of mangrove forests in some areas cannot be avoided as they are converted into shrimp farming or other developments.

Law on 32/2014 on Local Government

Local government plays an important role in climate change adaptation as they are the closest to the community to adequate adaptation measures. However, regarding addressing SLR, there is uncertainty and conflicts on coastal management. The conflict and uncertainty are increasing with the new regional law No 34/2014. According to this law, district and municipal level do not have any authority anymore to manage coastal areas. It is stated in article 14 (1) Law No 34/2014 that the authority and governance of forestry, marine, and energy is divided between central government level and provincial level. Previously, regency and municipal level has the authority to manage 0–4 mile and provincial level 4–12 mile. With the new law, currently, provincial level manages 0–12 mile of coastal areas. In the implementation, it is creating conflicts and uncertainty particularly because there is a significant change on coastal line due to erosion and abrasion and SLR. The provincial level claims that the permanently inundated land should be managed by provincial level Marine Affairs and Fisheries Agency, while the municipal level argued that this land should remain under their authority based on the previous map. In addition, there is uncertainty whether regency/district level can do mangrove planting program as now this area has become the responsibility of provincial level. Therefore, there is no effective response from the local municipal and district level as they wait for the initiative from provincial and central government levels due to the changing sectoral approach which becomes a major challenge in managing the coastal areas.

Case Study of SLR Impact on Human Security and Human Rights in Demak

Semarang and Demak have suffered serious problems with SLR and land subsidence for 20 years ago. Two case studies in Bedono village and Sri Wulan are highlighted to illustrate how climate change impacts human security and human rights. The adaptation conducted by local people is through mobility to a safe place, or some people are returning to their village as they have some effort from NGOs to plant the mangroves and use this mangrove as tourism to gain local community livelihood.

Bodono, Demak Central Java

Sea-level rise has caused two sub-districts to be permanently inundated in 1999. These two districts include Rejosari and Tambak Sari. According to the community, previously, Bedono is a welfare village, they produce rice and product from aquaculture. However, due to most of the agriculture land has been permanently inundated now the community finds other sources of livelihoods. People have been forcedly migrated due to this condition. In Tambak Sari, around 77 households and 200 households in Rejosari were migrated to the nearest areas. While some of the people from Rejosari were migrated to Purwosari and people from Rejosari were migrated to near villages in Gemulak and Sido Gemah (Nurochman 2018). They have occupied the irrigation land near the bank of the river managed by public work agency. Previously, all 2 sub-districts people who have been affected by SLR will be migrated by the government to outside, Java-like transmigration. However, before Day 1 of the transmigration program, people were rejected to be moved to Sumatra (Nurochman 2018). They preferred to move in near their village. However, the problems with new settlements in the bank of the river is that people are moved temporarily as it is government land for irrigation (Nurochman 2018). It should not be occupied as it serves as flood plain area. However, because there are no other options people live there without permanent housing and no electricity. It took long for the government to assist and respond to people's demonstrations to address the issue of environmental migration (Nurochman 2018). This is due to the current legislation where there is no procedure on how to address those issues. Therefore, there is uncertainty from the government to respond promptly to this problem. Appropriate legal and policy framework and operational measures are needed to help people move elsewhere in anticipation of harm, and to be protected and assisted if they are displaced. Below is the map of Bedono village. It can be seen that some of the areas have been inundated (Fig. 3.2).

The current adaptation options are conducted by NGOs such as Wetland International which assists the community to manage mangrove and make this mangrove as one of the tourist destination and source of community livelihood (Nurochman 2018). Other NGOs such as OISCA also help and assist the community to plant mangroves in the sub-districts that have been permanently inundated (Nurochman 2018). Currently, these mangroves are used as tourism areas. Local people who do not own boat will hire the boat to see the whole areas of mangroves that have already grown and see the previous sub-districts that have been permanently inundated and have become sea. The biodiversity of the areas now is back with the mangrove plantation such as birds and other local animals (Nurochman 2018).

Sri Wulan

Similar to Bedono, Sri Wulan village has also suffered from permanent coastal inundation due to sea-level rise. Four-hundred hectares of village areas have been permanently inundated, and most of the permanent inundated land is their farming land for agriculture (Hasan 2018). Therefore, most of the people now are working as laborers in industries in Semarang rather than farming or aquaculture (Hasan 2018). The livelihood has been shifted. According to the interview, the increase of SLR in Sri Wulan was as a result of reclamation that has been conducted by the government in Terboyo (Hasan 2018), Tanjung Mas Port and Airport in Semarang. Compared to other villages only in Sri Wulan there has not been many efforts that has been

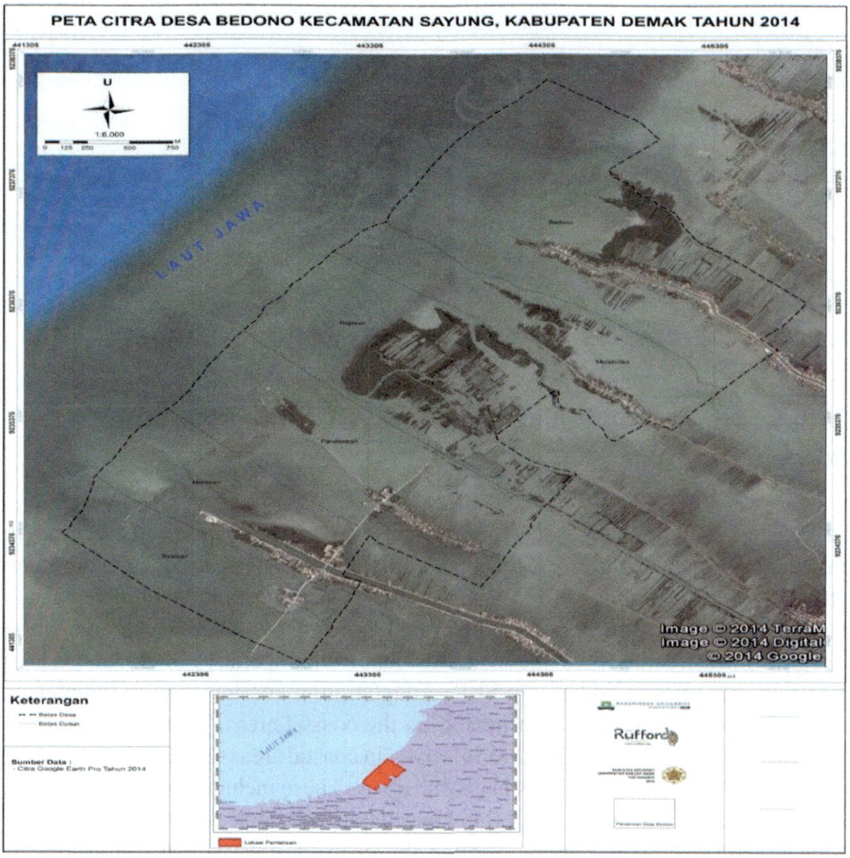

Fig. 3.2 Map of Bedono village

taken either from government and NGOs or industry to assist people in helping with reducing the risk of the impact of SLR. Hybrid engineering was not successful as the current is too strong, and there is not yet a sea wall and mangrove is also damaged and cannot grow due to strong sea current and waves (Hasan 2018). Below is the map of Sri Wulan village where half of the land for their farming and aquaculture has been permanently inundated. The only remaining land is their housing (Fig. 3.3).

The adaptation conducted by the community is finding livelihood in industries in Semarang. As there is no productive land in the village they choose to work as laborers in industries near their village. The problem of land subsidence and coastal inundation has made some people to move out from this village to Semarang or other villages.

Fig. 3.3 Map of Sri Wulan village

Climate Adaptation Policy to Address SLR in Semarang and Demak

The current climate change adaptation regime in Indonesia is more sectoral in approach. The increasing threat of climate change on human security and human rights in Indonesia requires the Indonesian government to improve climate change adaptation measures and needs an integrated and holistic approach to adaptation strategies. Infrastructure robustness and resilience are imperative in climate change adaptation measures in addressing SLR. Adaptation in coastal areas is complex as most of the population have been inhabiting the coastal areas. There is no uniform approach to climate change adaptation globally in coastal areas (Nicholas et al. 2007). There are three adaptation options in coastal areas. These include protection, accommodation, and retreat. All three options have environmental, economic, and social implications. The current adaptation option chosen by the government to address SLR in Semarang and Demak is building infrastructure. The Ministry of Public Works will establish the integration of sea wall and toll road from Semarang and Demak for 26 km length (PUPR Ministry 2018, Nurhidayah and McIlgorm 2019) (Fig. 3.4).

This sea wall is expected to protect coastal communities from sea-level rise impact and coastal flooding. However, some areas in Demak are excluded from the sea wall and projected to be suffering a lot from damage due to erosion because of the changes in the sea current, and due to this infrastructure some areas have been permanently inundated. Ideally, not only Semarang will be protected by the sea wall but also Demak.

Indeed, legislation and policy-makers have been slow to identify potential responses to address the impacts of sea-level rise. The establishment of infrastructures such as the sea wall is taking time and is expensive. With the length of coastal areas in Indonesia, it is costly to protect the whole of coastal line with sea wall. In addition, hard structures are likely to have an impact on local marine water circulations and ecosystems. Resettlement of populations can create major problems as the resettled people have difficulties in adjusting to their new environment and in finding new employment. Resettlement to the new location should remain as the last resort of the options to climate change adaptation.

Fig. 3.4 The proposed sea wall and toll road to be built in the Semarang and Demak areas. *Source* BBWS (2018)

Poor coastal communities for a long time has been suffering and are being marginalized by big businesses and industries. The privatization of coastal resources and their monopolization by business interests are a result of Law No 27/2007 on Coastal Area and Small Islands Management. The privatization (HP3) provisions have been revoked by Law No 1/2014 due to the protest of the fishermen community and suit to the Constitutional Court (Fig. 3.4).

Legislation plays a significant role in providing a basis for promoting climate change adaptation, improving social justice, and adaptive capacity. The existing legal framework on climate change adaptation tends to be more focused on structural measures, such as a hazard reduction, rather than non-structural measures, such as promoting resilience (Nurhidayah and McIlgorm 2019). For example, in addressing sea-level rise the government chooses to have a disaster risk reduction approach focusing on building infrastructure to reduce sea-level rise, as opposed to an ecological resilience based approach such as maintaining coastal mangrove forests (Nurhidayah and McIlgorm 2019). Mangrove forest has been clear to make way for other development either for port or aquaculture purposes.

Human security and Human Right Perspective on CCA: The Way Forward?

Despite sea-level rise has been considered as the threat of human security and human rights, slow response from the government has deteriorated the quality life of people who have suffered from the impact of SLR. While local community affected by SLR may claim from the government based on the violation of human rights to get remedies for the damages they suffer. The claim for human rights violation is difficult particularly if the community has illegally settled in the disaster-prone areas. So far the claim of the local community is more based on social and environmental violations conducted by the decision of the government. There is no such claim brought by local communities in Semarang and Demak as a result of government decision-making for infrastructure for CCA. However, there is a case in Jakarta. The Kiara on behalf of local fishermen brought the case to the court to stop the reclamation in the case of reclamation in Jakarta Bay., while reclamation has been considered as an integral part of giant sea wall project as adaptation options of Jakarta from the sinking by SLR. There was a rejection from small-scale fishers who live near Jakarta Bay. The small-scale fishers are concerned that their livelihood will be affected by the projects including the fishers in North Jakarta (Angke Bay, Marunda, and Cilincing) (Kiara 2014). Kiara claims that the project has not conducted an Environmental Impact Assessment (EIA). Kiara, a fisheries NGO estimated that 16,855 fishermen will be displaced by these projects (Elyda 2015). While this ambitious project will probably protect all the citizen's of Jakarta from SLR in long term; displacement of small-scale fisherman near the Jakarta Bay appears to be a short-term cost and raises serious valuation issues and social justice of climate change adaptation actions (Nurhidayah and McIlgorm 2019). As the result of this rejection of giant sea wall from the public, currently, the government of Jakarta proposes a new design to change the giant sea wall.

An integrated approach is needed to address the issue. ICZM which is incorporated into territorial planning could be an opportunity to minimize impacts and generate integration (Sierra-Correa and Kintz 2013). However, integration between sea zoning and territorial zoning is difficult to be implemented as a sectoral approach as sectoral ego between provincial level and municipal level are difficult to be resolved. For example, there is tension between provincial level and municipal level on the permanently inundated land whether it should become sea zoning and become the authority of provincial level or municipal level, which insists that the permanently inundated land is still subject to territorial planning. In addition, an integrated coastal zone management is difficult to be implemented in Indonesia. An adaptation approach particularly with the establishment of mega infrastructure is a more centralistic approach, which is the most notable feature of climate change adaptation particularly in terms of developing infrastructures in preventing coastal inundation. The centralistic approach is taken as there is a lack of local government funding. Most of the funding for these infrastructures is either loan or PPP (public–private partnership) and a small number of national and regional budget. While the centralistic approach is not bad at all; however, the centralistic approach is sometimes ignoring the voice and vulnerability of coastal communities at the local level.

Adaptation is a very important element for climate change and human security and human rights nexus. Uncertainty and risk of climate change are relevant to be considered in adaptation laws and policy. Therefore, it is essential that law and policy are incorporated for a long term by policy-makers and regulators to respond and adapt to climate change (Cassotta et al. 2016). As the way forward, amendment of Disaster Management Law, particularly, to include coastal flooding as the impact of SLR on the definition of flood or the establishment of a special law on climate change adaptation is needed. Currently, Indonesian law is not keeping up and laggard behind with the need for adaptation. There are key principles that need to be included to change the law and remove the barriers, to promote adaptation strategies, for example, Law No 32/2014 is creating uncertainty and conflict on the management of coastal areas between provincial level and municipal level; as a result, effective adaptation action is slow in response and not effective; to increase regulatory coordination among government bodies; to engage with more robust decision-making, for example, in SLR beside opting for expensive investment dikes and sea wall to hold back the sea. The government also needs to construct sewage treatment plant, drinking water treatment plant, hazardous waste treatment, and disposal facilities; and to take decisions whether to allow or not allow new residential and industrial settlement along disaster-prone areas and to sum any decision regarding long-term investment for climate change adaptation (Cassotta et al. 2016). IPCC has acknowledged the value of mix strategies including adaptation, technological development, and research on climate change adaptation (Cassotta et al. 2016).

From the case studies in Bedono Village and Sri wulan Village it can be seen that the government needs to address several issues including the protection of citizens against coastal flooding by establishing infrastructure such as sea wall or ecosystem-based approaches such as mangrove planting. In addition, the government needs to address the issue of internal environment forced migration. However, in the implementation, there is no adequate infrastructure and ecosystem-based approach that has been established by the government. Instead, the ecosystem-based approach such as mangrove plantation's much work is assisted by NGOs working with the community. Wetland International and OISCA international, for example, have been working toward mangrove plantation in Bedono Village. In addition, without adequate legal framework, it is difficult to address the issue of environment forced migration. The legal issue includes the certainty of people's property regarding the lost land because of SLR. There is uncertainty whether the community lost their land title because of their land has been permanently inundated. According to Ministry of Agrarian and Spatial Planning circulation letter the permanently inundated land will revert back to government land title.

There is a limitation in application of human security and human rights approach at the national level. The limitation is that both human security and human rights emphasize the safety and well-being of the individuals (Grunau, 2003), which makes the security of individuals secondary to that of the state. However, human security and human rights approach is a global normative framework that can provide a legal basis for the improvement of climate change adaptation conducted by the government. The great strength of human security as a political agenda for global governance seems to lie in the fact that implementing it "requires overcoming the compartmentalization of security, humanitarian, human rights, and development strategies by focusing on the protection and empowerment of people." (Oberleitner 2005).

The way forward, human security and human rights are both people centered. Both understand people to be "ends" and not means. Based on this principle there is an obligation that the state needs to protect and empower their citizen from the impact of climate change. There are several law and policy and issues that need to be improved in addressing SLR as follows:

1. Sectoral legislation hindered the effectiveness of efforts in addressing the problems of SLR and land subsidence. Therefore, it is difficult to achieve Integrated Coastal Zone Management (ICZM).
2. For local government, Law No 32/2014 has created more complexity and tension in the management of coastal areas. According to this law, the municipal and regency level does not have authority to manage coastal areas they previously managed at 0–4 nautical mile. Under this law, this authority is shifted to the provincial level at 0–12 nm. The tension lies particularly in the management of permanently inundated land. The provincial level uses the current coastal line which has changed as a baseline for ocean/coastal zoning plan while the municipal level object to this plan, instead preferring to use the previous coastal line which rejected the claim that permanently inundated land becomes the authority of provincial level.
3. Law No 32/2014 also creates uncertainty in the management of mangrove areas. Municipal level and regency level could not plant and rehabilitate mangrove as the authority has been shifted to the provincial level. Therefore, the municipal level argues that they can now only await and rely on the efforts from provincial and central governments to address the SLR and land subsidence problems.
4. Law No 24/2007 on disaster management does not define coastal flooding as a disaster, as this ongoing event occurs almost every day.
5. There is no special treatment afforded to poor people living in prone-disaster areas in coastal areas and other poor people living in safe areas. They are therefore burdened with the cost of rehabilitation of the damage of housing but could not afford to.
6. There is uncertainty as to the status of the land that has been permanently inundated due to SLR.
7. There is no special law to address necessitated environmental migration. There is uncertainty as to how to address the issue from a local government perspective.
8. The late response of local government efforts in addressing SLR and land subsidence has been filled by the effort of academics, NGOs, and INGOs and the corporate social responsibility (CSR) of industries near the coastal areas. These include mangrove planting and hybrid engineering programs.

Conclusion

Despite global recognition of the implication of climate change on human security and human rights, at the national level the protection and empowerment of local communities against SLR are hindered by overlapping laws, gaps, and laggard behind of legislation at national level to address SLR effectively. Paris Agreement requires all Parties, as appropriate, to engage in adaptation planning and implementation through, e.g., national adaptation plans, vulnerability assessments, monitoring and evaluation, and economic diversification. Improvements to the sectoral laws are also needed to incorporate adaptation laws. Given the uncertainty of climate change risk, the law should be prepared to cater for worst-case scenarios which may eventuate in the future. For example, there should be special laws to address the necessitated environmental migration. Currently, there is uncertainty as to how to address the issue from a local government perspective. The legislation may also need to include a funding mechanism for climate change adaptation. Such a funding mechanism is required for some public and private adaptation initiatives to offset the disproportionate economic burden of disadvantage on individuals or groups (McDonald 2010).

Human security and human rights emphasize on individuals. People are the center of approach in adaptation. The adaptation measures in developing countries such as Indonesia are not only hindered by sectoral and overlapping legislation but also financial resources for large-scale infrastructure, technology, knowledge mismatch, and human resources. The challenges of adaptation measures by the government have been shown in the implementation at the local level. Based on a case study in Bedono and Sri Wulan Villages, most adaptations are conducted by farmers, fishers, coastal dwellers; and residents of large cities will be autonomous and facilitated by their own social capital and resources. The adaptation measures by the government sometimes are taking a long time to be implemented.

Appropriate measures to improve the involvement of local people in climate change adaptation decision-making are needed. Policy options for adaptation are always consulted with the people affected by the decisions. This involvement will give a legitimate adaptation policy that has been accepted by local communities and improved social justice. For example, in the case of forced migration, the response should be informed and accepted by the whole community affected by this migration option. There should be an improvement in other existing laws to include and explicitly address the social justice implications of climate change impacts, through either better public consultation, or new forms of adaptation assessment process.

References

Barnett J (2001) Security and climate change. Tyndall Center for Climate Research

Cassotta S, Hossain K, Ren J, Goodsite ME (2016) Climate change and human securityin a regulatory multileve land multidisciplinary dimension: the caseof the arctic environmental ocean. In: Climate change adaptation, resilience and hazards climate change management, Springer International Publishing Switzerland

Chen CC, McCarl B, Chang CC (2012) Climate change, sea level rise and rice: global. Climate Change

Dedring J (2004) Human security and the UN Security Council. In: Shinoda H & Jeon H-W (eds) Conflict and human security: A search for new approaches of peace-building, Hiroshima: Institute for Peace Science, 45

Grunau (2003) The limits of human security: Canada in east timor 1:1 The Dispatch: Quarterly Review of the Canadian Defence & Foreign Affairs Institute https://d3n8a8pro7vhmx.cloudf

ront.net/cdfai/pages/322/attachments/original/1413011405/The_Limits_of_Human_Security.pdf?1413011405
Hasan, interview by Laely Nurhidayah (2018)
Human Development Report (1994) Oxford University Press, New York. http://hdr.undp.org/sites/default/files/reports/255/hdr_1994_en_complete_nostats.pd
ICCSR (2010) Indonesia climate change sectoral roadmap - ICCSR Scientific basis: Analysis and projection of sea level rise and extreme weather events
IPCC (2018) Global Warming of 1.5 0 C. UNEP
Mason M (2015) Climate change and human security: the international governance architectures. In: Redclift MR, Grasso M, Edward E (eds) Handbook on climate change and human security
McAdam J, Burson B, Kalin W, Weerasinghe S (2016) International law and sealevel rise: forced migration. University of New South Wales Research Series 2016:23
McDonald (2010) Mapping the legal landscape of climate change adaptation. In: Bonyhady T, Macintosh A and McDonald J (eds) Adaptation to climate change law and policy. 36, Federation Press
Nicholas R, Wong PP, Burkett V, Codignotto J (2007) Coastal system and low lying areas. Climate change 2007:impacts, adaptation and vulnerability, contribution of working group II to the fourth assessment report of the intergovernmental panel on climate change. Cambridge University Press 2007
Nurhidayah L, McIlgorm A (2019) Coastal adaptation laws and the social justice of policies to address sea level rise: an Indonesian insight. Ocean Coastal Manage 2019:11–18
Nurochman, interview by Laely Nurhidayah, 8 July 2018
Oberleitner G (2005) Human security: a challenge to international law? Global Governance
PUPR Ministry (2018) Paparan Tol Tanggul Laut: Jalan Tol Semarang Demak. Powerpoint presentation
Schofield C, Freeston D (2013) Options to protect coastlines and secure maritime. In: Gerrard MB and Wannier GE (eds) Threatened island nations legal implications of rising seas and a changing climate, 141–165
Southampton, University of (2015) Mangroves help protect against sea level rise.23 6. https://www.southampton.ac.uk/news/2015/07/mangroves-help-protectagainst-sea-level-rise
Security Implications of Climate Change: Sea Level Rise—Presentation by Prof. Walter Kaelin to the Security Council (2016) https://disasterdisplacement.org/staff-member/presentation-by-prof-walter-kaelin-to-the-security-council-security-implications-of-climate-change-sea-level-rise
Sierra-Correa PC, Kintz JRC (2013) Ecosystem-based adaptation for improving coastal planning. Marine Policy 2013: 385–393
UN Climate change news (2018) Global sea level rise Is accelerating—study. 2:13
UNDP (2018) https://hdr.undp.org/sites/default/files/human_security_guidance_note_r-nhdrs.pdf.
United Nations University (2008) Institute for environment and human security. Human Security, Climate Change
Wong, Losada IJ, Gatusso JP, Hinkel J, Khattabi A, McInnes KL, Saito Y and Sallenger A (2014) Coastal systems and low-lying areas. In: Climate change 2014: Impacts, adaptation, and vulnerability. Part A: Global and sectoral aspects. Contribution of working group II to the fifth assessment report of the intergovernmental panel on climate change. Cambridge University Press

Chapter 4
Integration into Development: Translating International Frameworks into Village-Level Adaptation

Skye Turner-Walker, Esti Anantasari, and Arry Retnowati

Abstract International climate change commitments have progressively prioritised addressing adaptation, particularly under the mechanisms of the United Nations Framework Convention on Climate Change (UNFCCC). In response, climate change adaptive capacity and resilience enhancing activities have been increasingly directed at local community or village scales. The aim of this chapter is to understand how conceptual underpinnings of adaptation are articulated into the programming of adaptation activities implemented at the local village levels under national adaptation planning in Indonesia. The local-level experiences of adaptation programming in implementation offer several insights to guide adaptation approaches and conceptions in informing national adaptation programming implementation at village levels. This chapter, therefore, situates the national adaptation programmes implemented in two case studies of agricultural areas in the Yogyakarta Region of Java, amidst Indonesia's country commitments and priorities on adaptation to village activity-level implementation. The chapter presents the local experiences of farmers in two key adaptation projects implemented at the village level as part of broader nation-wide programming. The results of which detail findings captured post-project implementation and provide examples of the implications of adaptation framing on programmatic outcomes over long-range timeframes. This chapter draws conclusions that significant gaps and disjuncture between the top-down nature of national adaptation programming directives exist in the translation of programming into village activities. In particular, in the ways in which local participants are engaged and have the agency to direct the activities of adaptation programming in a manner that is culturally and contextually appropriate, as well as locally specific, and able to be sustained in the uptake of the programme activities. These findings suggest that

S. Turner-Walker (✉)
Australian National University (ANU), Canberra, Australia
e-mail: skye.turner-walker@anu.edu.au

E. Anantasari · A. Retnowati
Universitas Gadjah Mada (UGM), Yogyakarta, Indonesia
e-mail: esti.anantasari@ugm.ac.id

A. Retnowati
e-mail: arry_retnowati@ugm.ac.id

R. Djalante et al. (eds.), *Climate Change Research, Policy and Actions in Indonesia*, Springer Climate, https://doi.org/10.1007/978-3-030-55536-8_4

limited engagement and meaningful inclusion with farmer communities in collaboratively developing programme adaptation activities in early design is still lacking in practice. When coupled with limited incorporation of farmer innovation, local knowledge and agency as part of the social processes key to long-term adaptation, overall, little impact or adaptive capacity is being fostered in the long-term in these cases of national adaptation activities.

Keywords Climate change adaptation · UNFCCC · Development programming · Village climate activities

Introduction

Internationally, limited attention has been given to the way climate finance within internationally framed climate adaptation programming is translated into interventions explicitly directed at instilling local community climate resilience and adaptive capacity (Taylor 2015). Less attention has been given to incorporating local and indigenous knowledge into formal climate change adaptation strategies, despite the interlinkages of integrating local and indigenous knowledge into science (Hiwasaki et al. 2014). Given that the factors that determine vulnerability tend to be context-specific, designing adaptation interventions to address such factors requires a locally tailored knowledge base (Few et al. 2007). However, much of this type of knowledge is not always in a format that is easily utilised (Schmuck-Widmann 2001) to climate change adaptation planners and planning processes. Particularly, as contrasts are often drawn between scientific epistemologies of climate change and those of local perspectives, regardless of a vast divergence of views (Bohensky et al. 2016). Capacity building remains a problem to which little attention has been paid so far in Indonesia, despite the importance of building local capacities (Yoseph-Paulus and Hindmarsh 2018). Geographically, research into the role and relevance of local context-specific knowledge and processes, both in understanding local applications of climate change and in contributing to adaptation strategies, have largely taken place in the Arctic and the Pacific (e.g. see Armitage et al. 2011; Alexander et al. 2011; Berkes et al. 2007; Cruickshank 2001, 2005; Krupnik and Ray 2007; Kuruppu 2009; Lefale 2010; McNamara and Prasad 2013; Weatherhead et al. 2010). Insufficient attention has been paid to the experiences of local farmers in the rendering of adaptation interventions in Indonesia, that are positioned within international development assistance and national commitments, to be translated into the practices of rural village communities in order to enhance climate resilience and adaptive capacity. This chapter, therefore, analyses the role of the conceptual underpinnings articulated into the programming of adaptation activities implemented at the local village levels in two comparative cases of programme implementation under national adaptation planning in Indonesia. The chapter describes the conceptual basis informing the programmatic approach to adaptation activities through which international climate frameworks are translated into national development planning for integration into local activities. Currently, Indonesia has several policy instruments and legal requirements that set out priorities and guidance for adaptation programming in the country, which are positioned within Indonesia's overall commitments under the UNFCCC process. From the articulation of international climate change adaptation frameworks into national planning for local activity implementation, we discuss here the policy and planning instruments that transfer the international conceptions of adaptation into national planning and local programming. We convey the local farmer experiences of the national adaptation programmes implemented in two case studies of agricultural areas in the Yogyakarta region of Java, amidst Indonesia's country commitments and priorities on adaptation to village activity-level implementation. These local-level experiences of adaptation programming in implementation offer several insights to guide adaptation approaches and conceptions in use for informing national adaptation programming implementation at village levels.

The chapter, firstly, outlines the methodology undertaken in gathering data through qualitative interviews undertaken with farmers, post-programme in two comparative case studies for investigation and as examples of "putting into practice" adaptation in the development sector through national programming. The chapter then reviews the literature on the conceptual underpinnings of adaptation in determining how adaptation programmes engage local farmers. Currently, Indonesia has several policy instruments and legal requirements that set out priorities and guidance for adaptation programming in the country, positioned within Indonesia's overall commitments under the UNFCCC process. Accordingly, we review the key policy structures for articulating international climate change adaptation frameworks into national planning and policy in Indonesia to situate the national adaptation programmes being implemented in two case studies, amidst Indonesia's national commitments and priorities on adaptation. We discuss here the positioning of climate finance and international commitments governing action on climate change adaptation, in order to understand the directives underpinning the National Action Plan on Climate Change Adaptation (Rencana Aksi Nasional Adaptasi Perubahan Iklim-RAN-API) for adaptation action planning and village-level targeted programmes. This section follows the implementation of these two case studies through the relevant line ministries to village-level implementation and links the approaches to adaptation activities to the literature on climate change adaptation, including the implications in integrating programming as local adaptation activities. Finally, the chapter documents the local-level and participant experiences in adaptive capacity and resilience enhancing activities undertaken under the RAN-API programming. The chapter follows with a series of considerations for how climate change adaptation is integrated into the development sector and implemented at local scales, with attention given explicitly to the ways that incorporating local knowledge and community culture fosters capacity and policy goals' effectiveness.

Case Study Approach

Two cases are drawn upon as comparative case studies for investigation and as examples of "putting into practice" adaptation in the development sector through national programming. This chapter provides a focus on the tensions between climate change adaptation programming globally or nationally influenced in design, yet dependent on implementation as activities at local village levels. In drawing on comparative findings between these two case studies, this research aimed to convey how attempts to increase adaptive capacity through adaptation interventions, intersect with the village-level locally determined strategies and systems for governing resources, and adaptive practices or strategies. In two coastal farming areas in the southern part of the Yogyakarta Region in Central Java, a series of interviews conducted over a duration of years, depict the community-level experiences of adaptation activities integrated into national development planning. The data collection phase of this research sought to elicit both community views regarding the interaction with climate change adaptation programming, as well as the views of implementers. However, presented in this chapter are only the common themes understood from semi-structured interviews taken with a sample of farmers across two sub-districts, directly on their views and experiences of these two programmes. The findings taken from the series of interviews and focus group discussions with farmers in the two case studies emphasise the ways that the local community were engaged during project implementation and responded to this programming. Semi-structured qualitative interviews for case study one were undertaken with households during 2019 selected on a rolling sample basis with both direct project participant and non-participant farmers from these villages and combined with earlier data captured in 2011 and 2013 (see Anantasari et al. 2011 and Retnowati et al. 2014) with follow-up interviews with the same participants conducted again in 2019. The interview questions focused on the impacts of climate change, the sociocultural processes and conceptions of land management adaptation over generations by the farming community, as well as interactions with programme activities.

The first case study in which findings are presented, describes findings of farmer experiences, of the village-level activities and processes of implementation and engagement undertaken in one sub-district area (consisting of three villages) of Kulon Progo District in Yogyakarta Region under a national climate adaptation programme. This programme was implemented under the Ministry of Marine Affairs and Fisheries (MMAF) through the local district government. The 'resilient coastal village development' programme (Pengembangan Desa Pesisir Tangguh or 'PDPT') was implemented nationally in Indonesia across a total of 66 coastal villages in 22 districts and cities during 2012–2014 (MMAF 2015). The second case study—describes community-level farmers in Gunung Kidul in the Yogyakarta region, and their interactions with climate change adaptation programming implemented through agricultural extension (initiated under the Ministry of Agriculture, and National Council on Climate Change) Climate Field School (CFS) programme experiences (Anantasari et al. 2011; Retnowati et al. 2014). The CFS programme was first held in Gunung Kidul District in 2007 with funding from the Ministry of Agriculture whose implementation was in collaboration with the Institute for Agricultural Technological Assessment (BPTP) in the Yogyakarta Region. These results discuss the CFS programme, which focused on teaching farmers to start recording data on a quantitative basis in order to transform agroclimatic tacit knowledge into explicit knowledge. Together, the sets of farmer experiences taken from both cases provide examples of how climate change adaptation programming activities, internationally framed, nationally programmed and then implemented at the village level, in practice combine the priorities of international climate policy and finance with the priorities and processes of village governance and community-devised strategies of adaptation. The longitudinal results of research conducted under the two case studies are used to show how current approaches to climate change adaptation interventions have influenced changes to adaptive capacity to climate change over time and their interactions with local village processes in doing so.

Conceptual Underpinnings in Determining the Approaches of Farmer Engagement in Adaptation Programming

This section of the chapter outlines a framework of adaptation and the key-related terms, to be able to describe how these concepts, central to the local adaptation activities in programming are developed and translate from universal principles under international climate change commitments, through national climate adaptation planning apparatus, into interpretations applied at the local village scale. We set out definitions of adaptation as an ongoing social process to discuss how interpretations of adaptation conceptually determine ways that programmes engage participants. We discuss adaptive capacity and how the processes of building capacity in the case studies, pivots on the assumptions of capacity being enhanced by assets,social capital, access, social processes, and agency. As vulnerability and resilience are discussed significantly in other chapters of this book (see Chap. 16), they are not included in the discussion here.

In a broad sense, vulnerability assessments have dominated as the method in which to establish climate change impacts identify target areas, sites or populations for designing adaptation strategies and programming (Downing et al. 2005). The assessments determine the outcomes in terms of the information generated and the options identified. However, the process of undertaking assessments in itself determines how this occurs and how strongly the assessments link with adaptation decision-making and actions (Miller and Bowen 2013). Top-down approaches are generally based on the projected future climate scenario impact projections and models developed on global or regional circulation models. They tend to become less accurate as they are downscaled to more detailed spatial scales, due to the uncertainties embedded in socio-economic factors, as well as in climate systems and biophysical and socio-economic impacts (Hinkel et al. 2010). These types of assessments are unable to account for microclimates, specific local environments and needs—and sometimes offer contradictory predictions when downscaled to the local level (see Ensor 2011; Dessai et al. 2009; Stainforth et al. 2007). Science data-driven or top-down climate vulnerability assessment processes (also termed biophysical or outcome vulnerability) are based on an analysis of climate change and its impacts, while participatory or bottom-up climate vulnerability assessment approaches (also termed social or context vulnerability) are based upon analysis of the people affected by climate change impacts (Dessai and Hume 2004; O'brien et al. 2007). The dynamics in linking scientific and local knowledge within climate change adaptation, remains challenging. Local knowledge tends to be relied upon to 'ground truth' climate modelling predictions and to 'fill gaps' in climate data (e.g. Reyes-Garcia 2015; Lefale 2010; Green et al. 2010; Green and Raygorodetsky 2010). Yet, local knowledge does not necessarily translate easily outside of the environments, cultures and people by which they are known. Nor do local knowledges compare or reduce well to the 'language of objectivism' (e.g. language that is objective, inductive and absolute rather than subjectivist).

At the centre of local-level targeted climate adaptation programming, is the need to facilitate and increase adaptive capacity, as a core activity (Jones et al.2010). When adaptation is perceived as a social process (Thornton and Manafsi 2010:134) of deliberate change in anticipation or reaction to external stressors (Nelson et al. 2007:387), the adaptation proces involves ongoing learning, as decision-making for adaptation efforts evolve and improve with newly emerging information and conditions (Tschakert and Dietrich 2010). This process of adaptation consists of an ongoing stream of activities, actions decisions and attitudes that inform decisions on all angles and aspects on life. This process is therefore also reflective of the existing social norms and processes of the (community or social) system (Adger et al. 2005) making local participation in early programme design crucial. In this sense, adaptation is seen as providing wider benefits, that are not just to specifically address and cope with climate change impacts but which are part of the development process (Apuuli et al. 2000). Adaptation actions are integrated into all aspects of life according to the demographic, cultural, geographical, ecological, economic, historical and technological contexts. Adaptation decisions are therefore also difficult to separate as adaptation decisions and actions from those of decisions and actions triggered by other social, cultural, political, economic or environmental events (Adger et al. 2005). At the centre of local-level targeted climate adaptation programming, is the need to facilitate and increase adaptive capacity, as a core activity (Jones et al. 2010).

Approaches to adaptation informing the implementation of adaptation activities have a basis in several conceptualisations. These range from the majority of adaptive capacity frameworks, which rely on assets and levels of capital as indicators for capacity. Building on from the term adaptation, adaptive capacity refers to the ability to adapt. 'Capacity is the "power to" do something' (Bebbington et al. 2006). Specifically, adaptive capacity is the capacity to adjust, modify or change characteristics or take actions that mitigate or moderate potential damage, as well as to take advantage of opportunities or cope with the consequences of shocks or stresses (Brookes 2003). Adaptive capacity includes "the conditions and characteristics that permit society at large (institutions, local groups, individuals, etc) access to and use of social, economic, psychological, cultural, and livelihood-related natural resources, as well as access to the information and the institutions of governance necessary to reduce vulnerability and deal with the consequences of disaster" (IPCC 2012). While widely contested on how the practical application of adaptive capacity translates, a key part is considered to be the active involvement of the individual, community or society involved in the process (Pettengell 2010). Access to resources is widely considered to be a key determinant of adaptive capacity (see Adger 2003b; Phillips 2003). However, it is also determined by societal decision-making structures (Adger et al. 2009), given that capacity resides in actors, both individual and collective (Bebbington et al. 2006). Factors such as the influence of other actors, social structures and institutions mirror wider societal debates between the dichotomy of agency and structures in determining actions (Giddens 1984, Bebbington 2000). Asset approaches assume that availability (of capital) automatically means usage and neglect to factor in the processes of decision-making and governance into adaptive capacity and other elements such as innovation, opportunities, beneficial exploitation and the sociocultural factors that both bound and provide beneficial institutional and cultural systems of governing (McGray 2009). Assets, are not only physical assets but account as sources of power via the way that people transform several types of assets or capitals (natural, human, financial, physical, cultural and social) into livelihood outcomes (Bebbington et al. 2006).

Other conceptualisations of adaptation seek to encompass beyond just what each system *has* to enable it to adapt—instead focusing on what a system *does* to enable it to adapt (McGray 2009). Conceptualisations based on increasing resilience, emphasise communities becoming more resilient and self-reliant, with the conceptualisation accentuating the agency of local people in making their own communities more resilient. Also understood as a (non-quantifiable) process, rather than a concept, 'resourcefulness', is an alternative term that has been picked up. The concept is poised as a response to the overly nebulous use of resilience framings, and in which to both problematise and redress 'recognition and redistribution', based on cultivating community conditions to conditions which develop social relations. Resourcefulness was proposed as a concept to transcend in practice, the applications of systems theory and resilience thinking with learning and mobilisation as local priorities and needs identified and developed by community actors (MacKinnon and Derickson 2012). Rather than being applied externally by agencies and experts (and donors), the application of resourcefulness emphasises learning and mobilisation as local priorities and needs that are both identified and developed by community activists (MacKinnon and Derickson 2012). Resourcefulness is applied by MacKinnon and Derickson (2012) to four key areas—(1) resources, (2) skills and technical knowledge, (3) indigenous and 'folk' knowledge/alternative ways of knowing and of knowledge production, and (4) recognition (in terms of rights, access and resources). Similar to adaptation, Mackinnon and Derickson (2012) conceive of the resourcefulness concept, as being understood as a process, rather than a condition that can be measured or quantified. Agency as a concept has received much more attention and application in practice than resourcefulness. The term is used widely in interlinkage with resilience, and social resilience, with agency-based resilience frameworks allowing for a reframing of resilience from origins in describing ecosystems, to a people-centred perspective (Bohle et al. 2009). Davidson (2010) suggests that successful social adaptation to climate change is more aptly described as a collective agency. Collective action is seen as the necessary nexus for adaptation and is key to many natural resource and related livelihood management decisions (Adger 2003a). Communities are often already characterised by collective action, social networks, experimentation and advocacy. Communities in local resource-dependent settings have also historically acted collectively to govern the livelihood resources (i.e. water, forest, fish, agriculture) that are seasonal, weather-dependent and variable (Adger 2003a). Such conceptsframe the ways in which adaptation programmes are developed and positioned as community activities. The following sections of this chapter discuss in application to the two case study findings, local experiences of adaptive capacity and resilience enhancing activities. The two case studies taken in point show that these conceptualisations of adaptation have had limited integration into informing programmatic planning and design under national adaptation planning actions to date.

The Translation of International Climate Change Adaptation Frameworks into National Planning and Climate Change Adaptation Programming in Indonesia

To understand the context shaping local experiences of adaptation programmes, this section of the chapter sets out Indonesia's country commitments and priorities on adaptation in which the programming for village activity-level implementation in the two case studies were developed. Under the United Nations Framework Convention on Climate Change (UNFCCC) Paris Agreement (2015), adaptation was defined as a global goal of "enhancing adaptive capacity, strengthening resilience and reducing vulnerability to climate change, with a view to contributing to sustainable development and ensuring an adequate adaptation response" (UNFCCC 2015). The responsibility of developed countries to provide support for adaptation and mitigation in developing countries is set out in Articles 3.1 and 4.4. During the UNFCCC 21st Conference of Parties (CoP) in Paris during 2015, ambitious targets for climate financing were set, and earlier commitments to pledging US $100 billion in global climate finance each year were reiterated (Scoville-Simonds 2016). The commitment pledged by developed countries to mobilise US $100 billion per year (minimum) in climate finance for developing countries, is shared between both mitigation and adaptation (Scoville-Simonds 2016). Adaptation finance was allocated based on both moral and legal grounds based on the notion that developed countries should provide assistance to developing ones (Füssel 2010), while adaptation allocations are given as differential distribution between (developed) countries' contributions to the causes of climate change through emissions, against the vulnerability to climate change impacts of developing countries. In 2010, the CoP adoption of the Cancun Agreement (UNFCCC 2011) had affirmed that adaptation must be addressed with the same level of priority as mitigation and that new and additional funding should be provided to developing countries, setting the background for the Paris Climate Accord. Under these international policy frameworks for adaptation, finance for adaptation was framed primarily concerning terms of shared responsibility. Adaptation gained greater international policy momentum under the 2007 Bali Action Plan, and the Copenhagen Accord of 2009 (Michaelowa and Stadelmann 2018:64). Following the IPCC's fourth assessment report, it was confirmed that more extensive adaptation than was happening, was required to reduce vulnerability to future climate change. The Conference of Parties (CoP) to the UNFCCC agreed to the Bali Action Plan in 2007 (UNFCCC 2008) and launched a comprehensive process to enable the full, effective and sustained implementation of the Convention through long-term cooperative action. The basis for the allocation of resources for adaptation being first enacted in the 1992 UN Framework Convention on Climate Change (UNFCCC), which stated that developed Parties 'shall also assist the developing country Parties that are particularly vulnerable to the adverse effects of climate change in meeting costs of adaptation' (UNFCCC 1992, Article 4, paragraph 4).

Since this time, globally, the rate of adaptation finance has grown steadily regarding support via projects, total commitments and as aid earmarked for adaptation (Michaelowa and Michaelowa 2012; Scoville-Simonds 2016). With future finance for climate activities expected to be of the same order of magnitude as that of development aid (Scoville-Simonds 2016). Currently, the majority of public adaptation finance flows between developed to developing countries as official development assistance (ODA) development aid, which is earmarked as adaptation or adaptation related (Scoville-Simonds 2016). With adaptation-ODA at over USD 10 billion annually, adaptation-ODA currently comprises the largest input to adaptation finance overall, by a significant margin (Scoville-Simonds 2016).

Indonesia's Climate Finance Landscape for Adaptation Programming

In 2016, global ODA to Indonesia allocations for adaptation (both loans and grants) were at US $2,253 million according to Organisation for Economic Co-operation and Development (OECD) data held on ODA (OECD 2018). This figure is more than double of what these amounts were 5 years earlier (Ampri et al. 2014). Although tracking the flow of climate finance into actions remains problematic (Ampri et al. 2014), according to reports analysing data from between 2012 and 2016, 68% of climate finance to Indonesia was directed to activities implemented on the ground, with 30% of those directed specifically to the capacity and knowledge building areas (Falconer and Glenday 2016; Ampri et al. 2014). The majority of international finance flows to central government ministries and agencies at 97% with only a small percentage going to local administrations, despite climate actions being implemented locally (Ampri et al. 2014). For adaptation flows, this allocation is largely directed on institutional strengthening, vulnerability assessments, and mainstreaming adaptation into government planning (Budiman et al. 2016). Notably, local administrations have limited financial and institutional capacity to implement adaptation activities, and a strong need remains for adaptation to be integrated into village-level activities. Conversely, figures for 2012, showed Indonesia to have received the highest amount of foreign direct investment (FDI) after India, aimed specifically at participatory climate change development (with US $6.7 million for participatory adaptation) (Dougherty et al. 2016). This landscape of climate adaptation finance governs how adaptation programming is shaped under different framings of adaptation and has implications that flow on to how climate adaptation development programming is approached and developed under national adaptation action planning and programmatic outcomes over longer timeframes. Although adaptation as an established goal has been galvanised under the international policy and climate financing mechanisms through the Paris Agreement, what comprises climate adaptation is unclear and without agreement for a particular standard, metric or universally accepted definition of what adaptation consists of in practice (Christiansen and Martinez 2018:8). How adaptation plays out in practice on the ground, or whether adaptation should be distinct from other traditional modes of development and development activities remains ambiguous (Sherman et al. 2016:708). Measures for evaluating climate adaptation actions are approached either in measuring a reduction in vulnerability, tracking and measuring the implementation of adaptation actions (e.g. spend, beneficiaries or infrastructure in place), and the effectiveness of adaptation (IPCC 2014). Overall, it is not well understood how effective long-term (post-programme implementation), international climate finance translated through national development programming into local village adaptation interventions in Indonesia, interact in the longer-term with village-level resource governance, agency and sociocultural knowledge for instilling enhanced adaptive capacity.

Adaptation Activities Under Indonesia's National Adaptation Planning

Under the UNFCCC commitment frameworks, countries are encouraged to develop national adaptation plans (NAPs) as a means of identifying medium and long-term adaptation needs, and the strategies and programmes to be developed and implemented as a means of addressing those needs. Although Indonesia is yet to submit a National Adaptation Plan (NAP) in this process to the UNFCCC, preparation is underway and planned for 2019/2020 (Sesotyaningtyas 2019). This NAP formulation will serve as a revision of Indonesia's previous National Action Plan on Climate Change Adaptation (Rencana Aksi Nasional Adaptasi Perubahan Iklim/RAN-API) formulated in 2012, and address inputs further into the national development plan (2020–2024). Alongside the RAN-API, Indonesia Climate Change Sectoral

Roadmap (ICCSR) by the National Development Planning Agency (Bappenas), and Indonesia Climate Change Adaptation Strategy under the National Council for Climate Change (DNPI) also set out adaptation programming. From 2007, Indonesia started formulating an Action Plan to Respond to Climate Change (Rencana Aksi Nasional Perubahan Iklim/RANPI) with which to mainstream climate change into the National Development Plan. In 2011, the government began to initiate the programme plan to address the effects of climate change. The RANPI was then updated in 2012 with the National Action Plan on Climate Change Adaptation (Rencana Aksi Nasional Adaptasi Perubahan Iklim/RAN-API) and forms the policy basis for national government working with local government, and international agencies working with local stakeholders to strengthen community resilience to climate change. The RAN-API mandates climate change adaptation strategies and planning sectorally across national ministries and agencies, and regional and local government priorities (Suarma et al. 2018).The National Climate Change Committee (NCCC) is tasked with integrating governmental climate change activities and involving relevant stakeholders, while the National Development Planning Agency (Bappenas) is responsible for implementing the national development plan, and also focuses on mainstreaming climate change actions into the development planning, as well as for aligning the Indonesian Climate Change Trust Fund with development priorities (Yoseph-Paulus and Hindmarsh 2018). The RAN-API mandates climate change adaptation strategies and planning sectorally across national ministries and agencies, and regional and local government priorities (Suarma et al. 2018).

Under the 2012 RAN-API, activities for adaptation strategies are targeted in five key sectoral areas—agriculture, marine and fisheries, coastal areas and small islands, infrastructure and health. Within the five sectoral areas, RAN-API sets out adaptation options, scoping actions and locations and institutions for these under target clusters, and subclusters allocated within: (1) economic resilience (food security, energy); (2) livelihood resilience (health, settlements, infrastructure); (3) resilience of ecosystems and environmental services; and priority areas (urban area, coastal areas and small islands) (Kawanishi et al. 2016). RAN-API provided short-term adaptation programmes and activities (2014), as well as setting out medium (2015–2019) and long-term (2020–2025) programme and activity goals for climate change adaptation in Indonesia. While at the same time, the Indonesia Climate Change Sectoral Road Map guides how climate change activities are programmed more broadly. The Indonesia Climate Change Sectoral Roadmap (ICCSR) provide inputs into national development planning, including the preparation of action programming for the Road Map on climate change adaptation in coastal areas and the agriculture sector (Bappenas 2010). For example, including the prioritisation of capacity building strategies and vulnerability assessments as key actions (Yoseph-Paulus and Hindmarsh 2018). Under the RAN-API climate change impact, responsive activities with vulnerability assessments have been carried out at a national scale for provinces, regencies and cities, with capacity building and adaptation actions at community-level facilitated predominately under NGO and donor agency programming, alongside government programming allocated under sectoral agencies.

The Resilient Coastal Villages and Climate Field School Programmes Implementing National Adaptation Priorities

Forming part of the programmatic activities for the National Action Plan Addressing Climate Change (2012), are, under the Ministry of Maritime Affairs and Fisheries, the Resilient Coastal Villages Programme (Pengembangan Desa Pesisir Tangguh or 'PDPT') and under the Ministry of Agriculture, the Climate Field School (CFS) programme. In 2012–2014, the resilient coastal village development (PDPT) programme was implemented across a total of 66 coastal villages in 22 districts and cities throughout Indonesia (MAAF 2015). While addressing the vulnerability of the agricultural sector, the CFS programme was undertaken across 33 provinces of Indonesia, under the Ministry of Agriculture and BMKG. Although both programmes target villages engaged in agricultural livelihoods, the programmes are administered under different Ministries according to the geographic and sectoral affiliation, and the targeting of vulnerable coastal areas and small islands and agriculture as strategic clusters of vulnerability. The area geographically located on the coastline in case study one was allotted under the Ministry of Marine and Fisheries, whereas the area for case study two (which is also a coastal area) allotted under the Ministry of Agriculture. Although both areas are primarily agricultural livelihood areas, fishing livelihoods are reportedly more common in the second case study under the Ministry of Agriculture programming. In case study one, farmers state that there are limited or supplementary fishing livelihoods in their villages. The CFS programming falls under Cluster 1: Adjustment of the food production system to climate change variation—(1) adaptation of food production systems to climate variations and change through the development of types of crops, planting patterns and cultivation technology that are more resilient to extreme climate variation occurrences. (2) Development of climate information system, an integrated crop calendar system and an early warning system in terms of threats of droughts and floods as well as of organic disturbances that adversely affect crops, livestock and fish. Various adaptation and mitigation activities have been conducted by the Ministry of Agriculture for this, which include the development of CFS and the adjustment of planting calendars (Suroso et al. 2009). Coastal areas as noted, are covered by a target for the resilience of specific areas (Suroso et al. 2009) and programming for case study areas was developed accordingly. Strategically these seek to ensure the stability of livelihoods of coastal areas against the threat of climate change, increase the environmental quality of coastal areas, develop adaptation structures, adjust spatial plans and develop and optimise research and information systems to build resilience to climate change in coastal areas (Suroso et al. 2009). The Pengembangan Desa Pesisir Tangguh (PDPT) programme nationally was based on building stability against the threat of climate change, planning for adaptation to climate change in identified villages, and adjusting relevant spatial plans to include climate change threats. Lastly, focus included building ongoing development and optimal use of climate change research and information systems in these target areas. Under a further environmental and ecosystem management cluster of programming in the national plan of adaptation action, efforts were directed at developing the Coastal Resilient Village (CRV) (Bappenas 2014; Darajati 2012).

Climate vulnerability and risk assessments have been serving as an entry point for initiating adaptation planning in Indonesia, at local (village) and regional district levels, and determine much of the climate change adaptation strategies and planning process that takes place thereafter, such as in case study one. Based on assessments of exposure, sensitivity and adaptability, the PDPT programmatic concept as an example, was based on enhancing the resilience of the 'climate disaster resilient village' (CDRV) by increasing adaptive capacity to mitigate the risks of disasters driven by climate change. However, criticisms of the programme emerged that the approach lacked the participation of those with the experience and best practices for including key local knowledge in the design phase, such as those of the local stakeholders on their local resources and environment for transformative participation to occur (Suarma et al. 2018). The national-scale top-down vulnerability assessments determining targets for PDPT had little focus for the everyday adaptation strategies needed for the agricultural livelihoods to be climate-resilient, and focused on readiness for significant but less frequent events than the impacts of climate variability, seasonal changes and instability, increased pest incidence, drought wind and flood damage to agriculture and farmers' everyday livelihoods and livelihood security. Based on the premise that climate change responses should be more sensitive to the needs and risks faced by more vulnerable communities, community approaches to adaptation and implementation at village levels are supposed to build on the proposition that local communities have the skills, experience, local knowledge and networks to increase their own resilience (Forsyth 2013). Building local knowledge, skills and experiences into programatic approaches further could greatly benefit the outcomes for the activities of the RAN-API in achieving meaningful and sustained adaptation.

The coastal string of agricultural-livelihood dependent villages along the beach area of Kulon Progo are susceptible to high winds and flooding and subject to hot, dry periods in summer. Yet, this area also is vulnerable to coastal storms, inundation and has high tsunami potential. With current predictions suggesting that the area is highly vulnerable to a tsunami if a significant earthquake occurs in the sea in the south of Java (BMKG 2019a). The resilient coastal village development programme (Pengembangan Desa Pesisir Tangguh or PDPT) was implemented in three villages in the Panjatan sub-district of Kulon Progo District. These three villages were targeted in the programme to become "resilient coastal villages" following a series of national-level climate vulnerability assessments undertaken across Indonesia and used to identify specific coastal areas of high vulnerability as villages to target for the programme. The Ministry of Maritime Affairs and Fisheries through the Directorate General of Marine, Coastal and Small Islands formulated the national programming of the PDPT under a five-phased process to roll out the programming in each target area. Phase one was a preparation stage for coordination of the drafting team, relevant stakeholders, methodology and preparation of work plans, which was then followed by the collection of secondary data across government agencies. Stage three then involved field surveys to ground truth information gathered from government agencies for data accuracy and then in stage four identify potential areas for activities based on indicators of social and population factors, assets and infrastructure, economy, institutions, natural resources, environmental conditions, and coastal natural disasters. Finally, under phase five, coastal village cluster priorities were determined through the ranking and scoring of these results against a set of seven indicators in order to identify the final target village areas of the programme (MMAF 2016).

The CFS programme taken here as the second case study was first held in Gunung Kidul District in 2007 with funding from the Ministry of Agriculture whose implementation was in collaboration with the Ketindan Center for Agricultural Training (BBPP) and the Yogyakarta Special Region of Agriculture. The Agency for Meteorology, Climatology and Geophysics (BMKG), under the Ministry of Research and Technology, collects weather, climate and earthquake data from 200 observation stations placed across the country, and provides information to the public on these events. Part of their mandate includes carrying out drought prediction in Indonesia based upon satellite rainfall data. One key aspect under international collaboration is the development of the Drought Early Warning System (DEWS) within the Climatological Early Warning System. The Indonesian National Disaster Management Agency (Badan Nasional Penanggulangan Bencana/BNPB) manages the local coordination of the DEWS data and forecasts, working with extension officers at the local levels with farmers. At an operational level, coordination for the climate field in agriculture is through both the Regional Board of Disaster Management (BPBD) and district agricultural extension officers. The Department of Agriculture is the leading national body for coordinating the use of climate information for agricultural planning, in particular for local farmers, through the National Center for Agricultural Extension Development, with a large portion of agricultural management referred through the Kabupaten (district) level of government. The CFS programme was initially established via collaboration between the Seasonal Drought Early Warning System (SDEWS) and CFS initiatives under the Department of Agriculture and the National Council on Climate Change (DNPI). Imbedded in a methodology inherited from the original Farmers Field Schools, established under the UN Food and Agricultural Organisation (FAO) from the 1980s, the basis for Farmer Field Schools upon which Climate Field Schools are based is on a methodology consistent of (voluntary) farmer groups and a basis of learning principles for adoption and adapting and applying these in the field. The basis for farmer field schools uses principles of learner-centred education based on demonstration plots that allow for experimentation, participation and a comparison of differences so that farmers can assess for themselves the effectiveness of adopting certain practices. The approach was usually also organised through farmer or extension worker facilitators (Tarrant 2014). Adopting this methodology for climate adaptation was intended to establish an approach to responding to climate change that is farmer-driven.

These two programmes highlight the different approaches being implemented via government agencies under the national adaptation planning actions (RAN-API). Explicit differences in the approaches and focus of programming are evident between the two case studies according to the agencies in which they are administered, and the distinctions between climate change adaptation and disaster risk reduction as key areas in which resilience or adaptive capacity activities were developed. As the programme for the Gunung Kidul area is under the Ministry of Agriculture, the focus is on limiting the impact of climate change on agricultural practices and livelihoods. However, as the programme for Kulon Progo is allocated under the Ministry for Marine Affairs and Fisheries, the programmatic emphasis is on limiting the impacts of coastal storms, tsunami preparation, coastal environmental protection/coastal buffering and marine and coastal-based livelihoods (primarily disaster risk reduction focused), rather than emphasising the inclusion of climate change adaptation in agricultural livelihoods such as slow-onset impacts of drought, water scarcity, increased heat or flood, or climate variability such as changes in seasonal patterns affecting agricultural productivity.

According to the findings in the two case studies discussed, adaptation to climate change has been approached in terms of reduction to risk and vulnerability but with little regard and incorporation of farmers' social processes, the agency in innovation and local knowledge within the engagement approach and design of the adaptation programmes for village implementation. Findings from the case studies found that programme activities were largely pre-determined rather than responsive to local farmers' processes, needs. As well as neglecting to position early engagement as a starting point for designing the programmatic approach. In PDPT, for example, programme objectives were focused on physical conditions and technically based (Suarma et al. 2018:128).

The Experiences of Local Villages in Adaptive Capacity and Resilience Enhancing Activities

The first case study in which findings are presented in this section describe the experiences of community-level farmers targeted for the Kulon Progo implementation of the PDPT producing vegetable and fruit crops in the coastal area. Targeted in the programme to become a "resilient coastal village" following a series of climate vulnerability assessments undertaken throughout the country used to identify project target areas, the village communities according to the interviews undertaken are primarily engaged and dependent on farming livelihoods. Prior to the implementation of the PDPT programme, farmers in the area actively engaged in determining strategies for retaining land access and strengthening farming livelihood options in a harsh coastal environment. These villages are more widely known for the strategies employed in farming unfavourable agricultural land due to coastal conditions, and engaging active strategies in the face of land conflict pressures to retain the area for sustaining farming livelihoods. Previous agricultural studies undertaken in the case study area in Kulon Progo found highly innovative adaptive farming techniques employed (Dinarti et al. 2013; Supriyanto et al. 2012). In particular, innovative agricultural livelihood diversification based around collective management systems (Raya 2014). These systems emphasise social capital and social relations as being conducive to ongoing adaptation process in land management, that are also applicable to climate change (Supriyanto et al. 2011, 2012). The latter study particularly highlighting farmers' ability to control environmental, physical, economic, human and political forms of capital to build up an active, innovative and adaptive farming community in coastal area. Notably, these system have enabled an agricultural livelihood-based community to flourish in a coastal sandy environment utilised as highly productive agricultural land.

The programme activities implemented under the project according to each village were for each of the 3 target villages: 1) the purchasing of water pumps, and three-wheeled motorbike for transport of goods under business development, and construction of community building facilities, revegetation of the sand dune and coastal area, loudspeaker and communication system for the disaster warning system, construction of a tsunami evacuation route. 2) The construction of joint business kiosks for business development, construction of a road and tsunami evacuation route and revegetation of the dune and coastal area. 3) Business development, including the purchase of a communal three-wheeled motorbike, purchase of water pumps, revegetation of the river area and beach area, construction of a road and tsunami evacuation route, as well as a loudspeaker and communication system for disaster warning. These activities were implemented over a 2-year period. The PDPT programme was implemented within each village through a facilitator team from across the district, a village empowerment team (of male and female representatives for each village), as well as a technical team (headed by representatives from local and district government agencies). The programme across the three villages also had several coordinators and assistant staff employed to assist the programme, and a further team of staff appointed to the programme in Kulon Progo from the Ministry of Marine Affairs and Fisheries (KPP) offices, both from Kulon Progo and Jakarta. However, although the PDPT programme was implemented through several layers of coordination through which to engage local residents, interviews in 2019 across the three villages of Bugel, Karang Sewu and Banaran found that there was no awareness of the programme by most residents interviewed, and where there was awareness, only limited knowledge of the activities or the reasons for these (with residents only having heard the programme name). Tsunami evacuation routes built under the programme have, however, been well utilised for access to farming areas and transportation of produce and supplies on a daily basis. Those respondents that knew about the programme and had been involved in the programme—reported that they initially only heard about the programme by chance, rather than through an approach that socialised and engaged local residents with the programme's objectives. Key community heads reportedly also remained largely unaware of the PDPT programme and, in general, little knowledge of what the programme entailed where it was known. Overall, findings from the interviews fundamentally highlighted a clear need for programme activities to have focused on agricultural-related impacts of climate change on land rather than being focused on fishing livelihoods and marine impacts. Communities report in 2019, years on from the implementation of the PDPT commenced in 2012 and finished in 2016, that the activities to protect against climate change, such as the protection and conservation of coastal areas against winds—remain community organised and driven. The experience of the PDPT programme by villagers (where they are aware that the programme took place) is that the programme was concerned with large significant disaster events rather than the day-to-day impacts that relate directly to their experiences.

Farmers in the area for decades have been adapting plant varieties to the conditions of the area in order to produce successful yields, innovating water efficiency solutions, windbreaks, inter-cropping, agroforestry, soil fertility measures and essentially implementing climate-smart agriculture in a harsh coastal (sand) area through their own techniques of innovation, community learning and adaptation. Farmers continue to maintain their resilience through practices of livelihood and crop diversification. Farmers interviewed, in general, were very aware of increased climate variability, and impacts, particularly in rainfall changes, increased heat, coastal winds, and ocean inundation events. The variability in seasons and growing conditions related to the unpredictability from previous experiences of past growing conditions were of great concern (altering the use of local knowledge and traditional practices like use of the Javanese calendar *Pranoto Mongso*). Variations in planting times and pest incidences are now very hard to predict and have resulted in substantial crop losses. Along with increased coastal wind impacts damaging crops with ocean salt spray and adding to soil salinity. Farmers have had to adapt watering practices according to increased heat. The processes of farmers' learning and innovation (learning from neighbours' successes, and by individual trial and error—then picked up by others if successful), were not incorporated into the programme. Nor were the social sharing practices and cooperative agreements in managing the area included (e.g. agreed strategies for crop dates to prevent pest incidence, negotiate market prices and so on). Revegetation activities for conserving the coastal sand areas and adding ecosystem-based protection, for example, were undertaken under the PDPT programme. However, non-local revegetation varieties were originally introduced under the programme for the first year (and failed), until local farmers' knowledge and suggestions for local species were incorporated into programme activities. These activites were then reportedly successful in the second year of the programme.

The PDPT programme also connects with the *musyawarah perencanaan dan pengembangan* or Musrenbang process. The Musrenbang process is intended to address and bring discussion about local development needs. Overall, the process should provide local desa (villages) with access to the budget for their needs. If communities come together and decide upon priorities for short-term improvements, then submit these priorities to the local government planning department, then resources should be assigned according to these priorities' (according to the availability of funds). However, in reality, most farmer households were unaware of the opportunities and processes in which to participate and link in with local government planning. This has implications to the longer-term sustainability and maintenance of the infrastructure and activities carried out under the PDPT programme, and the programme outcomes in the long-term.

The second case study—describes community-level farmers in Gunung Kidul in the Yogyakarta Region of southern Java, and their interactions with climate change adaptation programming in development, implemented through agricultural extension Climate Field School (CFS) programme experiences (Anantasari et al. 2011;

Retnowati et al. 2014). Observations, as well as both ethnographic and participatory research, were conducted to understand how farmers interpret and use the information on seasonal rainfall, rainfall variability, distribution and agricultural impacts. These results discuss a CFS programme that taught farmers to start recording data on a quantitative basis and transform tacit knowledge into explicit knowledge of agro climate factors. The CFS programme was first held in Gunung Kidul District in 2007 with funding from the Ministry of Agriculture whose implementation was in collaboration with the Institute for Agricultural Training Assessment (BPTP). The Farmers Group in Wareng IV Hamlet (Wareng Village, Wonosari Sub-district), were chosen to be one of the pilot projects for implementing the CFS programme in Yogyakarta. The Indonesian government CFSs are in several places in Java. As the first CFS, designed by the Ministry of Agriculture, Indonesian scholars and foreign counterparts, were introduced in Indramayu District, West Java Province in 2003, developing the CFS practices to help improve analysis and anticipation of changes in weather conditions. Meanwhile, from 2011 to 2018, the CFS has been implemented in 316 locations spread across 33 provinces in Indonesia with a total of 8,000 participants (BMKG 2019a). During 2012 and 2013 focus group discussion (FGD) with 200 representatives of farmers and sub-districts officials in karst area ofGunung Kidul, there were not yet specific climate change adaptation programmes introduced to the farmers (groups) through sub-district (local) government. Climate information had only been introduced within the agriculture and environmental agency and the BPBD recently in the last decade, and discussion of climate change has also taken place at the regional level. The Regional Board of Disaster Management deal with potential disaster events due to climatic change, but the incorporation of climate change information falls within the jurisdiction of BMKG. Field instructors and officers for plant pest observation and control were charged with the role of agricultural extension workers transferring climate knowledge and measures to the farmers. Hence, some programmes led by BMKG on climate field school targeted these agents. Farmers that had no involvement with the CFS continued to develop climate adaptation strategies based on their experiences, observation and knowledge of local environment without this information. Therefore, the agriculture programme introduced by the Regional and Local Government was adjusted based on the farmers' understanding of their surrounding ecosystem.

The national government through BMKG put higher consideration on climatic factors following the impact of tropical cyclone *Cempaka* which hit the southern part of Java and Bali during 22 November 2017–1 December 2017. Following the cyclone impacts, the CFS also began targeting fishermen as well as the farmers. Information from farmers and Observers of Plant Pest Organisms (POPT) officers stated that CFS, which had the cost of assistance covered by the central government, had been carried out in several villages in Gunung Kidul District. BMKG has implemented the CFS programme, which became known as the CFS-BMKG from 2011 to 2018 with the CFS being one of BMKG's efforts to improve climate literacy and dissemination of climate information for agriculture. This falls under accordance with Presidential Instruction Number 5 of 2011, namely Safeguarding National Rice Production in the Face of Extreme Climate Conditions. CFS is also in line with the

Government's Nawacita programme, which is the seventh Nawacita programme to enhance economic sustainability. The BMKG states that yields from CFS Phase 3 activities show an increase in production of up to 30% compared to its production average, showing that the presence of CFS activities on farmer groups can increase agricultural productivity. Information from the Gunung Kidul District Agriculture Service states that the last CFS activities were carried out in Gunung Kidul District in 2016/2017 with facilitation funds from the World Bank. The implementation of CFS consists of three stages, namely CFS stage 1/Agro-climate Socialization (CFS-1) for Local Governments/Related Services, CFS stage 2 (CFS-2) for Agricultural Extension (PPL/POPT) and CFS stage 3 (CFS-3) for Farmers and Farmers Groups. In CFS-1 there were 1386 participants, 551 CFS-2 participants and 1023 CFS-3 participants, who joined in 40 farmer groups. Nowadays 33 provinces have participated in the CFS.

Those interviewed felt the positive benefits of activities facilitated, including involving staff and researchers from the IAARD through the Hydrological Research Institute and Agro-climate (Balitklimat). Yet, the CFS had a number of programmatic design issues that prevented the effectiveness and ability of the programme to properly engage with participants. The timeframes of the CFS did not allow the farmers' knowledge and practices to adapt to the CFS and incorporate the CFS practices to climate change. Firstly, rather than being held in the rainy season, CFS activities were held during the dry season, when there were limited rainfall and so fewer agricultural activities taking place. Ultimately, failing to allow any demonstration to do what 'in season' should take place according to the farmer field school approach. At the same time, in 2018, there was no allocation of funds from the Gunung Kidul District Budget for the facilitation of the CFS Programme. The absence of this fund allocation is due to the CFS programme no longer being prioritised. At the time of writing—no further information about plans to continue or follow-up of the CFS at this site since were planned to follow on from the 2010 implementation in Gunung Kidul.

The farmers' learning process also terminated once the CFS ended as potential understanding was lost in the CFS learning only being introduced as a once-off whereby farmers were unable to make further use of such 'once-introduced learning'. Whereas climate variability and more severe extreme events, as well as their impacts on plants and soil, require ongoing detailed observation, and depend on farmer confidence gained through years of experience. This meant that sustaining continuity in the programme maintained at each site was crucial. Information on projected rainfall was disseminated by the BMKG. However, the farmers complained that they did not receive any locally valid information in a timely fashion, or in forms that they could easily understand, to serve as the basis for their actions. Farmers in Gunung Kidul widely use *Pranoto Mongso* (see Retnowati et al. 2014). *Pranoto Mongso* also used in Kulon Progo) is an inherited traditional knowledge seasonal calendar which has been acknowledged and considered as part of cultural identity in central Javanese agricultural-based communities. Part of which includes reading signs from nature, known as "*titen*", and built into farmers' process of observing agro-meteorological changes by observing, recording, analysing and evaluating existing knowledge for adaptation. Interpreting the signs from the surrounding environment, people in Karst Gunung Kidul adopt ways of living with their environment in a way to conserve, preserve and maintain sustainability (Retnowati et al. 2014). Incorporation of local knowledge and conceptions for undertaking land management such as *Pranoto Mongso* are therefore crucial to determine their norms, values and belief in regards to their activities, particularly in the area of water and land management and farmers' responses to climate change or variability. In Karst regions like Gunung Kidul, timing and crop variety are considered important to adjust within a given characteristic of the physical setting. By reading the signs for the rainy season, farmers start to plough and plant seeds known as "*awu-awu*", even if the rain has not started yet. In this uncertain period of environmental changes, those who still practice *awu-awu* are taking risk of failing cultivation. Other farmers, mostly young generation farmers, have started to cultivate when the rain falls to avoid the risk (see Retnowati et al. 2014). For CFS in Gunung Kidul, however, interviews noted that the only medium used by farmers conveying the knowledge from BMKG, was television, which released mostly general weather forecasts, and climate information based on large-scale scientific prediction and evaluation. This information was also not downscaled to be of use locally by the farmers.

Conclusion

The concepts laid out in this chapter defining adaptation as an ongoing social process, are shown as key in informing programmatic approaches adaptation that integrate international climate frameworks through national development planning into local-level. In the ways that programming engages, is participatory in design, and incorporates the ongoing nature of social processes into timeframe designs. Adaptive capacity has not been enabled in the timeframes and programatic design of the two cases taken as examples of national adaptation planning activites in implementation. Neither programme focused on the long-term community agency aspects of adaptive capacity, as communities did not report engagement in the early conceptions of programme design. The findings show that missed opportunities to engage fully, those targetted for adaptation activites, have meant that local-level contextual interpretations, knowledge and ways of responding and adapting to change as an ongoing process found that in application, programme adaptation activities have had limited outcomes in the longer-term. The participant experiences and interpretations in the case studies strengthens the assumption that the intervention of any new ideas to communities (such as climate change information and adaptation response strategies) requires contextual time framing with both local knowledge and state policy, that fosters community culture and incorporates community capacity. Whereas the implementation of activities might result in information and training being received, longer term uptake and continued engagement of strategies involve mutually conceived strategies built on local experiences that current national adaptation planning programme design timeframes have been unable to account for and effectively respond to. The findings show that the approaches used in programming, particularly in engaging communities, largely determine the success of programming outcomes, particularly in the longer term. How programming fosters local engagement in to early design phases, prior to determining the overall approach to programming, is shown as a pre-cursor in these cases for how successfully a programme addresses the adaptation needs, and community-directed ways of employing future strategies of adaptation. A greater building of local knowledge, skills and experiences into programmatic approaches could further benefit the outcomes for the activities of the RAN-API in achieving meaningful, sustained, ongoing adaptation.

References

Adger WN (2003a) Social capital, collective action, and adaptation to climate change. Econ Geogr 79(4):387–404

Adger WN (2003b) Social aspects of adaptive capacity. In: Smit JB, Klein R, Huq S (eds) 2003, Climate change, adaptive capacity and development. Imperial college press

Adger WN, Hughes T, Folke C, Carpenter S, Rockstrom J (2005) Social-ecological resilience to coastal disasters. Soc Ecol Resilience Coast Disasters 309(5737):1036–1039

Alexander C, Bynum N, Johnson E, King U, Mustonen P, Neofotis P et al (2011) Linking Indigenous and scientific knowledge of climate change'. Bio Sci 61(6):477–484

Ampri I, Falconer A, Wahyudi N, Rosenberg A, Ampera B, Tuwo A, Glenday S, Wilkinson J (2014) The landscape of public climate finance in Indonesia. An Indonesian ministry of finance and CPI report. Available at: https://climatepolicyinitiative.org/publication/landscape-of-public-climate-financein-indonesia-3/

Anantasari E, Winarto YT, Stigter K (2011) Climate Field School: Learning and understanding some scientific knowledge of climate. In: Winarto YT, Stigter K (eds) Agrometeorological learning: coping better with climate change. Deutschland: LAP LAMBERT Academic Publishing GmbH and Co. KG, pp 44–85

Apuuli B, Wright J, Elias C, Burton I (2000) Reconciling monitoring and assessment. Environ Monit Assess 61(1):145–159

Armitage D, Berkes F, Dale A, Kocho-Schellenberg E, Patton E (2011) Co-management and the co-production of knowledge: learning to adapt in Canada's arctic. Glob Environ Change 21:995–1004

Bappenas (2010) Indonesia climate change sectoral roadmap. Jakarta: ICCSR, Republic of Indonesia (Peta Jalan Sektoral Perubahan Iklim Indonesia, Badan Perencanaan Pembangunan Sasional, Pemerintah Indonesia)

Bappenas (2012) National action plan for climate change adaptation (RAN-API), Synthesis report, ministry of national development planning/national development planning agency, Government of Indonesia (Rencana Aksi Nasional untuk Adaptasi Perubahan Iklim (RAN-API), laporan sintesis, Kementerian Perencanaan Pembangunan Nasional/Badan Perencanaan Pembangunan Nasional, Pemerintah Indonesia)

Bappenas (2014) The national action plan for climate change adaptation (RAN-API), Ministry of national development planning/national development planning agency (Rencana nasional adaptasi perubahan Iklim (RAN-API), Kementerian perencanaan pembangunan nasional/badan perencanaan pembangunan nasional (Bappenas))

Bebbington AJ (2000) Re-encountering development: livelihood transitions and place transformations in the Andes. Ann Assoc Am Geogr 90(3)

Bebbington A, Dharmawan L, Fahmi E, Guggenheim S (2006) Local capacity, village governance, and the political economy of rural development in Indonesia. World Dev 34(11):1958–1976

BMKG (2019a) Climate Field School collaboration between BMKG and the community to increase Indonesian food production. Agency for Meteorology Climatology and Geophysics (Sekolah lapang iklim kolaborasi bmkg dan masyarakat tingkatkan produksi pangan Indonesia, Badan Meteorologi, Klimatolgi dan Geofisika BMKG), online: https://www.bmkg.go.id/berita/?p=sekolah-lapang-iklim-kolaborasi-bmkg-dan-masyarakat-tingkatkan-produksi-pangan-indonesia&lang=ID&tag=berita-utamaAccessed 9 Mar 2019
BMKG (2019b) Official BMKG Press Release Regarding the Potential of the 8.8 SR Earthquake and the 20m Tsunami in the South of Java, Press Release GF. 00.02/004/DG/VII/2019, Sadly, M. Agency for Meteorology, Climatology and Geophysics (BMKG), 21 July 2019 (Respon BMKG terhadap isu gempabumi m 8,8 dan tsunami di pantai selatan jawa, badan meteorologi, klimatologi, dan geofisika BMKG, Jakarta, 21 Juli 2019) online: https://dlh.kulonprogokab.go.id/detil/483/press-release-resmi-bmkg-terkait-potensi-gempa-bumi-88-sr-dan-tsunami-20-m-di-selatan-pulau-jawa. Accessed 28 July 2019
Bohle H, Etzold B, Keck M (2009) Resilience as agency. University of Bonn, Geography, Germany
Bohensky EL, Kirono DGC, Butler JRA, Rochester W, Habibi P, Handayani T, Yanuartati Y (2016) Climate knowledge cultures: stakeholder perspectives on change and adaptation in Nusa Tenggara Barat, Indonesia. Climate Risk Manage. 12:17–31
Budiman I, Takama T, Pratiwi L, Soeprastowo E (2016) Role of microfinance to support agricultural climate change adaptation in Indonesia: encouraging private sector participation in climate finance. Future of Food: J Food, Agric Soc 4(3):55–68
Christiansen L, Martinez G, Naswa P (eds) (2018) Adaptation metrics: perspectives on measuring, aggregating and comparing adaptation results. UNEP DTU Partnership, Copenhagen
Cruickshank J (2001) Glaciers and climate change: perspectives from oral tradition. Arctic 54(4):377–393
Cruickshank J (2005) Do glaciers listen? Local knowledge, colonial encounters, and social imagination. UBC Press, Vancouver
Davidson DJ (2010) The applicability of the concept of resilience to social systems: some sources of optimism and nagging doubts. Soc Nat Resour 23:1135–1149
Darajati W (2012) National workshop on the tangguh coastal village development program (PDPT), National action plan for climate change adaptation (RAN-API), Presentation, Jakarta, December 13, 2012 (Lokakarya nasional program pengembangan desa pesisir tangguh (PDPT), Rencana aksi nasional adaptasi perubahan Iklim (RAN-API), Presentasi, Jakarta, 13 Desember 2012)
Dessai S, Hulme M Lempert R, Pielke R (2009) Climate prediction: a limit to adaptation? In: Adger WN, Lorenzoni KL, O'Brien KL (eds) Adapting to climate change: Thresholds, values, governance, Cambridge University Press, Cambridge pp 64–78
Dessai S, Hulme M (2004) Does climate adaptation policy need probabilities? Clim Policy 4:107–128
Dinarti SI, Wastutiningsih SP, Supriyanto S (2013) Collective innovation of coastal agricultural farmers in Panjatan Subdistrict of Kulon Progo. Agro Ekon 24(1) (Kecerdasan kolektif petani di lahan pertanian pasir pantai Kecamatan Panjatan Kabupaten Kulon Progo. Agro Ekonomi 24(1), Juni)
Dougherty A, Taylor J, Hidayani R, Fatimah D (2016) Climate change vulnerability assessments in Indonesia: where are the women's perspectives? Asian cities climate resilience working paper series, vol 37
Downing T, Patwardham A, Klein R, Mukhala E, Linda S, Winograd M, Ziervogel G (2005) Assessing vulnerability for climate adaptation. In: Lim B, Spanger-Siegfried E (eds) 2005, Adaptation policy frameworks for climate change: developing strategies, policies and measures, chapter: 3, Cambridge University Press, pp 67–89
Ensor J (2011) Uncertain futures: adapting development to a climate change. Practical Action, UK
Falconer A, Glenday S (2016) Taking stock of international contributions to low carbon, climate resilient land use in Indonesia, Working paper, Climate Policy Initiative, online: https://climatepolicyinitiative.org/wp-content/uploads/2016/02/Taking-Stock-of-International-Contributions-to-Low-Carbon-Climate-Resilient-Land-Use-in-Indonesia-Full-Study.pdf. Accessed 10 Jan 2018

Few R, Brown K, Tompkins E (2007) Public participation and climate change adaptation: avoiding the illusion of inclusion. Climate Policy 7(1):46–59
Forsyth T (2013) Community-based adaptation: a review of past and future challenges. Clim Change 4(5):439–446
Füssel H-M (2010) How inequitable is the global distribution of responsibility, capability, and vulnerability to climate change: a comprehensive indicator-based assessment. Glob Environ Change 20(4):597–611
Giddens A (1984) The constitution of society: outline of the theory of structuration. Polity Press, Cambridge
Green D, Billy J, Tapim A (2010) Indigenous Australians knowledge of weather and climate. Clim Change
Green D, Raygorodetsky G (2010) Indigenous knowledge of a changing climate. Clim Change 100:239–242
Hinkel J, Nicholls RJ, Vafeidis A, Tol RSJ, Avagianou T (2010) Assessing risk of and adaptation to sea-level rise in the European union: an application of DIVA. Mitig Adapt Strat Glob Change 5(7):1–17
Hiwasaki L, Luna E, Syamsidik, Shaw R (2014) Process for integrating local and indigenous knowledge with science for hydro-meteorological disaster risk reduction and climate change adaptation in coastal and small island communities. Int J Disaster Risk Red 10:15–27
IPCC (2012) Managing the risks of extreme events and disasters to advance climate change adaptation. A special report of working groups I and II of the intergovernmental panel on climate change. In: Field CB, Barros V, Stocker TF, Qin D, Dokken DJ, Ebi KL, Mastrandrea MD, Mach KJ, Plattner GK, Allen SK, Tignor M, Midgley PM (eds) Cambridge University Press, Cambridge, UK, and New York, NY, USA
IPCC (2014) Noble I, Huq S, Anokhin Y, Carmin J, Goudou D, Lansigan F, Osman-Elasha B, Villamizar A (2014) Adaptation needs and options. Final draft. IPCC working group II chapter 14. https://ipcc-wg2.gov/AR5/images/uploads/WGIIAR5-Chap14_FGDall.pdf
Jones L, Ludi E, Levine S (2010) Towards a characterisation of adaptive capacity: a framework for analysing adaptive capacity at the local level. Overseas Development Institute, London, UK
Kawanishi M, Preston BL, Ridwan NA (2016) Evaluation of national adaptation planning: a case study in Indonesia. In: Kaneko S, Kawanishi M (eds) Climate change policies and challenges in Indonesia. Springer, Tokyo
Krupnik I, Ray GC (2007) Pacific walruses, indigenous hunters, and climate change: bridging scientific and indigenous knowledge. Deep Sea Res Part II: Top Study Oceanogr 54(23–26):2946–2957
Kuruppu N (2009) Adapting water resources to climate change in Kiribati: the importance of cultural values and meanings. Environ Sci Policy 12:799–809
Lefale PF (2010) *Uáafa le aso* stormy weather today: traditional ecological knowledge of weather and climate. The Samoa experience. Clim Change 100(2):317–335
MacKinnon D, Derickson KD (2012) From resilience to resourcefulness: a critique of resilience policy and activism. Prog Hum Geogr 37(2):253–270
McGray H (2009) The national adaptive capacity framework: key institutional functions for a changing climate. Pilot Draft. World Resources Institute, Washington, DC
McNamara KE, Prasad SS (2013) Valuing indigenous knowledge for climate change adaptation planning in Fiji and Vanuatu. UNU-IAS TKI
Michaelowa K, Michaelowa A (2012) Development cooperation and climate change—political economic determinants of adaptation aid. In: Michaelowa A (ed) Carbon markets or climate finance? Routledge, Abingdon, pp 39–52
Michealowa A, Sradelmann M (2018) Development of universal metrics for adaptation effectiveness. In: Christiansen L, Martinez G, Naswa P (eds) Adaptation metrics: perspectives on measuring, aggregating and comparing adaptation results. UNEP DTU Partnership, Copenhagen
Miller F, Bowen K (2013) Questioning the assumptions: the role of vulnerability assessments in climate change adaptation. Impact Assess Proj Appraisal 13(3)

Möhner A (2018) The evolution of adaptation metrics under the UNFCCC and its Paris agreement. In: Christiansen L, Martinez G, Naswa P (eds) Adaptation metrics: perspectives on measuring, aggregating and comparing adaptation results. UNEP DTU Partnership, Copenhagen

Moloney G, Leviston Z, Lynam T, Price J, Stone-Jovicich S, Blair D (2014) Using social representations theory to make sense of climate change: what scientists and nonscientists in Australia think. Ecol Soc 19(3)

MMAF (2015) Strategic plan of the directorate general of marine area management 2015–2019, directorate general of marine area management, number 2/Per-DJPRL/2016, ministry of maritime affairs and fisheries (MMAF) (Rencana Strategis Direktorat Jenderal Pengelolaan Ruang Laut tahun 2015–2019, Direktorat Jenderal Pengelolaan Ruanglaut, Nomor 2/Per-DJPRL/2016, Kementarian Kelautan dan Penikaan (KKP))

MMAF (2016) Coastal village development plan (PDPT) Realizing the improvement and expansion of the pro-people program, ministry of maritime affairs and fisheries (Pengembangan desa pesisir tangguh (PDPT) Mewujudkan peningkatan dan perluasan program pro-rakyat, Kementarian Kelautan dan Perikanan (KKP)

Neil Adger W, Dessai S, Goulden M, Hulme M, Lorenzoni I, Nelson DR, Naess LO, Wolf J, Wreford A (2009) Are there social limits to adaptation to climate change?. Climatic Change 93(3–4):335–354

Nelson DR, Adger WN, Brown K (2007) Adaptation to environmental change: contributions of a resilience framework. Ann Rev Environ Res 32(1):395–419

O'Brien K, Sygna L, Leichenko R, Adger W, Barnett J, Mitchell T, Schipper L, Tanner T, Vogel C, Mprtreux C (2007) Disaster risk reduction, climate change adaptation and human security. GECHS Report 2008:3

OECD (2018) ODA disbursements to countries and regions, OECD Stat, Organisation for Economic Co-operation and Development (OECD), online: https://stats.oecd.org/Index.aspx?DataSetCode=Table2AAccessed 10 Oct 2018

Pettengell C (2010) Climate change adaptation: enabling people living in poverty to adapt. Oxfam GB, Oxford

Phillips J (2003) Determinants of forest use among communal farmers in Zimbabwe. In O'Brien K, Vogel C (eds) Coping with climate variability: the use of seasonal climate forecasts in South Africa. Ashgate

Raya AB (2014) Farmer group performance of collective chilli marketing on sandy land area of Yogyakarta Province Indonesia. Asian Soc Sci 10(10)

Retnowati A, Anantasari E, Marfai M, Dittmann A (2014) Environmental ethics in local knowledge responding to climate change: an understanding of seasonal traditional calendar *Pranoto Mongso* and its phenology in karst area of Gunung Kidul, Yogyakarta, Indonesia, 4th International Conference on Sustainable Futures for Human Security, Sustain 2013, Procedia environmental sciences, 20:785–794

Reyes-Garcia V, Fernandez-Llamazares A, Gueze M, Garces A, Mallo M, Vila-Gomez M, Vilaseca M (2015) Local indicators of climate change. WIREs Clim Change

Schmuck-Widmann H (2001) Facing the Jamuna river: indigenous and engineering knowledge in Bangladesh. Bangladesh Resource Centre for Indigenous Knowledge, Dhaka, Bangladesh

Scoville-Simonds M (2016) The governance of climate change adaptation finance—an overview and critique. Int Dev Policy 7:1

Sesotyaningtyas M (2019) The national climate change adaptation policy workshop, June 24 2019, secretariat of the national climate change adaptation plan of actions (Rencana aksi nasional adaptasi perubahan iklim sekretariat RAN API gelar workshop 24 Juni 2019, kebijakan adaptasi perubahan iklim nasional sekretariat rencana aksi nasional adaptasi perubahan kilim)

Sherman M, Berrang-Ford L, Lwasa S, Ford J, Namanya DB, Llanos-Cuentas A, Maillet M, Harper SL, IHACC research team. (2016). Drawing the line between adaptation and development: a systematic literature review of planned adaptation in developing countries. WIREs Clim Change 2016

Stainforth S, Allen M, Tredger E, Smith L (2007) Confidence, uncertainty and decision-support relevance in climate predictions. Philos Trans R Soc A 365:2145–2161

Suarma U, Hizbaron D, Sudibyakto S, Nurjani E (2018) Participatory implementation with climate change related policies in urbanised area of Indonesia. Indonesian J Geogr 50(2)

Supriyanto S, Herianto AS, Wastutiningsih SP (2011) Social community model in the coastal sand area of Panjatan Subdistrict, Kulon Progo Regency, proceedings of the national seminar on the research results of the social economics department in Yogyakarta, 8 Dec 2011. (Supriyanto S, Herianto AS, Wastutiningsih SP (2011) Modal social komunitas petanti lahan pasir pantai Kecamatan Panjatan Kabupaten Kulon Progo', Prosiding Seminar Nasional Hasil Peneltian Jurusan Sosial Ekonomi Pertanian di Yogyakarta, 8 Desember 2011)

Supriyanto I, Subejo S, Herianto SH, Wastutuningsih SP, Untari DW, Dinarti SI, Winanti DL (2012) Farmer adaptation strategies to climate change. Faculty of Agriculture, University of Gadjah Mada, Yogyakarta. (Strategi adaptasi petani terhadap perubahan iklim, Fakultas Pertanian, Universitas Gadjah Mada). Pintal Publishing, Yogyakarta

Suroso DSA, Hadi TW, Salim W (2009) Indonesia climate change sectoral roadmap (ICCSR) Synthesis report. Technical report, Republic of Indonesia, Jakarta

Tarrant J (2014) Mid-term evaluation of the climate services supporting adaptation in the Indonesian food crops system, technical report, USAID, engility/international resources group (IRG), climate services partnership mid-level evaluations at the international institute for climate and society at Columbia university. online: https://www.climate-services.org/wp-content/uploads/2016/04/Indonesia-Evaluation.pdf

Taylor M (2015) The political ecology of climate change adaptation. Routledge, UK

Thornton T, Manafsi N (2010) Adaptation: genuine and spurious: demystifying adaptation processes in relation to climate change. Environment and society: advances in research 1. Berghahn Books, pp 132–155

Tschakert P, Dietrich KA (2010) Anticipatory learning for climate change adaptation and resilience. Ecology and Soc 15(2)

UNFCCC (1992) 1992 UN framework convention on climate change (UNFCCC), international environmental treaty, adopted on 9 May 1992, Rio de Janeiro, Brazil online: https://unfccc.int/resource/docs/convkp/conveng.pdf

UNFCCC (2008) Report of the conference of the parties on its thirteenth session, Bali 3 - 15 December 2007, UN Framework Convention on Climate Change (UNFCCC), 14 March 2008 https://unfccc.int/resource/docs/2007/cop13/eng/06a01.pdf

UNFCCC (2015) Paris Agreement, FCCC/CP/2015/L.9/Rev.1 UNFCCC Secretariat

Weatherhead E, Gearheard S, Barry RG (2010) Changes in weather persistence: insight from Inuit knowledge. Glob Environ Change 20:523–528

Yoseph-Paulus R, Hindmarsh R (2018) Addressing inadequacies of sectoral coordination and local capacity building in Indonesia for effective climate change adaptation. Clim Dev 10(1):35–48

Chapter 5
Company–Community Partnership and Climate Change Adaptation Practices: The Case of Smallholders Coffee Farmers in Lampung, Indonesia

Ayu Pratiwi, Guenwoo Lee, and Aya Suzuki

Abstract Climate change affects agricultural production system and the livelihood of the smallholder producers. To help farmers increase their adaptive capacity, private sector may serve as an alternative to the government through company–community partnership, which aims to directly acquire production from the farmers while simultaneously increasing their capacity through knowledge exchange, market access, and social capital. Despite their pivotal roles, how the private sectors operate in promoting climate change adaptation and mitigation strategies to their smallholder clienteles are still understudied. This chapter examines the effects of company–community partnership upon the climate change adaptation and mitigation practices of the smallholders' coffee producers in Lampung, Indonesia. Using propensity score matching (PSM) and inverse probability weighting regression (IPWR), we found that farmers possessing motorized vehicle, larger farm size, and more active networks inside their locality are more likely to join the partnership. Such partnership also positively affects the adaptive capacity of smallholder farmers in two ways. First, it improves their farm income, thus reducing their income vulnerabilities; and second, through increased propensity to adopt resource-conserving and agroforestry techniques due to more opportunity for knowledge exchange and a higher degree of social capital.

Keywords Agricultural technology · Climate change · Adaptation · Mitigation

---

A. Pratiwi (✉)
Turku School of Economics, Economic Geography Unit, Turku, Finland
e-mail: ayu.pratiwi@utu.fi

G. Lee
Institute of Economic Research, Hitotsubashi University, Kunitachi, Japan

A. Suzuki
Graduate School of Frontier Sciences, Department of International Studies, The University of Tokyo, Bunkyo City, Japan

R. Djalante et al. (eds.), *Climate Change Research, Policy and Actions in Indonesia*, Springer Climate, https://doi.org/10.1007/978-3-030-55536-8_5

Introduction

Climate change poses challenges on the future of agriculture and the livelihood of the global population. Assessment report developed by the IPCC's Working Group II in 2014 focused on two challenging topics, namely agriculture and food security (Field 2014). Chapter 7 specifically outlined the current effects of climate change, the projected decline in crop production by 2030 that may follow, and what farmers can do to mitigate some of the negative impacts of climate change. The report concludes that adaptation of current crops production and farming techniques will be key to a sustainable future, and mitigating the effects on agriculture will require concerted efforts from intersectoral stakeholders.

In Indonesia, climate change presents significant risks to meet the demand of production and distribution of food. The country's population has increased at an average rate of 1.49% during the period of 2000–2010 (Firman 2016), causing pressures on Indonesia's already limited resources to ensure food security of its growing population. Poverty alleviation too remains a challenge with 10.96% of Indonesia's population still living in poverty in 2014 (Indonesia 2015). At the same time, deforestation, forest fires, and degradation of peat land have placed the country as the world's fifth largest emitter of greenhouse gases and the largest contributor of forest-based emission (Wijaya et al. 2017). Apart from the deteriorating environment, climate change will increase economic vulnerability of the smallholder producers due to significant projected decrease in yields. More intense rainfall is predicted, which can result in about 2–3% more rainfall each year, in addition to decreased soil fertility by 2–8%, leading to the declining share of rice yields by 4% per year and maize by 50% per year (Sari et al. 2007).

It is therefore critical for Indonesia to achieve its national climate pledges in keeping the earth temperature rise well below the 2 °C threshold called for in the Paris Agreement. As outlined in the Nationally Determined Contribution (NDC) document to the Paris Accord, Indonesia has set ambitious goals related to production and consumption of food, through integrated plans to build resilient food system through the enhanced actions on sustainable agriculture and plantation (Indonesia 2015). One imperative measure is to introduce short-term climate adaptation and long-term mitigation strategies to the actors involved in the food systems, who are primarily agricultural producers. However, governments often fail to provide incentives to increase the demand for adaptation and to implement efficient strategies for the agricultural producers who are mostly the smallholders. Past studies have also shown that barriers to adaptation and mitigation practices are high, primarily due to technical, information, and cultural constraints. For instance, "Climate Field School" in Indramayu region in 2003 was carried out to promote adaptive application of climate forecasts to crop selection decisions, but after 5 years, adoption is virtually non-existent (Siregar and Crane 2011). This is due to perceived risks associated with such practices and lack of understanding of the benefits. In Papua, farmers perceived that climate change has little impact to their livelihood due to more apparent important factors such as logging, mining, infrastructure development, and political decentralization (Boissière et al. 2013), which explained why the programs on climate change have little to no effects.

The role of private sectors is therefore indispensable to fill the vacuum in pushing the adaptation and mitigation strategies among the smallholders. Recently, voluntary sustainability standards such as Rainforest Alliance and UTZ, and company–community partnership, have been introduced as an alternative channel to increase adaptation intensity. The main purpose for private sectors, particularly multinational companies (MNCs) to engage in the company–community partnership, is to procure commodities directly from the farmers, resulting in more effective supply chains and consequently increased demands of production from their main export markets (Narrod et al. 2009). These partnerships have strong potential to improve the farmers' adaptation through several mechanisms, such as by offering training and support and microfinancing. Technical trainings are often implemented to increase farmers' knowledge of climate change impacts and potential adaptation technologies to generate better quality of produce and protect the environment. Sometimes supports in the form of a microfinance scheme was promoted to provide farmers with the possibility to finance adaptation measures, and more efficient adaptation technologies were distributed over the initiative's network (Borsky and Spata 2018).

Despite their major roles in shaping the global agricultural supply chain and possibly strong roles in enforcing adaptation and mitigation strategies, little attention has been given to examine to what extent private sectors manage to increase their smallholder clienteles' adaptive capacity. This chapter speaks to fill those gaps by addressing the following research questions: first, what are the determinants of participation in company–community partnership initiated by the private sector; and second, to what extent this partnership helps the smallholders increase their farm income and climate change adaptation strategy simultaneously through the adoption of resource-conserving agricultural practices, crop diversification, and forest tree planting.

The rest of the chapter is organized as follows: the Literature Review section provides a theoretical ground of the company–company partnership and how such initiatives were carried out in the study area; the Methodology outlines the details of the survey undertaken between 2012 and 2014, with respondents' profile comprises coffee farmers with and without partnership status in almost equal numbers; the Estimation Strategies describe the regression techniques employed in this study, namely the propensity score matching (PSM) and inverse probability weighting regression (IPWR); the Results section elaborates the estimation results; and finally the last section concludes with conclusion and the broader discussion upon this study's relevance to the Paris agreement.

Literature Review

Company–Community Partnership

Partnership between community and companies has been hailed as one of the key mechanisms toward sustainable development in a global scale at the Earth Summit in Johannesburg in 2002 (Vermeulen et al. 2003). Involvement of the private sectors in the natural resources management and governance has garnered both optimism and skepticism. The proponents suggest that partnership with companies may encourage more opportunities for local actors to exploit their comparative advantages while at the same time obtain technology transfer and more effective market access (Vermeulen and Mayers 2006). Conversely, critics widespread that partnership may replace instead of complementing the roles initially carried out by the government, thus may overlook the bigger issues such as climate change and poverty alleviation. For instance, Bitzer et al. (2008) found that partnerships are unable to turn the coffee chain into a sustainable chain due to imbalance in actor involvement.

In Indonesia, most of coffee growers are smallholder farmers who generate majority of the country's output, overpowering big state plantation. The general coffee supply chain is shown in Fig. 5.1. The ever-increasing global coffee demand necessitates multinational companies (MNCs) to procure coffee beans directly from the farmers. Traditionally, firms' linkages with smallholders have been solely through procurement from farmer groups. This system is quite common as a means of reducing the transaction costs of dealing with many smallholders individually (Chowdhury et al. 2005). In 1994, multinational companies (MNCs) through the domestic roasters as their subsidiaries started to provide access to markets for the coffee farmers community in Lampung through the establishment of the cooperatives, which features to include the ROSCA[1] function, within the farmers group.

In an effort to establish the partnership agreement, company officers work with extension official to locate farmers group producing the better yields, which imply that either the farmers group selected for partnership performed relatively better than the others or the members are influential enough to gain access for the whole group via extension official. However, other farmer groups having intention to join the partnership may apply for the certification to the company representatives, which will then conduct a series of inspection upon the quality of coffee production and yield to determine the access to the partnership. The partnership is agreed on group basis, and not on individual levels.[2] The partnership agreement primarily gives its members the rights to directly sell the coffee beans to the MNCs or domestic roasters with premium price. However, the farmers should adhere to the several mandatory quality

[1]Rotating Savings and Credits Association, a peer-to-peer lending/banking system by the group of individuals who agreed to meet for a defined period (usually every month) to save and borrow together, also known as *Arisan*.

[2]This means the treatment group is assigned at farmers group level, but analysis is done at individual level. As a result, there may be a group effects heterogeneity that affected the treatment assignment that may not be accounted for in the analysis.

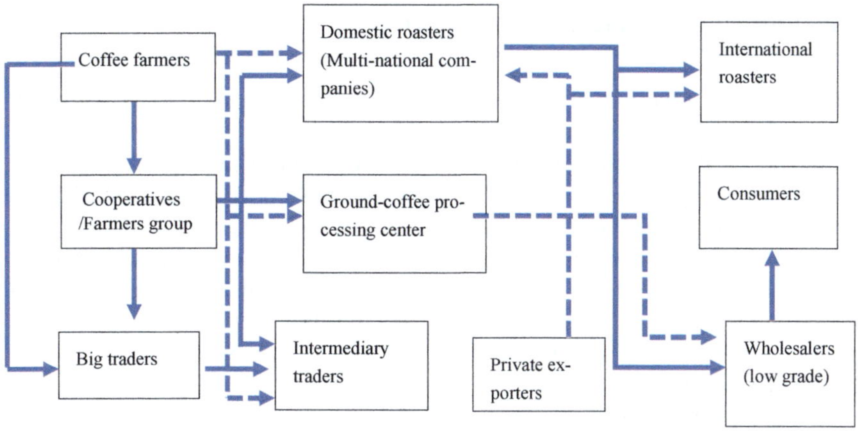

Fig. 5.1 Coffee supply chain in Lampung

checks for production, namely the zero usage of chemical pesticides, resulting in export quality of coffee produce. The MNCs examine the quality of the coffee beans after they brought them to the action center every time they purchase. If the coffee beans passed the quality control, then MNCs will buy those beans and transfer the money to the farmers. The process usually takes two days to one week for inspection until farmers obtain their money. According to the interviewed farmers and extension officials in the region, MNCs usually set a higher price of 10–20% more than the market price for the beans procured via the partnership.

What makes the practice different with "contract farming" is that MNCs did not specify the quantity and the timing to sell so farmers may individually sell to the MNCs if they can show the agreement certificate. If they did not meet the quality standard, MNCs will return the produce and farmers may sell to non-MNCs or local traders. Furthermore, MNCs have sometimes channeled their corporate social responsibility (CSR) initiatives into them, and even empowered the farmers with access to financial cooperatives, as the partnership agreement certificate may serve as collateral. As the focus of MNCs is predominantly on the organic production, the MNCs sometimes carried out various training to thousands of Indonesian coffee farmers on the use of organic substances and to improve yields.[3] While the use of organic farming is encouraged by the MNCs, the usage of resource-conserving and agroforestry techniques as a means to mitigate climate variabilities are often overlooked. Farmers both in partnership and non-partnership have the same access to agricultural extension officials as their primary sources of agricultural information, which may provide them with environmental conservation techniques that the MNCs are lacking. This is why farmers in both groups may have the same opportunity for the adoption of these techniques, making their cases comparable.

[3]For instance, see https://swa.co.id/swa/listed-articles/pusat-pelatihan-untuk-petani-kopi (accessed 02/04/2019) and https://food.detik.com/berita-boga/d-2565092/cofffee-made-happy-program-pelatihan-kopi-akan-dimulai-di-lampung (accessed 02/04/2019).

Agricultural Characteristics in Lampung

Coffee and cocoa have been major export commodities from Lampung province, most notably in Tanggamus and West Lampung districts. Climate and soil conditions are suitable for the growth and production of coffee plants. Coffee popularity in the region is driven by the relatively high-selling and stable price since 2007. Climate change may affect the production of coffee beans due to changes in rainfall patterns and increases in air temperature. The prolonged dry season may decrease the quality of coffee beans due to the increasing number of empty fruits. Extreme changes in air temperatures can modify the biochemical composition and the flavor (Carr 2001; Silva et al. 2005); it can also cause stressed plant growth and abnormalities (DaMatta and Ramalho 2006).

Efforts to address climate change on coffee plants can be done through the application of cultivation technology, both in terms of adaptation and mitigation to climate change. We review the climate-related environmental changes, their impacts, and the coping mechanism in Table 5.1. As outlined in Table 5.1, efforts to tackle the climate effects from the perspective of smallholder producers can be done through the application of appropriate cultivation strategies. The short-term adaptation strategies include planting and protecting crops and trees through the management of shading tree planting; managing soil nutrients and erosion through the application of soil and water conservation techniques; crop diversification to reduce income volatility through incorporation of more perennial crops; and switching to more resilient crops or strengthening the physicality of the crops from salinity, heat, and pest, through application of grafting methods. One of the long-term adaptation strategies for the smallholders comprises capturing more carbon from the atmosphere through agroforestry systems with forest tree planting.

A wide range of studies (Albrecht and Kandji 2003; Palm et al. 2005) has substantiated the fact that agroforestry systems, even if they are not primarily designed for carbon sequestration, present a unique opportunity to increase carbon stocks in the terrestrial biosphere. Agroforestry trees play important roles to cycle nutrient and improve the microclimate (Mbow et al. 2014). In the surveyed area, farmers retained hardwood trees in their farmland, such as Teak *(Tectona grandis)*, Albasia *(Albizia Falcata)*, Jabon *(Anthocephalus cadamba)*, Banyan *(Ficus microcarpa)*, Rattan (*Calamus manan)*, and Bamboo *(Bambusoideae)*. These hardwood trees are seen as monetary investments due to the high-valued wood price while at the same time served as shading tree for cocoa, fruits, and other perennial crops. Simple agroforestry practices to incorporate more diverse perennial crops are also found to particularly benefit the low-income farmers in reducing their income volatility and chances of complete crop failures (Pratiwi and Suzuki 2019).

Climate change results in erratic rainfall and high- and low-temperature spells, making the plant survival as well as food and nutritional security under threat because fruits and vegetable crops are found to be sensitive to environmental extremes like drought and flooding (Bhatt et al. 2013). One of the methods is to enhance or improve the genetic tolerance to such stresses by developing tolerant varieties through grafting

Table 5.1 Climate change adaptation and mitigation strategies, barriers to adaptation, and possible intervention

Climate change adaptation and mitigation strategies, barriers to adaptation and possible intervention							
Short-term adaptation strategies	Possible solution	Long-term mitigation strategy	Possible solution	Barriers to adaptation	Possible solution	Possible intervention	Possible solution
Planting and protecting crops and trees[a]	Shade-trees/hardwood cultivation	Sequestering CO_2 from the atmosphere[a]	Forest-tree planting agroforestry system	Technical constraints[a]	Technical counselling	Technical assistance[a]	Frequent meeting with formal social networks i.e. company representative and agricultural extension official
Managing soil nutrients and erosion[a]	Soil and water conservation technique			Information constraints[a]	Frequent transmission of agricultural information	Better output—market integration[a]	Better product quality and better access to improved market through partnership with companies
Crop diversification[a]	Perennial-crops agroforestry			Financial and resources constraints		Social empowerment[a]	Frequent meeting with informal social networks i.e. fellow farmers, farmer group members

(continued)

Table 5.1 (continued)

Climate change adaptation and mitigation strategies, barriers to adaptation and possible intervention							
Switching to varieties tolerant to heat, drought or salinity[a]	Bud-grafting techniques	Reducing carbon emission		Institutional constraints		Replanting and conservation	
Diversifying farm operation (combination of livestock, aquaculture, non-farm)				Social and cultural constraints		Funding, aid, financial assistance	
Water management, optimizing irrigation						Policies, regulations, and law enforcement, and infrastructure improvements	

Note those with[a] are the adaptation and mitigation strategies we specifically try to cover in this chapter. Adapted and synthesized from various sources, i.e., Bezner-Kerr et al. (2012), Beyerl et al. (2018), Albrecht and Kandji (2003), and Mbow et al. (2014)

over selected vigorous rootstocks. The application of bud grafting techniques for coffee plants is encouraged by the agricultural extension official in the region. Coffee plants developed through the bud grafting technique are faster to grow and yield better harvest. At the age of one year since the connection is made, harvest can be obtained. Selection of budding candidates by farmers is generally very simple, which is based on the criteria of high fruit production and large seed size (Randriani et al. 2014).

If dry season is prolonged for more than three months consecutively, a traditional conservation method called *rorak* (soil and water conservation method) is needed. *Rorak* is made with length 75–100 cm, width 30–40 cm, and depth 40–60 cm, with distance to coffee plants around 60–100 cm apart (Supriadi 2015). The amount of *rorak* should be 50% of the total coffee plant per hectare. To improve the structure and porosity of the soil and increase the ability to bind water, the rorak must be filled with mulch or leaf litter. In Lampung region, *rorak* is suitable in the coffee plantation with the slope level of 8–15%, especially in landscape that is sensitive to erosion (Mulyoutami et al. 2004).

Methodologies

We administered the survey to 16 randomly selected farmers groups out of 36 listed groups based on regional census in 2008 in Sumberejo and Pulau Panggung sub-district in Tanggamus district, which listed 240 coffee farmers as their members in 2012. Of 16 groups, seven groups belong to the company partnership. Data collection spans three years period, from September 2012 to 2014. The partnership agreement for all surveyed groups started before our baseline survey. We collected information on household demographic, various socioeconomic characteristics, and the knowledge and adoption of resource-conserving agricultural practices, namely water and soil conservation practices, grafting methods, agroforestry, crop diversification, and income generated from farming activities, as our outcome variables.

Table 5.2 shows the summary statistics of the variables examined in this study. Farmers in the company partnership hire more labor and own larger farmland; they are also more connected through transportation means such as car and motorbike. In terms of network, farmers in the partnership are better connected, and particularly have intensive interaction with fellow group members. No significant differences are found in terms of education and age between the farmers who belong to the partnership versus non-partnership. However, results on the outcome variables show significant differences between these groups, with farmers who belong to the partnership tend to have higher farm income, adopt conservation and grafting techniques, practice crop diversification, and forest tree based agroforestry.

Table 5.2 Summary statistics

Variables	Partnership		Non-partnership		*p*
	Mean	SD	Mean	SD	
Household education					
Age of head	45.608	11.637	45.610	11.561	0.999
Years of schooling of head	8.285	3.394	8.172	3.199	0.654
Years of schooling of most educated person	10.462	2.812	10.855	2.585	0.056
Years of experience as coffee farmers	17.433	11.915	17.008	10.504	0.612
Household composition					
Ratio of males in the household	0.507	0.162	0.549	0.175	0.001
No. of household members	3.856	1.166	4.072	1.367	0.027
No of adults in the household (15–70 years old)	2.777	1.013	2.826	1.125	0.555
Migration status					
Lampung native	0.091	0.288	0.067	0.250	0.227
Second-generation migrant	0.645	0.479	0.575	0.495	0.055
Farm characteristics					
No. Of hired labor	3.651	2.709	2.962	2.453	0.004
Owned farmland (ha)	1.375	1.031	1.083	0.825	0.000
Cultivated farmland (ha)	1.342	0.962	1.064	0.809	0.000
No of cattle owned	2.000	1.275	2.107	0.994	0.714
Access					
Own mobile phone (=1 if yes)	0.860	0.348	0.836	0.371	0.382
Own motorbike (=1 if yes)	0.917	0.276	0.831	0.376	0.000
Own car (=1 if yes)	0.096	0.296	0.039	0.194	0.002
Network information					
No. Of names household seek for farming advice	5.200	2.995	4.337	2.967	0.000
No. Of names household seek for farming advice from within farmers group	3.273	2.824	2.654	2.650	0.003
No. Of names household seek for farming advice from outside farmers group	1.544	1.694	1.340	1.598	0.102
Know extension agent (=1 if yes)	0.895	0.307	0.889	0.315	0.781
Outcome variables					

(continued)

Table 5.2 (continued)

Variables	Partnership		Non-partnership		p
	Mean	SD	Mean	SD	
Log of total farm income	16.466	1.137	16.017	1.144	0.000
Adopting grafting (=1 if yes)	0.945	0.228	0.872	0.334	0.001
Adopting conservation techniques (=1 if yes)	0.873	0.334	0.825	0.380	0.079
Planting spice crops (=1 if yes)	0.708	0.455	0.611	0.488	0.006
Planting hardwood (=1 if yes)	0.190	0.393	0.128	0.334	0.022
Observations	363		360		

Estimation Strategy

We conduct estimation in two phases. The first stage of regression estimates the determinants of joining the company partnership, and the second estimates whether company partnership substantially drives the adoption of sustainable practices, i.e., the soil and water conservation and grafting methods, and plant diversification, and improves farm income.

For the first step of regression, we evaluate the marginal probability effects (MPE) of explanatory variables to be in a partnership with companies based on probit and logit models. Through the result of this analysis, we intend to define what factors determine the partnership with companies. We estimate

$$\frac{\partial P(\text{Dummy Partnership}_i = 1)}{\partial x_{it,j}} = \emptyset\left(x_{it}^{'}\beta\right)\beta_j \quad \text{MPE in Probit} \tag{5.1}$$

$$\frac{\partial P(\text{Dummy Partnership}_i = 1)}{\partial x_{it,j}} = \Lambda\left(x_{it}^{'}\beta\right)\left(1 - \Lambda\left(x_{it}^{'}\beta\right)\right)\beta_j \quad \text{MPE in Logit}$$

where t implies the year of 2012, 2013, and 2014; Dummy Partnership = 1 denotes membership in company partnership and Dummy Partnership = 0 denotes the non-membership; $x_{it,j}$ represents household characteristics relevant to the membership in the partnership in each year of 2012, 2013, and 2014; β is the vector of coefficients to be estimated; and $\varnothing\left(x_{it}^{'}\beta\right)$ is the value of the standard normal probability density function at $x_{it}^{'}\beta$.

In the second step, we examine the effects of having these partnerships with MNCs on farmers' income and climate change adaptation status. However, individuals may self-select into treatment based on their anticipated monetary and nonmonetary benefits, resulting in possible reverse causality (endogeneity) between the partnership status and the outcome variables. Due to the absence of baseline data, we could not employ a more rigorous statistical model to eliminate endogeneity. With this limitation, the optimum way is to compare the farmers with partnership status and without. Thus, we employed propensity score matching (PSM) and inverse propensity score weighted regression (IPWR) to reduce the differences that were due to observable characteristics between farmers in partnership and non-partnership. While PSM and IPWR can control for the observable effects related to the partnership agreement, they cannot exogenize the unobserved effects such as willingness to work hard. In conducting PSM, we ensure that the covariates are adequately balanced between treatment and control individual to remove the observable bias between the two groups. Hence, the systematic difference between farmers with partnership and non-partnership with the same values of covariates can be attributable to the partnership.

While we conduct various algorithm methods for the matching, the kernel matching methods are reported here, because they gave the lowest mean bias estimates. Kernel matching (KM) are nonparametric matching estimators that use weighted averages of all individuals in the control group to construct the counterfactual outcome. Thus, one major advantage of these approaches is the lower variance which is achieved because more information is used (Caliendo and Kopeinig 2005). We then estimated the average treatment effect on the treated (ATT) using bootstrapping methods.

$$\begin{aligned} \text{ATT} &= E\big(Y_1 - Y_0 | \text{Dummy Partnership}_i = 1\big) \\ &= E\big(Y_1 | \text{Dummy Partnership}_i = 1\big) \\ &\quad - E\big(Y_0 | \text{Dummy Partnership}_i = 1\big) \end{aligned} \tag{5.2}$$

where $E\big(Y_1 - Y_0 | \text{Dummy Partnership}_i = 1\big)$ is the expected causal effect of the partnership for farmers joining the partnership. $E\big(Y_1 | \text{Dummy Partnership}_i = 1\big)$ is the observed outcome, while $E\big(Y_0 | \text{Dummy Partnership}_i = 1\big)$ represents the counterfactual outcome.

Although the PSM is able to remove the systematic observable difference between users and non-users, it reduces the efficiency of the estimation (Heckman et al. 1998). Therefore, we further conduct an inverse propensity score weighting regression to achieve consistency in estimates in a process known as "doubly robust" estimation. This method uses the inverse of the propensity score as weights to run a linear regression of the outcome variables, as proposed by Robins and Rotnitzky (1995) and as later improved by Hirano et al. (2003). Here, the weight is equal to one for observations that are in the partnership, and $px(x)/(1 - ps(x))$ where px is the propensity score for non-partnership observations (Hirano and Imbens 2001).

Result

Determinants of Participation in Company–Community Partnership

We estimate the determinants of participation in the company partnership and show the results in Table 5.3. Farmers with larger farm size, possession of motorbike and car as well as larger networks in their farmers group appear to have been positively associated with the membership in company partnership. Mobile phone possession,

Table 5.3 Determinants of company–community partnership: average marginal probability effects from the pooled probit and logistics model

Variables	(1)	(2)
	Probit marginal effects	Logit marginal effects
Age of household head	−0.0003	−0.0004
	(0.002)	(0.002)
Years of education of household head	−0.005	−0.005
	(0.006)	(0.006)
Log of cultivated farmland	0.077***	0.0762***
	(0.029)	(0.029)
Mobile phone possession (=1 if yes)	−0.111*	−0.112*
	(0.066)	(0.067)
Motorbike possession (=1 if yes)	0.159**	0.160**
	(0.075)	(0.077)
Car possession (=1 if yes)	0.162**	0.165**
	(0.082)	(0.083)
No. of advice networks who are farmers group members	0.019***	0.020***
	(0.007)	(0.008)
No. of advice networks who are not farmers group members	0.014	0.015
	(0.012)	(0.013)
Know extension officials (=1 if yes)	0.024	0.0231
	(0.064)	(0.065)
Years of experience of coffee farming	−0.0002	−0.0002
	(0.002)	(0.002)
Observations	655	655
Pseudo R-squared	0.034	0.034
F test of joint significance	0.002	0.003
chi^2	28.08	26.46

Robust standard errors in parentheses, $^{***}p < 0.01$, $^{**}p < 0.05$, $^{*}p < 0.1$

however, is found to be negatively correlated with the partnership. Thus, it can be inferred that relatively wealthy farmers with larger farm size, with motorized vehicles possession, are more influential and connected, hence the higher likelihood to join the company partnership.

Several possible mechanisms arise to explain the results. In our surveyed region which may reflect the Indonesian rural situation in general, farmers value face-to-face meeting for information sharing and tend to seek it primarily using motorbike due to the distance and contour of the rural road (Matous et al. 2015). These farmers with high capacity for mobility are more likely to be exposed with important informal channels for information sharing about resource-conserving practices (Bodin and Crona 2009); thus higher likelihood to get connected with extension officials or company officers who then refer them to join company partnership. While mobile phone is important for farmers to contact their peers, they mostly used the text message to ensure that the person they would like to meet is available at that time (Matous et al. 2015). Mobile phone serves as a catalyst of information seeking, not the primary tools.

However, only knowing such opportunity is not enough to warrant the membership, as they have to maintain the quality of their production in addition to using non-chemical pesticide. Even if only one farmer in the group did not meet the production standard, all members will be given a warning to adhere to the requirements. This is when the strong network intensity of the group helps to push the members to meet certain production qualities as set by the MNCs because the group partnership status relies on individual compliance. These peers networks may lead them to keep the constant quality of their produce, thus maintaining their membership over the time. This corroborates the results of Pratiwi and Suzuki (2017, 2020), who found that farmers who have more active networks inside their locality tend to be more innovative and resourceful in their farming practices.

Effects of Company–Community Partnership on Adaptation and Mitigation Practices

We then employ logit to match the partnership group with the non-partnership group using kernel-based matching method and estimate the propensity score of whether a farmer will join the partnership. Figures 5.2 and 5.3 show an overlap of treated and untreated cases distribution and matching situation. Through the matching, average bias decreases from 10.4 to 5.9%, and 291% untreated and 319% treated cases are located in the region of common support. The balancing test result shows that four variables were significant before the matching, but no more significant after matching, implying desirable matching quality.

Using the matching, we obtained the average treatment effect on the treated (ATT) of the company partnership in Table 5.4. We show the ATT of the company partnership upon the adoption of soil and conservation techniques, grafting methods,

Table 5.4 Effects of company–community partnership on climate change adaptation and mitigation practices: PSM Kernel matching using bootstrapping

Variable	Farm income (in Rupiah)	Log of farm income	Adopting water and soil conservation technique	Adopting grafting technique	Planting spice crops (= 1 if yes)	Planting hardwood trees (=1 if yes)
ATT	5,490,000**	0.253***	0.048*	0.060***	0.077**	0.065***
	(2,200,000)	(0.083)	(0.031)	(0.021)	(0.038)	(0.025)
Observations	650	650	650	650	650	650

Bootstrapped standard errors in parentheses, $^{***}p < 0.01$, $^{**}p < 0.05$, $^{*}p < 0.1$. Bootstrap analysis was repeated 50 times

agroforestry technique, crop diversification, and farm income. Farmers who belong to the partnership are found to significantly generate more annual farm income by 5.5 million Rp (less than US$500) compared to those who do not. This is probably contributed by the higher prices of coffee beans procured by the MNCs relative to the market price, in addition to higher production yields resulting from crop diversification compared to the control group.

While these conservation practices are not mandated by the MNCs, farmers in partnership are more likely to adopt conservation and grafting techniques compared to the control group by 5 and 6% points. They also have higher propensity to diversify with spice crops and plant hardwood trees compared to the non-partnership group by 8 and 7% points. While farmers belong to the partnership may be more

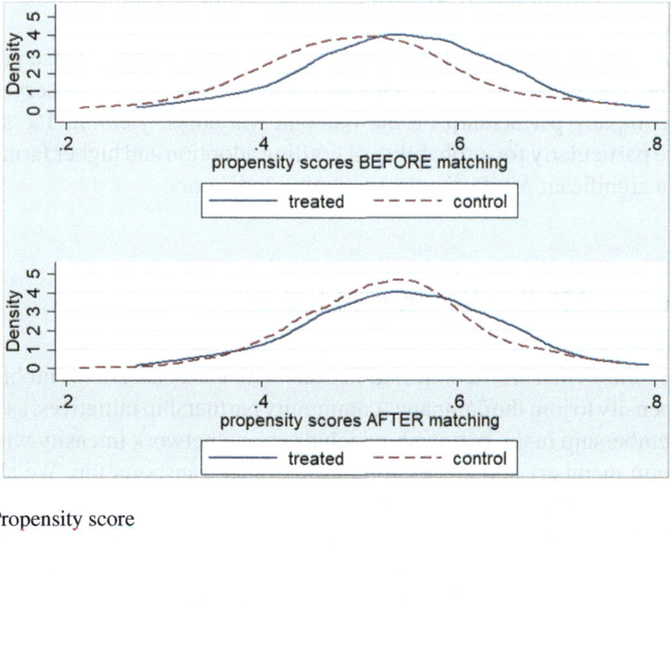

Fig. 5.2 Propensity score

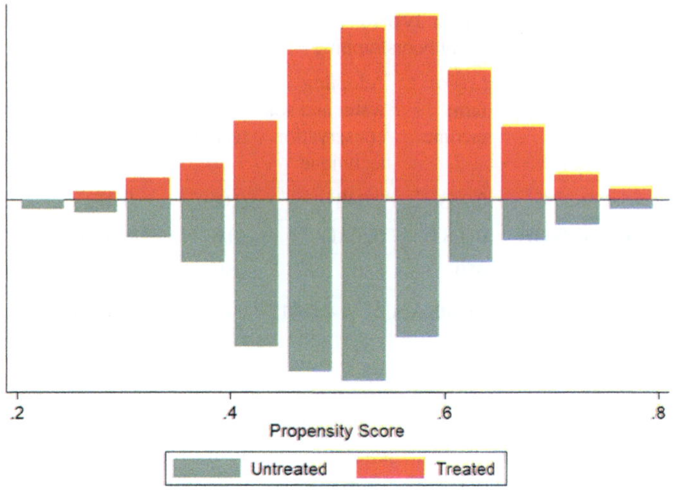

Fig. 5.3 Matching result

exposed to more information channels compared to those who do not, having information may not necessarily lead to adoption due to perceived risks and benefits associated with the technologies. However, farmers in partnership may have changed their expected returns of using such technologies, including the diversification, after obtaining significantly more income from joining the partnership. Furthermore, their strong reinforcement from the network inside their locality or farmers group may have augmented the popularity of such technologies, resulting in continuous usage of such practices.

To further enhance the robustness of our analysis, we conduct inverse propensity score weighted regression on each of the outcome in Table 5.5. We observe positive effects of company partnership on the outcome variables, yielding 1% statistical significance particularly for probability of grafting adoption and higher farm income, while retain significant results for the remaining indicators.

Discussion and Policy Recommendation

This chapter shows that well-connected farmers with better access for mobility have higher propensity to join the company–community partnership initiatives, as determinants of membership in the partnership include strong network intensity with fellow farmers group members and possession of motorized transportation. We also show that company–community partnership managed to increase the adaptive capacity of their smallholder clienteles through two channels. First is through improved farm incomes resulted from premium coffee selling prices, which is the primary feature of the partnership agreement. And second is through the usage of resource-conserving

Table 5.5 Impact of company partnership on resource-conserving technology adoption and income (Inverse Propensity Score Weighting Regression)

Variables	(1)	(2)	(3)	(4)	(5)
	Log of farm income	Adopting water and soil conservation technique	Adopting grafting technique	Planting spices crop (=1 if yes)	Planting hardwood trees (=1 if yes)
Company partnership (=1 if yes)	0.271***	0.0570*	0.0673***	0.0933**	0.0639**
	(0.0800)	(0.0313)	(0.0222)	(0.0385)	(0.0295)
Age of household head	−0.00612	0.00311*	0.000143	−0.00653**	0.00242
	(0.00624)	(0.00175)	(0.00114)	(0.00274)	(0.00240)
Years of education of household head	0.00158	0.00630	−0.00506*	−0.0145*	0.00404
	(0.0176)	(0.00557)	(0.00280)	(0.00764)	(0.00712)
Log of cultivated farm	0.873***	−0.0255	0.0328*	0.147***	−0.00893
	(0.0908)	(0.0281)	(0.0186)	(0.0384)	(0.0334)
Mobile phone possession (=1 if yes)	−0.387**	0.169**	0.0720	0.0483	−0.0124
	(0.183)	(0.0720)	(0.0520)	(0.0878)	(0.0709)
Motorbike possession (=1 if yes)	0.268	−0.105***	−0.0617**	−0.212**	−0.0779
	(0.180)	(0.0398)	(0.0266)	(0.0952)	(0.0966)
Car possession (=1 if yes)	0.441**	0.0113	0.0277**	0.166**	−0.0898
	(0.178)	(0.0602)	(0.0140)	(0.0707)	(0.0683)
No. of advice network in farmers group	0.0590***	-0.00408	0.00342	0.0103	0.00271
	(0.0142)	(0.00750)	(0.00271)	(0.00793)	(0.00752)
No. of advice network outside farmers group	0.0765***	0.00331	0.0134**	0.0156	0.00678
	(0.0282)	(0.0107)	(0.00522)	(0.0129)	(0.0124)
Knowing extension official (=1 if yes)	0.257	-0.0216	0.0542	0.0355	0.103
	(0.172)	(0.0574)	(0.0473)	(0.0842)	(0.0627)
Years of experience in coffee farming	0.00253	0.00205	−0.000193	0.00502**	−0.00462**
	(0.00490)	(0.00152)	(0.000918)	(0.00213)	(0.00187)
Living near extension official 15 min walking	0.0730	0.0386	0.0252*	−0.0114	−0.0249
(=1 if yes)	(0.110)	(0.0476)	(0.0129)	(0.0678)	(0.0598)
Native of Lampung	0.154	0.0196	−0.0363	0.130	0.0288
(=1 if yes)	(0.171)	(0.0674)	(0.0375)	(0.0896)	(0.0795)

(continued)

Table 5.5 (continued)

Variables	(1)	(2)	(3)	(4)	(5)
	Log of farm income	Adopting water and soil conservation technique	Adopting grafting technique	Planting spices crop (=1 if yes)	Planting hardwood trees (=1 if yes)
Second-generation migrant to Lampung	−0.112	−0.00541	−0.0131	0.0692	0.0719
(=1 if yes)	(0.117)	(0.0446)	(0.0200)	(0.0635)	(0.0481)
Constant	16.15^{***}	0.619^{***}	0.921^{***}	1.021^{***}	0.0614
	(0.471)	(0.140)	(0.0875)	(0.242)	(0.204)
Observations	618	646	655	655	655

agricultural techniques and crop diversification practices as short-term adaptation, and forest tree planting as long-term mitigation strategies. As these practices are not required by the partnership agreements, farmers in partnership may have obtained the information somewhere else, probably through informal interactions with market practitioners such as traders and company representatives, and agricultural extension officials. While farmers have no pressure to incorporate these techniques into their practices, they still performed them relative to the non-partnership groups. This indicates that the connections with more information source may have enabled them to revise their expected returns of using such practices through improved incomes resulted from the partnership agreements, and strong reinforcement from peer farmers to constantly use the technologies.

While climate change adaptation has yet to be the main concerns of the company–community partnership, they somehow managed to encourage adoption of adaptation and mitigation practices to their smallholder clienteles to some extent. This spillover from company partnership indicates that adoption of sustainable practices come from various channels, from local government initiatives to the private sectors. Thus, roles that were traditionally played by the public sector (for example, extension services) can be transferred to, or shared with, the private sector (Borsky and Spata 2018). However, as connections, possession of motorized transportation, and larger farm size seem to be a prerequisite of joining this partnership, this initiative may exclude the poorest of the farmers. Despite the fact that they may be the ones who suffer the most from climate change due to capital constraints, the poorest of the farmers may be worsened by the non-inclusivity nature of the partnership. Company–community partnership should consider more affirmative action initiatives to reach these poorest communities, with stronger engagement and involvement from the local government to ensure the inclusivity.

Although adaptation issues have garnered more focus during the formulation of the Paris Accord, adaptation has typically received lower levels of global support and has mobilized less action from the private sector, with only 16% of global finance dedicated toward climate adaptation in 2014 (OECD 2015). Hall and Persson (2018) also found that climate change adaptation governance is low in precision and obligation, which is partly explained by the notion that adaptation is perceived as a contested public good. While the public sectors, particularly the governments, must take the lead in financing climate change action, more active participation of the private sector in a global scale is critical to achieve the scale needed to transition to climate-resilient economies.

References

Albrecht A, Kandji ST (2003) Carbon sequestration in tropical agroforestry systems. Agr Ecosyst Environ 99:15–27

Beyerl K, Mieg HA, Weber E (2018) Comparing perceived effects of climate-related environmental change and adaptation strategies for the Pacific small island states of Tuvalu, Samoa, and Tonga. Island Stud J 13

Bezner-Kerr R, McGuire KL, Nigh R, Rocheleau D, Soluri J, Perfecto I (2012) Effects of industrial agriculture on climate change and the mitigation potential of small-scale agro-ecological farms'. Animal Sci Rev 2011:69

Bhatt S, Gething PW, Brady OJ, Messina JP, Farlow AW, Moyes CL, Drake JM, Brownstein JS, Hoen AG, Sankoh O (2013) The global distribution and burden of dengue. Nature 496:504

Bitzer V, Francken M, Glasbergen P (2008) Intersectoral partnerships for a sustainable coffee chain: Really addressing sustainability or just picking (coffee) cherries? Global Environmental Change 18:271–284

Bodin Ö, Crona BI (2009) The role of social networks in natural resource governance: what relational patterns make a difference? Glob Environ Change 19

Boissière M, Locatelli B, Sheil D, Padmanaba M, Sadjudin E (2013) Local perceptions of climate variability and change in tropical forests of Papua. Ecology and society, Indonesia, p 18

Borsky S, Spata M (2018) The impact of fair trade on smallholders' capacity to adapt to climate change. Sustain Dev 26:379–398

Caliendo M, Kopeinig S (2005) Some practical guidelines for the implementation of Propensity Score Matching. IZA–Institute for the study of labor, Discussion Paper

Carr M (2001) The water relations and irrigation requirements of coffee. Exp Agric 37:1–36

Chowdhury SK, Gulati A, Gumbira-Said E (2005) High value products, supermarkets and vertical arrangements in Indonesia.

DaMatta FM, Ramalho JDC (2006) Impacts of drought and temperature stress on coffee physiology and production: a review. Braz J Plant Physiol 18:55–81

Field CB (2014) Climate change 2014–Impacts, adaptation and vulnerability: regional aspects. Cambridge University Press

Firman T (2016) Demographic patterns of Indonesia's urbanization, 2000–2010: continuity and change at the macro level. Contemporary demographic transformations in China, India and Indonesia, Springer, pp 255–269

Hall N, Persson Å (2018) Global climate adaptation governance: why is it not legally binding? European J Int Relations 24:540–566

Heckman JJ, Ichimura H, Todd P (1998) Matching as an econometric evaluation estimator. Rev Econ Stud 65:261–294

Hirano K, Imbens GW (2001) Estimation of causal effects using propensity score weighting: an application to data on right heart catheterization. Health Serv Outcomes Res Method 2:259–278

Hirano K, Imbens GW, Ridder G (2003) Efficient estimation of average treatment effects using the estimated propensity score. Econometrica 71:1161–1189

Indonesia Ro (2015) Intended nationally determined contribution

Matous P, Todo Y, Pratiwi A (2015) The role of motorized transport and mobile phones in the diffusion of agricultural information in Tanggamus Regency, Indonesia. Transportation 42:771–790

Mbow C, Smith P, Skole D, Duguma L, Bustamante M (2014) Achieving mitigation and adaptation to climate change through sustainable agroforestry practices in Africa. Curr Opinion Environ Sustain 6:8–14

Mulyoutami E, Stefanus E, Schalenbourg W, Rahayu S, Joshi L (2004) Pengetahuan lokal petani dan inovasi ekologi dalam konservasi dan pengolahan tanah pada pertanian berbasis kopi di Sumberjaya, Lampung Barat. Agrivita 26:98–107

Narrod C, Roy D, Okello J, Avendaño B, Rich K, Thorat A (2009) Public–private partnerships and collective action in high value fruit and vegetable supply chains. Food Policy 34:8–15

OECD O (2015) Climate finance in 2013–14 and the USD 100 billion goal. Organisation for Economic Cooperation and Development (OECD) in collaboration with Climate Policy Initiative (CPI)

Palm CA, van Noordwijk M, Woomer PL, Alegre J, Arévalo L, Castilla C, Cordeiro DG, Hairiah K, Kotto-Same J, Moukam A (2005) Carbon losses and sequestration following land use change in the humid tropics. Slash and Burn: the Search for Alternatives Columbia University Press, New York (USA) pp 41–63

Pratiwi A, Suzuki A (2017) Effects of farmers' social networks on knowledge acquisition: lessons from agricultural training in rural Indonesia. J Econ Struct 6:8

Pratiwi A, Suzuki A (2019) Reducing agricultural income vulnerabilities through agroforestry training: evidence from a randomised field experiment in Indonesia. Bulletin Indonesian Econ Stud 55:83–116

Randriani E, Dani D, Tresniawati C, Syafaruddin S (2014) Hubungan Antar Karakter Vegetatif, Komponen Hasil, dan Daya Hasil Kopi Robusta Asal Sambung Tunas Plagiotrop. Jurnal Tanaman Industri dan Penyegar 1:109–116

Robins JM, Rotnitzky A (1995) Semiparametric efficiency in multivariate regression models with missing data. J American Statistical Assoc 90:122–129

Pratiwi A, Suzuki A (2020) Does training location matter? Evidence from a randomized field experiment in Rural Indonesia. Agric Food Econ 8(1):1–23

Sari A, Maulidya M, Butarbutar R, Sari R, Rusmantoro W (2007) Executive summary: Indonesia and climate change. World Bank Working Paper on Current Status and Policies, Department for International Development Available at https://siteresourcesworldbankorg/INTINDONESIA/Resources/226271-1170911056314/3428109-1174614780539/PEACE-ClimateChangepdf

Silva EAd, Mazzafera P, Brunini O, Sakai E, Arruda FB, Mattoso LHC, Carvalho CR, Pires RCM (2005) The influence of water management and environmental conditions on the chemical composition and beverage quality of coffee beans. Braz J Plant Physiol 17:229–238

Siregar PR, Crane TA (2011) Climate information and agricultural practice in adaptation to climate variability: the case of climate field schools in Indramayu, Indonesia. Culture Agric Food Environ 33:55–69

Supriadi H (2015) Budidaya tanaman kopi untuk adaptasi dan mitigasi perubahan iklim. Perspektif 13:35–48

Vermeulen S, Mayers J (2006) Partnerships between forestry companies and local communities: mechanisms for efficiency, equity, resilience and accountability. Partnerships in Sustainable Forest Resource Management: Learning from Latin America: 127

Vermeulen S, Nawir AA, Mayers J (2003) Better livelihoods through partnership? A review of the impacts of deals between communities and forestry companies on local development. International Conference on Rural Livelihoods, Forests and Biodiversity, pp 19–23

Wijaya A, Chrysolite H, Ge M, Wibowo CK, Pradana A, Utami A, Austin K (2017) How can Indonesia achieve its climate change mitigation goal? An analysis of potential emissions reductions from energy and land-use policies. World Resources Institute World Resour Inst Work Pap: 1–36

Chapter 6
Engagement of Small and Medium-Sized Manufacturing Enterprises in Individual Flood Adaptation in Indonesian Coastal Cities—Implications for Flood Governance

Thomas Neise, Matthias Garschagen, and Javier Revilla Diez

Abstract Flood risk in urban coastal areas is on the rise due to increased population, urban growth, environmental degradation, and climate change impacts. State authorities are often overstrained to provide adequate flood risk reduction, including for small and medium-sized manufacturing enterprises (SMEs), the backbone of urban economies in Indonesia. SMEs are oftentimes located in coastal areas or along rivers and are frequently affected by flooding. Hence, they are forced to adapt individually in order to protect their investment goods, maintain their production processes, and secure livelihoods of owner and employees. This chapter investigates how SMEs in urban areas are exposed to and respond to floods. We analyze which kind of flood adaptation measures SMEs implemented and why. The paper also examines the enablers and barriers to convene SMEs around collective flood adaptation measures in order to alleviate flood risk more substantially. The study uses a mixed-method approach and takes Jakarta and Semarang as examples for the empirical research. Interviews with representatives of firms, business associations, industrial parks management, and public authorities give an understanding of SMEs' adaptation challenges. A quantitative survey of 120 participants presents an overview of chosen firms' individual flood adaptation measures. The analysis shows that SMEs rely on incremental flood adaptation measures (e.g., elevating, pumps) and barely engage in more substantive or transformative risk reduction measures. The SMEs hesitate to engage in collective flood adaptation strategies due to a lack of joint governance mechanisms and sparse support from public authorities and managers

T. Neise (✉)
Institute of Geography, Osnabrück University, Seminarstr. 19 a/b, 49074 Osnabrück, Germany
e-mail: thomas.neise@uni-osnabrueck.de

M. Garschagen
Department of Geography, Ludwig-Maximilians-University of Munich, Luisenstraße 37, 80333 Munich, Germany
e-mail: m.garschagen@lmu.de

J. Revilla Diez
Institute of Geography, University of Cologne, Albertus-Magnus-Platz, 50923 Cologne, Germany
e-mail: j.revilladiez@uni-koeln.de

R. Djalante et al. (eds.), *Climate Change Research, Policy and Actions in Indonesia*, Springer Climate, https://doi.org/10.1007/978-3-030-55536-8_6

of industrial parks. In sum, the chapter provides relevant results for policy design, suggesting a stronger integration of SMEs into integrative flood adaptation policy in Indonesian coastal cities.

Keywords Jakarta · Semarang · Flood · Manufacturing firms · Governance

Introduction

In many countries of the Global South, small- and medium-sized manufacturing enterprises[1] (SMEs) are exposed to floods. This is also true in Indonesian coastal cities where a high number of firms collocate due to an abundant number of workers, a high market potential, and a better infrastructure. At the same time, the global proliferation of flood hazards is becoming increasingly visible (EM-DAT 2018) and can further be expected for the future (IPCC 2018). An important reason is that climate change tends to raise the sea-level and the magnitude of storm surges (IPCC 2018). Furthermore, high urbanization in sensitive low-lying flood-prone areas causes the reduction of green space and land subsidence due to groundwater extraction (e.g., Garschagen et al. 2018; Garschagen and Romero-Lankao 2015; Garschagen 2015; Sudmeier-Rieux et al. 2015). All these aspects accelerate the flood-proneness of SMEs.

So far, public authorities in Indonesia inadequately address the flood risk of SMEs and/or fail to include SMEs into state-led—often insufficient—flood mitigation policies in coastal cities (Neise et al. 2018). This is a remarkable deficit that hampers effective and more comprehensive flood risk reduction. A central task is, therefore, to better understand the factors that hinder an integrative flood adaptation in which all actors involved (e.g., residents, public authorities, the civil society) jointly organize and implement large-scale adaptation measures, such as a polder or pump systems. In theory, it is to be expected that such large-scale measure can be more effective than individual ones and can realize a higher investment efficiency as costs can be distributed among all actors.

Small and medium-sized manufacturing firms are the backbone of Indonesia's economy, especially to alleviate poverty and to provide jobs for the lower skilled population (Pakpahan 2013; Sandee 2009). Flood hazards induce business disruptions, such as destroyed raw materials or power outages that result in lower outputs, worker layoffs, and degrading livelihoods (Hallegatte 2014). Hence, the susceptibility of SMEs to flood hazards also leads to vulnerabilities for the residents and the economic development within Indonesian coastal cities at large.

Research with a focus on firms is burgeoning but rather focusses on estimating losses and damages provoked by natural disasters or survival rates of firms (e.g., Bahinipati et al. 2017; Pathak and Ahmad 2016). Lately, the literature is highlighting how enterprises in industrialized countries adapt to floods to sustain their business

[1]Our definition of SMEs relies on the frequently applied definition by the European Commission (2015) that call all firms with fewer than 50 employees as small enterprises and enterprises with fewer than 250 employees are defined as medium-sized enterprises.

(e.g., Berkhout et al. 2006). However, a comprehensive understanding of firms' individual competencies to adapt to natural hazards is still lacking, particularly in hazard-prone emerging and developing countries, such as Indonesia.

Also, within the debate of flood governance, SMEs play a minor role in Indonesia and at the global level. The debate around adaptation governance rather focuses on state actors, individual households, or groups of households in community-based adaptation approaches (e.g., Garschagen 2013; Garschagen 2016; Birkmann et al. 2010, 2014; DiGregorio et al. 2016). Yet, since public authorities are in many countries overstrained to provide an adequate flood risk reduction particularly also for enterprises, it seems to be necessary to include firms as a key stakeholder in adaptation governance considerations. In fact, the question arises how SMEs can enhance the performance and effectiveness of flood governance arrangements.

Thus, we argue that it is crucial to gain at first an understanding of firms' own activities on adaptation. Accordingly, in a first step, the chapter is to analyze how SMEs develop individual flood adaptation strategies by utilizing their dynamic capabilities (i.e., resources and competencies) and to identify deficiencies. Since the SMEs cannot often reduce their flood exposure substantially, we discuss in a second step how flood risk reduction can be better addressed by an integrative approach, i.e., in a joint effort of firms, together with public authorities, residents, and the civil society institutions. By doing so, the chapter complements the literature by discussing how SMEs can be vital players of flood governance in coastal cities.

The remainder of the chapter is structured as follows: Sect. "Flood Risk and SMEs' Individual Adaptation" provides a conceptual frame for the analysis of how SMEs develop their individual adaptation by utilizing their routines and dynamic capabilities. Section "The Role of SMEs on Broader Flood Governance in Indonesia" describes the methodological approach and provides a data overview. The fourth section presents the magnitude of flood exposure of SMEs and the empirical results of the chosen individual adaptation by SMEs in Jakarta and Semarang. Section "The Role of SMEs on Broader Flood Governance in Indonesia" elaborates the formal perspective on flood governance in Indonesia. It also presents examples of SMEs' engagement on flood governance and highlights deficits. Based on these results, we suggest recommendations for a more integrative flood governance in Sect. "Discussion: Proposal for a More Integrative Flood Governance". The last section provides key conclusions.

SMEs' Dynamic Capabilities on Flood Risk Reduction

Numerous papers estimated firms' loss and damage from flood events. Also, research has dealt with the recovery time and/or survival rate of firms (e.g., Bahinipati et al. 2017; de Mel et al. 2012; Sydnor et al. 2017). Other literature has focused on prevention activities by firms. For instance, Kreibich et al. (2007) discuss the preparedness and precaution measures of firms in the context of floods in Saxony, Germany 2002. A survey of small businesses and their willingness on flood risk mitigation strategies in the United Kingdom revealed that they prefer community-level based flood adaptation and would consequently waive individual adaptation (Thurston et al. 2008). Wedawatta and Ingirige (2012) conducted four case studies with SMEs in London and state that the SMEs responded to floods by protection measures (e.g., flood gates), enhancing business continuity management and usage of insurance coverage.

Existing literature deals particularly with disaster events, but recurrent and aggravating flood risks are less researched. Also, research on the individual adaptation activities of SMEs remains sparse in comparison to the plethora of research on communities and households. Moreover, this research—especially in hazard research—lacks strong theoretical conceptualization to grasp (in-)action of individual adaptation. Promising approaches to conceptualize firms' individual adaptation emerged in organization studies and evolutionary economic geography, namely the idea of routines (cf. Nelson and Winter 1982) and the dynamic capabilities framework (cf. Teece et al. 1997).

Firms' routines enable to understand firms' behavior to constantly adapt to changing conditions caused by market dynamics and/or natural hazards such as floods. Routines describe firms' behavioral patterns (e.g., standard operating procedures, organizational structures) that describe how firms carry out their regular business activities in ordinary situations (Berkhout et al. 2006; Nelson and Winter 1982; Zollo and Winter 2002). Empirical evidence shows that firms differ in the ability to adapt, i.e., some firms performing better than others due to their better routines that enable more effective adaptation to business impacts such as floods (Cyert and March 1963).

When firms face changes of high magnitude, speed or unpredictability, and their established routines become ineffective, they are required to modify their routines to facilitate an effective adaptation (Gavetti and Levinthal 2000; Nelson and Winter 1982; Zollo and Winter 2002). For instance, when confronted with a significant increase in flood exposure, firms need to start to search for new adaptation measures such as new dikes. At the same time, the dynamic capability framework stresses that the ability to introduce new routines depends heavily on firm-specific organizational and strategic abilities (Teece et al. 1997; Eisenhardt and Martin 2000). "Dynamic capabilities", therefore, refer to higher level competencies that facilitate firms to develop, reconfigure, and integrate internal and external resources and routines to address changing economic, social, and ecological circumstances. They encompass learned bundle of competencies (e.g., skills, technologies, and knowledge) that ensure firms to develop, integrate, or modify resources and routines (Teece et al. 1997).

We understand firms' individual adaptation as the mobilization of resources and learned competencies to develop, integrate, reconfigure and finally utilize firms' routines to respond effectively to flood hazards or to waive routines that have been proven as an insufficient basis for adaptation (Eisenhardt and Martin 2000; Teece et al. 1997; Zollo and Winter 2002). However, maladaptation can occur if firms possess inadequate dynamic capabilities or the efficacy of the competencies is under- or overassessed (Barnett and O'Neill 2010; Grothmann and Patt 2005). In such situations, adaptation turns into maladaptation if responses are performed that lead to an increase rather than a reduction in susceptibility and risk to floods. Also, Berkhout et al. (2006) note that companies often question the economic benefits of adaptation compared to the needed investments. Hence, the companies might waive adaptive actions or postpone activities of adaptation until the flood exposure causes severe damages and losses.

In order to relate adaptation measures to the necessary input of dynamic capabilities and their efficacy, we distinguish between soft and hard measures. Adaptation regarding organizational measures, for instance, back-up plans for production or shipping, are understood as soft adaptation. The enterprises need to learn appropriate competencies to implement and adapt them flexibly. But soft measures such as an emergency response team require continual funding and/or deployment of manpower. Soft (i.e., organizational) measures are very suitable if the flood risk can be anticipated by experience or forecast systems. Technological adaptation and investment in the built infrastructure are mostly considered as hard adaptation measures. Simple hard measures such as elevating the plant level only require minor skills and technologies. The SMEs can rely on their own competencies and implement them on their own or use small financial resources to hire a local construction company. Regarding more advanced hard (i.e., technologically based, capital-intensive) measures, e.g., a polder system, the enterprises typically need to consult and hire engineering companies. The enterprises are forced to acquire external competencies. In addition, these measures involve high investment costs (Agrawala et al. 2011; Hallegatte 2009, 2014).

Method and Data

The empirical research is based on a comparative case study approach (cf. Yin 2014) that compares SMEs in Jakarta and Semarang and regarding their dynamic capabilities. The research followed a mixed-method approach. In a first step, we assessed the flood exposure of SMEs by spatial analysis of secondary data provided by the Central Agency on Statistics of Indonesia (BPS). In a second step, qualitative data material was collected from 48 in-depth interviews with business owners or directors of small ($n = 25$) or medium-sized ($n = 23$) manufacturing firms. The interviews aimed to provide an understanding of the disruptions of firms' businesses processes caused by floods and the determinants (e.g., resources, skills and knowledge) of firms' rationales to implement adaptation measures or the inability to do so.

In addition, 21 expert interviews were conducted with other key informants. The interviews with representatives of business associations and industrial park management authorities dealt with the cooperation between the enterprises and the interviewed stakeholder within an industrial park or among firms within the business sector. The expert interviews with local public authorities (e.g., local disaster management authorities, BAPPEDA Kota Semarang, and Public Works DKI Jakarta) gave an understanding of the flood mitigation policy in Jakarta and Semarang. In addition, the interviews centered on the cooperation of public authorities with SMEs on flood adaptation initiatives and the respective regulations that facilitate or hamper public-private partnerships. In a third step, a survey with 120 SMEs collected quantitative data of which adaptation forms the firms are using.

The gathered information will be presented in a triangulated form to provide a comprehensive insight on SMEs' engagement on individual adaptation and to identify potential deficiencies. The empirical research was conducted between August 2015 and April 2017 in Indonesia. All interviews were conducted together with researchers from the Indonesian Institute of Sciences (LIPI).

The in-depth firms' interviews were analyzed following the principles of the structured qualitative content analysis (e.g., Mayring 2015). The content analysis aimed to identify the similarities and dissimilarities of firms' individual adaptation strategies regarding their location, business size, resources, and flood experience. Also, both case study regions allowed to examine the influence of different regional governance systems.

In order to preserve the anonymity of the enterprises, the interviews quotes are coded with a two-digit code. The first letter refers to the case study area (J = Jakarta; S = Semarang). The second letter indicates the business size of the interviewed enterprise (M = medium-sized; S = small).

Flood Risk and SMEs' Individual Adaptation

Flood Risk of SMEs in Coastal Cities

A large amount of Indonesia's population and of its economic activities collocate in low-elevation coastal zones (LECZ) with high exposure to floods (McGranahan et al. 2007; Neumann et al. 2015). In these flood risk areas also a high density of small- and medium-sized manufacturing firms can be found (cf. Fig. 6.1).

Figure 6.1 displays the number of SMEs located in flood-prone districts in Jakarta and Semarang. The number of small enterprises with no more than 20 employees was extracted from the Village Potential Statistics (PODES) 2014 (BPS Indonesia 2015b). Firms with a number of employees between 20 and 249 were collected from the directory of medium-sized and large manufacturing firms 2015 (BPS Indonesia 2015a). This information was joined with data from InaRisk (2018) that assessed the flood risk for entire Indonesia.

The analysis demonstrates that in both cities the highest risk of floods exists at the LECZ and along rivers (e.g., the Ciliwung river in Jakarta). So, 58.6% (6667 units) of all enterprises is located in high flood risk districts in Jakarta (e.g., Kali Deres, Cakung, Penjaringan). In Semarang, the share of SMEs located in flood risk areas, such as Genuk, Pedurungan, or Semarang Utara, comprise 54.4% (1814 units).

The SMEs face different characteristics of flood hazards in both cities. The respondents in Jakarta explained during the in-depth interviews that their businesses have been affected particularly by extreme flood events, which occurs when high rainfall

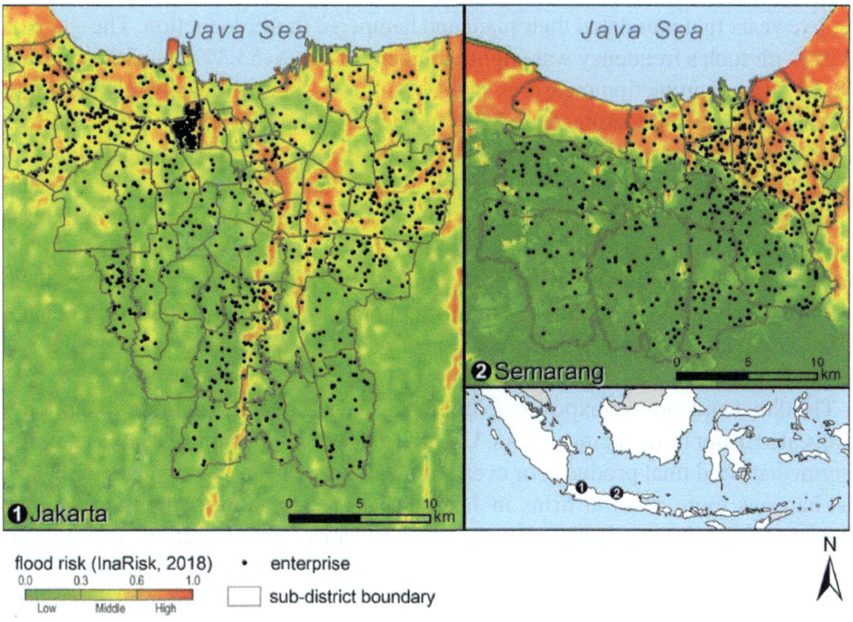

Fig. 6.1 Flood risk for SMEs in Jakarta and Semarang. *Data* BPS Indonesia 2015a, b; InaRisk 2018; design: Thomas Neise

coincides with high tides, e.g., in 2002, 2007, and 2012. Firms near the river were also annually exposed to fluvial (i.e., riverine) floods (e.g., JS-7; JS-10) and the ones, close to the coast were plagued by tidal floods several times per years (e.g., JM-3; JM-5). The origin of floods differs in Semarang. Some firms (e.g., SM-5; SM-9) were exposed to large flood events, such as in 2011 and 2014, but most of the firms experienced small, predominately tidal flooding (e.g., SM-14).

The survey shows that enterprises are more often affected by indirect flood exposure than direct impacts on their business (cf. Fig. 6.2). From 120 SMEs every second enterprise (50.8%) experienced between one and five direct flood events within the

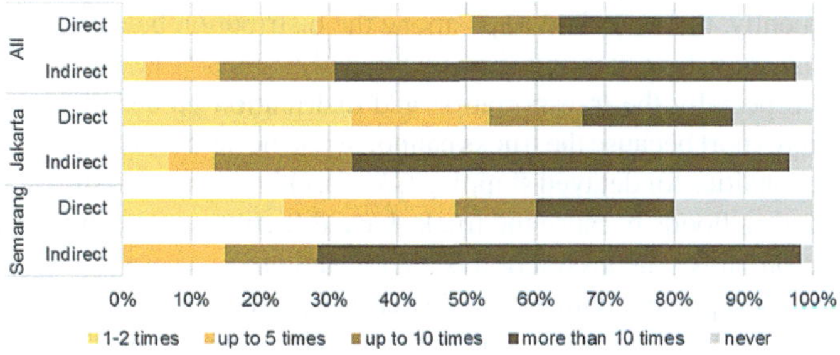

Fig. 6.2 Direct and indirect exposure to floods in Jakarta and Semarang ($n = 120$)

last five years that inundated their plant and hampered their production. The share of SMEs with such a frequency was slightly higher in Jakarta (53.3%) than in Semarang (48.3%). Usual protection measures such as pumps or sandbags can barely block the high water level inundating the SME's premises. More severe were the indirect consequences for the SMEs, meaning that they were exposed to inundated access routes, disrupted delivery by suppliers, or power outages. Two-thirds of all enterprises stated that floods indirectly exposed them more than ten times within the last five years. Although the SMEs established adaptation measures to protect their plant, they must face the consequences of inadequate flood risk reduction measures in their neighborhood. As Semarang faces more frequently tidal floods, 70% of the enterprises experienced indirect consequences caused by floods. In Jakarta, the share was 63.3%.

The direct and indirect exposure to floods caused a variety of business disruptions, as revealed in our in-depth interviews. Usually, the inundation damages raw materials, intermediate and final products, or even the machines. Due to the higher water level, this happens more often at firms in Jakarta than in Semarang. About one-quarter of the SMEs in Jakarta had to throw away and replace raw materials or products. Likewise, one-quarter of the small-scale firms in Semarang experienced losses due to damaged raw materials or products. For instance, wood panels of a furniture firm were taken by a flash flood, and a high amount of the remaining wood panels got wet and became unusable (SS-10). However, the SMEs face the highest financial losses by halted production and the inability to sell their products. Typically, the demand for, e.g., food products decreases during flood events. Market traders purchase fewer products because local markets are also inundated, or their customers cannot reach the markets due to floods (JS-7).

Generally, enterprises more commonly experienced indirect losses in both cities. Power outages belong to the most occurring problem. Either the power suppliers shut down the electricity network in order to prevent electrocution, or the power supply is just collapsing during flooding. Therefore, the enterprises rely on generators, but they can only bypass short power cuts. Enterprises with a high dependency on machinery must reduce the production rate or suspend the production process temporarily. Especially medium-sized firms are affected by power outages in Jakarta and Semarang alike (e.g., JM-3; SM-7; SM-10). Disrupted distribution channels affect primarily medium-sized firms that distribute their products regionally or exports their products. Consequently, the firms had to postpone the distribution because access roads are impassable. A fish processing firm mentioned in Jakarta (JM-5): "Everything slows down, especially the transportation and distribution process." The company had to delay its export because the trucks cannot reach the harbor. Therefore, the firms paid contract penalties for delayed shipping (JM-5). Other firms also stop the delivery completely during floods because the trucks have difficulties to reach the customers or transportation hubs. The production is halted, and products are stored to a higher level of the plant to prevent damages or decreasing product quality due to the floodwater (e.g., JM-3). The enterprises stated that usually, the customers understand the reasons for delayed shipping but enterprises, for instance, in the garment sector must pay monetary penalties. Since the enterprises compete globally, the enterprises are afraid to lose their customers to competitors in China or Vietnam. Disrupted distribution channels concern particularly firms in Jakarta because of more severe flood events than in Semarang where the shipping is mostly practicable during tidal floods. However, the salty water from the tidal floods corrodes the vehicles or machines of firms more often in Semarang (e.g., SM-1). Hence, enterprises face additional costs by replacing or repairing their assets.

SMEs' Individual Adaptation Measures

This section presents the chosen individual adaptation measures and highlights the deficiencies that the SMEs have to face. The analysis triangulates the insights from the in-depth interviews and the survey. The results of the survey give a general overview of the different forms of adaptation measures, while the interviews provide deeper insights on lacking resources and competencies to implement an effective flood risk reduction.

The survey shows that the vast majority of enterprises (80.8%) applies adaptation measures. The share of adapting enterprises is considerably higher in Semarang (85.0%) than in Jakarta (76.7%). Taking a closer look at the type of adaptation, it can be shown that elevating the plant or the access street of the facilities is by far the most widespread measure (80.6%). Due to a higher exposure to land subsidence and tidal floods, firms in Semarang more often (86.5%) elevated their plant than the ones in Jakarta (73.9%). Interestingly, community-based clean-up activities, as a soft measure, is the second most applied measure. The share of enterprises doing community-based cleaning is considerably higher in Jakarta (50.0%) than in Semarang (34.6%). In the interviews, the enterprises stated that cleaning the neighborhood is far more often undertaken after a large flood event took place than regularly to prevent clogged rivers or canals. This happens particularly in Jakarta, where riverine flooding causes severe inundation. As displayed in Fig. 6.3, all other options

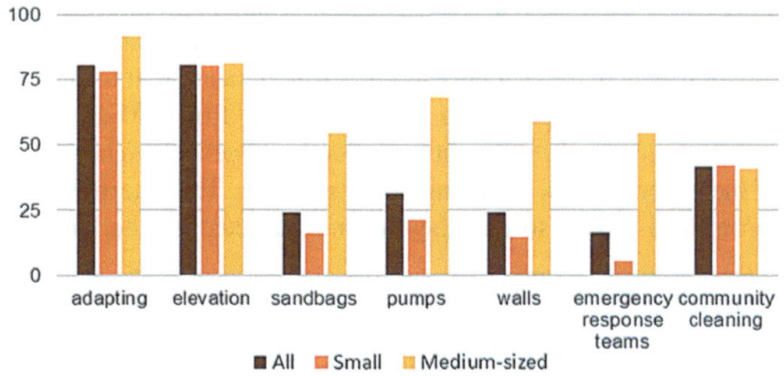

Fig. 6.3 Adaptation measures of small and medium-sized enterprises ($n = 120$). *Source* own survey

are far less implemented by the SMEs, and the share of implementation is relatively similar in Jakarta and Semarang.

Remarkably, more than every second enterprise (58.2%) that is adapting to floods ($n = 97$) uses more than one adaptation form. Both in Jakarta and Semarang, the enterprises apply on average two adaptation forms (2.2). Small firms implement on average 1.8 adaptation measures ($n = 75$; SD 0.94), while the number of adaptation measures of medium-sized firms is twice as high (3.6; $n = 22$; SD: 1.79). The higher variety of adaptation measures applied by medium-sized enterprises can be explained by the fact that the firms possess more financial resources and skills to organize a more diverse adaptation strategy that includes soft options, such as an emergency response team but also measures strengthening the built infrastructure (e.g., protective walls) or technological solutions, such as pumps. While small enterprises need to channel their scarce financial resources to fewer adaptation measures.

The interviews show that the SMEs both in Jakarta and Semarang predominately rely on their long-established routines and hardly develop new competencies of adaptation. The flood risk is relatively predictable, and the enterprises mentioned that they could handle small flooding through elevation and other hard measures, such as pumps. The firms do not see any need to extend their adaptation measures. The firms barely invest more resources in adaptation or develop new skills to find other, more effective measures. In Semarang, enterprises stated that they elevated their plant several times and plan to continue doing so (e.g., SS-1; SS-7; SM-7). However, the deficit of this adaptation choice became particularly apparent during large flood events in Jakarta (e.g., 2007, 2012) and Semarang (2012). The enterprises mainly pursued a 'wait-and-see'-strategy because their preferred adaptation measures, such as elevation of the plant or pumps, were largely ineffective against the high inundation level over weeks.

Certainly, the elevation reduces the flood risk as it prevents the water from flowing into the firm's property or the access roads from being flooded. However, the concrete used for the elevation increases the rate of land subsidence due to soil surface stress and thus increases the long-term flood risk. The SMEs always fall back on their established routine, i.e., alleviating the flood risk by elevation. For small surface and tidal floods, the elevation is effective and can protect the plant in the medium-term. However, the plant must be elevated again due to the continuous land subsidence, the elevation of main streets or of adjacent premises which means that the water can flow into their own plant. An exception is a food-processing firm (SM-14) at the harbor of Semarang that anticipated its exposure to floods proactively. The company took advantage of the expertise of a hired engineering officer. The enterprise constructed all buildings on deep pillars so that they are not affected by the land subsidence anymore.

A major reason that SMEs cannot develop new competencies or generate resources for adaptation is their weak financial conditions. Especially, small enterprises that produce metal, or plastic intermediate products face several business disadvantages that also hamper their investment budget for adaptation. They have often not recovered from the Asian financial crisis in 1997, suffer from the weak value of the rupiah that increases the prices for raw materials, and they have difficulties to compete with the quality and price of imported products, for instance, from China (JS-3; JS-4; JS-9). Their competitiveness continues to deteriorate due to business disruption by floods. Therefore, the already narrow profit margins are further hampered, thus impeding investment in more productive machines or in adaptation measures (e.g., JS-8; JS-12). Consequently, small enterprises typically stop their production and distribution when they are exposed to floods and continue after the water has receded (e.g., JDS-10). Medium-sized firms that oftentimes possess more resources and are in a better competitive situation can develop a more sophisticated flood adaptation. Thus, the enterprises can finance, for instance, special staff for emergence response or a flood risk monitoring. The survey shows that more than every second medium-sized firm established an emergency response team, while only 5.3% of all small firms integrate this soft measure in their portfolio. Medium-sized enterprises can invest their resources to train their staff and implement new organizational procedures to alleviate the impact of floods on their production processes.

The engagement of SMEs in individual adaptation is reaching its limit, especially in the case of indirect flood exposure. As mentioned before, the SMEs are more exposed to indirect flood exposure. The firms can only mitigate to certain extent side effects of flooding, such as power outages or inundated streets by generators or stopped delivery. Especially in Semarang, the respondents stated that their pumps become useless. When the rivers reach a high water level or the sea has high tide the firms cannot drain the water from their premises (SMG-7; 23; 24). The respondents complained about clogged or too narrow drainage canals (e.g., SMG-26). Moreover, the SMEs are not in a position to fix the insufficient drainage system because it is public infrastructure, and not all firms engage in cleaning clogged canals. Another issue is the unpassable streets that cause indirect business disruptions. The street network in the industrial parks has poor quality, especially in Semarang. Only the main streets are often paved, and the other streets are gravel roads with many potholes. Rain and tidal flooding convert the gravel roads into muddy ones full of puddles complicating that trucks can pass through. Main streets to the harbor or to the customer are inundated and clogged by slower traffic speed causing delayed shipping and higher transportation costs. Overall, the SMEs face additional costs and disrupted supply chains.

Although most of the SMEs adapt individually to their flood risk, their adaptation remains insufficient, thus they are still exposed to floods. Two major constraints underline insufficient flood risk reduction. First, the individual adaptation measures orientate to a large extent on hard measures (i.e., built infrastructure) that have a reactive character. The SMEs utilize their long-established competencies on flood risk reduction, even though the flood risk cannot be reduced substantially in the long run. Second, the dynamic capabilities of the SMEs are not enough to coordinate and fund larger scale flood adaptation strategies, such as an expanded drainage canal or a better street network. In order to solve these issues completely, collective approaches and the support from crucial stakeholders, such as the governmental authorities or the managers of industrial parks, are needed to cover the high costs of larger scale adaptation measures. Therefore, the remainder of the chapter discusses and analyzes whether the integration of the SMEs' engagement on flood adaptation should be rather integrated into collective adaptation in order to raise the efficacy of flood risk reduction. To do so, we review the formal perspective of flood governance in Indonesia and then present empirical results of limited flood governance.

The Role of SMEs on Broader Flood Governance in Indonesia

Formal Regulations on Flood Governance

The idea of adaptive governance is based on a collaboration of all affected stakeholders (e.g., state authorities, firms, civil society, and residents) collaborate in order to achieve adaptation to natural hazards (e.g., Archer et al. 2014; Berkes and Ross 2013). Governance is ideally characterized by polycentric and multi-layered institutions, participative and collaborative relationships, self-organization, and networks (Djalante et al. 2011; Folke et al. 2005; Pahl-Wostl et al. 2013). With a particular focus on the complexities of cities, "adaptive urban governance" argues that governance requires the consideration of the interrelationship between formal (e.g., laws and regulations) and informal institutions (e.g., customs, social capital). Therefore, effective adaptation needs structural measures (e.g., flood protection systems) as well as the integration and, if possible, the modifications or transformations of multi-level governance structures and disaster management planning systems (Birkmann et al. 2010; Solecki et al. 2017).

In Indonesia, a considerable effort has been done through revising legal frameworks of the disaster management policy accentuated especially by the passing of law No. 24/2007 on disaster management and the establishment of national and local disaster management agencies (BNPB and BPBDs) (Djalante and Garschagen 2017; Lassa 2013). The new agencies share the responsibility on disaster risk management policy and thus receive technical and financial resources to develop and implement disaster risk reduction strategies such as a hazard early warning system (BNPB 2015; Das and Luthfi 2017). Although the BPBD is assigned as the coordinator to implement disaster risk management policies, it has to coordinate its plans with numerous ministries (e.g., the Ministry of Public Work and Housing, the Ministry of Home Affairs and the Ministry of National Development Planning) as well as with the local disaster management agencies (Mardiah et al. 2017). Furthermore, the BNPB and BPBDs should respect the needs of the stakeholders. Therefore, the BNPB and the provincial BPBDs are encouraged to organize disaster risk reduction platforms that facilitate the coordination of risk reduction activities among different public authorities, the civil society, NGOs, and the private sector (i.e., firms) (Djalante and Garschagen 2017; Lassa 2013; UNISDR 2007).

In respect to the private sector, Indonesia's legal framework states that all companies should support disaster risk reduction. Formally, the contribution and responsibilities of firms are mentioned in the disaster management law (24/2007). According to article 16, paragraph 3 preparedness activities shall be the responsibility of the national and regional government, but the implementation should be jointly undertaken with the communities and business institutions[2] (Government of Indonesia 2007). Moreover, business institutions are encouraged to implement disaster management plans regularly on their own and in cooperation with other institutions (article 28 law 24/2007). Business institutions are also supposed to support the governmental disaster management policy as a form of corporate social responsibility (article 29) (Government of Indonesia 2007). This legal obligation formally applies for all firms. In reality, it is above all large, state-owned, and multi-national firms engage in disaster risk reduction (cf. Rembeth 2015). Our own research also found, for instance, that large domestic and multi-national firms financed and executed a large-scale pumping system and a drainage canal in one of Jakarta's largest industrial areas in Cakung (cf. Neise et al. 2017).

A major shortcoming concerns the potential of public-private partnerships. Respondents from public authorities explained that local regulations forbid a co-funding of large-scale adaptation measures from the public and the private sector (Interviews Bappeda DKI Jakarta 2017; Public Works DKI Jakarta 2017). Hence, public-private partnerships regarding flood risk reduction are not existing. However, within the realms of Corporate Social Responsibility (CSR), companies can finance large adaptation measures themselves, or the government can force companies to build infrastructure to compensate for obtaining a building permit (Interview Public Works DKI Jakarta 2017).

SMEs' Engagement on Flood Governance

This section presents empirical results on whether SMEs benefit from adaptation activities supported by governmental authorities and/or the management of the industrial park where many firms are tenants. We demonstrate that flood governance is often poorly established, hampering effective flood risk reduction for SMEs.

Regarding the support of the government, an ambivalent picture can be drawn. Both in Jakarta and Semarang, a few enterprises benefit from large-scale adaptation measures, such as an enlarged flood canal or a water basin built by the government. However, only enterprises in the vicinity experience fewer flooding. Many enterprises stated that they miss any action by the government alleviating their flood exposure (e.g., JM-4; JS-3; JS-14; SM-5; SM-9; SS-1; SS-11). Besides the few large infrastructure projects on flood mitigation, the government in Jakarta and Semarang typically elevated the main roads due to the ongoing land subsidence. Consequently, the enterprises need to uplift their plant level likewise in order to prevent that the rainwater flows into their property (e.g., JS-2; JS-10; SM-7; SM-11; SS-9). The

[2]The law 24/2007 defines business institution as any legal institution including state-owned enterprises, regional-owned enterprises, cooperatives, or private enterprises domicile within the territory of the Republic of Indonesia (Government of Indonesia 2007).

enterprises are forced to invest their scarce resources into an adaptation form (i.e., elevation) that barely reduce the flood exposure:

> Apparently, the government uplifted the road because they want to have a project. So, they could get money. Actually, we are sad because not all people here can afford to elevate their property. So, that their property is sinking and the people and businesses becoming the victim (SS-2).

In Semarang, the industrial park "Lingkungan Industri Kaligawe" (LIK) with a high number of SMEs in the eastern district Genuk provides an example of lacking support for the SMEs. For instance, the business director of a food-processing enterprise (SM-8) mentioned

> That's what I am always asking, why the government neglects the LIK area? The government is always focusing on developing fancy areas in the city center or new big industrial parks. Although we have to pay tax every year, the government never maintains this industrial park well.

Enterprises expressed the poor conditions in LIK. Streets are damaged, the water canals are clogged, and the access street is often impassable due to tidal floods. The respondents also stated that neighboring firms went bankrupt due to the flood-proneness and the general weak economic conditions (e.g., SM-1; SM-5; SS-8). The enterprises also do not expect any support from the management of the industrial park. The management collects money from the tenants, but there is a lack of investment in the maintenance of the industrial park (e.g., SM-1; SM-4; SM-8). Therefore, the firms demand support from the government to eliminate the poor infrastructure and flood exposure.

Usually, the lacking support for flood risk reduction mainly addresses the industrial park management authorities. Like the case of LIK, enterprises located in the industrial park Terboyo in Genuk complained about the weak action on flood risk reduction measures. The industrial park authority, under the supervision of the district government, attempted to reduce the tidal flood exposure by a pump house and a higher river embankment after several complaints of the enterprises. Also, the main access road was elevated. However, enterprises are still exposed to tidal floods. Therefore, some enterprises collaborated and elevated the street in front of the property by their own although they are tenants paying for the maintenance of the industrial park (e.g., SM-5; SM-13; SS-5). Regarding the bad condition in Genuk the representatives from the spatial planning authority (Interview BAPPEDA Semarang 2017) mentioned that the governmental authority only cares for the main access road of the industrial parks in Genuk, but they feel not accountable for the area within the industrial parks.

The situation in Genuk demonstrates that the SMEs do not receive sufficient support for flood risk reduction. Instead, they have to rely on their own engagement in adaptation, preferably as a collaboration on elevating the street level between adjacent firms. Sound flood governance between the SMEs, the industrial park management authorities, and the city authority is missing, mainly due to formal regulation on co-funding of flood mitigation that prevents collaboration between governmental authorities and the companies (cf. Sect. "Formal Regulations on Flood Governance").

A different picture can be drawn for the industrial park in Muara Baru, Jakarta. The industrial park nearby the sea predominately consists of fish processing medium-sized firms and small metal producers. The SMEs benefit from state-led adaptation measures. Nearby two large pump houses and a water basin were constructed. Also, the access street was enlarged and elevated. Lately, the protective seawall was heightened (Interviews Fisherman Authority 2017; Public Works DKI Jakarta 2017). In addition to the improved infrastructure, the enterprises appreciate the proper support from the industrial park management named PERUM and the local office of the Fisherman Authority. PERUM is proactively cleaning the area and maintaining the infrastructure. The management held regular meetings with the business owners to talk about flood problems or lacking infrastructure. For small projects, such as repairing potholes or damaged wall parts, the enterprises are providing the funding. Regarding large projects, PERUM forwards the request of the SMEs to the governmental authorities (JM-3; JM-5). This case demonstrates that close dialogue between the enterprises and the industrial park management enables the willingness of the SMEs to engage in joint maintenance flood adaptation measures. In addition, the enterprises are represented by powerful institutions (i.e., PERUM and Fishermen authority) that push the SMEs' demand for supportive flood risk reduction measures provided by the government.

However, such sound flood governance remains an exception in Jakarta, as well. For instance, enterprises in the industrial areas in the northwest of Jakarta (e.g., Kali Deres, Cengkareng) experienced little interest from the government regarding their flood exposure. A plastic products business owner (JS-1) in Kali Deres mentioned: "Apparently, the government forgets the businesses and residents here because we are too far away from the city center." The industrial parks are not managed by any authority and lack of collaboration. Also here, the government elevated the main roads and the enterprises jointly clean sporadic the rivers and canals (JM-5; JS-2; JS-5; JS-14). But for the main activities, the enterprises do not collaborate, also not with governmental authorities or within business associations. The enterprises individually elevate their plant and rely on their own pumps. The lack of engagement of governmental authorities in flood risk reduction originates due to the spatial plans for these industrial areas as a representative from BAPPEDA DKI Jakarta (2017) stated

> Based on detail spatial planning, industrial areas which are located in North Jakarta and East Jakarta are highly exposed to rob [tidal flooding], floods, and land subsidence. We see some firms in the flood-prone areas as a threat. They use groundwater extraction. […] Together with the industrial agency, we are trying to reduce the number of manufacturing firms. We are promoting the existing industries or new investments to move to Cikarang and Karawang.

Similarly, especially small enterprises at the riverbanks at Jakarta' major rivers (e.g., in Mampang Prapatan and Pejagalan) do not receive any support on flood adaptation. Since the government dredged the river and enlarged the riverbank, the enterprise will be evicted jointly with the residents (e.g., JS-8; JS-11; JS-12). Due to the eviction policy[3] (cf. Leitner and Sheppard 2017; Garschagen and Surtiari 2018), the entrepreneurs that mostly work in a semi-formal or informal business environment and within a very local market (i.e., neighborhood) lose their livelihood. At their new home, primarily social housing complex, they are not allowed to reopen their business, for instance, of food production. Overall, the Jakarta government focuses on finishing the large-scale flood mitigation project funded by the World Bank that includes increasing the water capacity of the main rivers, building large pumping stations at the coast, and heightening the coastal flood protection wall (Interviews BBWSCC 2017; Public Works 2017). However, this program has two major shortcomings. First, the severities that small enterprises will lose their business, and their customer is largely ignored. Second, the measures have reduced the magnitude of large floods especially in the central business district of Jakarta but lacks to combat inundations in settlements of lower income residents with a substantial number of SMEs and even tend toward an intensification of flooding (Garschagen et al. 2018).

Discussion: Proposal for a More Integrative Flood Governance

The empirical research shows that SMEs concentrate on individual adaptation options that have a very limited long-term impact. Especially small enterprises lack to perform sophisticated, more effective measures due to their inadequate competencies and limited financial resources. At the same time, cooperation between the enterprises and even more collaboration with governmental authorities and the industrial park managers lacks to a large extent. Also, it became clear that state-led adaptation measures are insufficient to alleviate the flood risk for SMEs and do not take place in line with the adaptation efforts of SMEs. Until now, the flood mitigation policies in Jakarta and Semarang do not seem to be sufficient, as measures by public authorities, companies, and communities are performed side by side in an uncoordinated manner (cf. Neise et al. 2018).

Given the overarching goal to reduce flood risk for the two cities as a whole, our analysis suggests that the flood mitigation policy in both cities should be enhanced toward a more integrative flood governance approach, including SMEs and utilizing their individual capacities. Such an agenda enables to gain more efficacy by synchronized flood adaptation measures and higher investment efficiency. The costs of measures can be shared, and a higher efficacy can be achieved if rather large-scale measures (e.g., a polder or pump house) are implemented. More far-reaching adaptation measures not only have a positive effect on reducing the flood risk for the enterprises but also for the inhabitants, who often live in the immediate vicinity of the manufacturing firms. Overall, the urban infrastructure is improving, and more

[3]Since 2014, mainly informal settlements at the banks of the Ciliwung river and the Pluit retention pond have been forcedly evicted from the Jakarta administration. The eviction aims to dredge the river and retention pond as well as to reinforce the embankment (Leitner and Sheppard 2017).

effective flood adaptation measures are being introduced, so that businesses remain important for local employers and taxpayers (Linnenluecke and Griffiths 2015; Neise et al. 2017). In order to demonstrate the potential of SMEs as a crucial player on collaborate flood risk reduction Neise et al. (2019) found that micro-, small- and medium-sized enterprises in Jakarta and Semarang are willing to contribute to join large-scale flood risk reduction measures (e.g., a polder system) if they receive support from public authorities and the community. Hence, the enterprises are not purely driven by their self-interest but rather are interested in participating in joint activities that can substantially reduce the flood risk for all affected actors.

In the following, we propose policy instruments that can help to unfold the potential of SMEs in supporting the state authorities and the community on adaptation to floods. Such integration could strengthen the flood governance in flood-prone coastal cities.

First, the analysis has shown that particularly the SMEs can rarely adapt effectively to flood events due to limited financial resources and strong competitive pressure. Therefore, dynamic capabilities could be strengthened by the implementation of business continuity management within small firms. Business continuity management encompasses managerial processes to identify risks that can impact critical business processes. Based on this assessment—optimally—preventative measures, preparedness activities, and response solutions are carried out (Disaster Recovery Institute 2016; International Labour Organization 2011). Studies stress that SMEs commonly lack to develop business continuity management independently (Disaster Recovery Institute 2016; Kato and Charoenrat 2018). Therefore, it is suggested that state authorities, business associations, or the chambers of commerce offer training programs and raise the awareness of business continuity planning tools. This might encourage SMEs to implement a business continuity planning in order to pursue a more proactive flood adaptation strategy. Business continuity planning can be also beneficial for the SMEs in order to improve their usually low level of competitiveness.

Second, the collaboration between SMEs and large enterprises should be encouraged. Usually, large firms own better dynamic capabilities (i.e., competencies and resources) to develop sophisticated adaptation measures (Neise and Revilla Diez 2019). Therefore, it is recommended that these firms share their knowledge with smaller firms. On the one hand, this might enhance the flood adaptation strategy of SMEs and consequently, their viability. On the other hand, large firms can take more attention to SMEs as critical suppliers. Consequently, the entire manufacturing sector and even small firms could reduce their flood risk more effectively, through an improved synchronization of individual measures. If large companies and SMEs collaborate, it is possible that large-scale, more effective adaptation measures (e.g., a polder system) can be implemented, for instance, in an industrial park. Moreover, the manufacturing sector would become more competitive and create more jobs and tax revenues; thus, could contribute to a more advanced regional development with lower flood risk for the entire region (Neise et al. 2018). As stated above, the Indonesian disaster management law already encourages corporate social responsibility activities by companies on disaster mitigation. This formal regulation should be enhanced while large companies collaborate with local SMEs on flood adaptation.

Third, we detected that the collaboration between state authorities, the community, civil society, and firms is only weakly developed; although formal regulations and laws on collaborative disaster management are articulated. Thus, it is suggested to improve law enforcement and develop a more advanced multi-level flood governance in which firms, particularly SMEs, play an active part. The interviews with political stakeholders allow drawing the conclusion that engagement on flood adaptation is requested only by large firms because they are evaluated as profitable and are motivated to express corporate social responsibility. The group of small enterprises is oftentimes outside of the debate on flood governance. Hence, it is recommended that smaller enterprises are more integrated into state programs on collective adaptation to floods. This is in line with contemporary reconfigurations of flood risk management that emphasizes the active engagement of residents in the conceptualization of flood risk reduction measures (Tseng and Penning-Rowsell 2012). For instance, in the United Kingdom, "Local National Partnerships" were established that bring together interested people, organizations, and firms that seek to improve the local adaptation to environmental hazards (Carter et al. 2015; Department for Environment, Food and Rural Affairs 2017). In this respect, small enterprises might be recruited as local partners that execute and maintain jointly with the citizen local adaptation measures, such as a pump house.

Fourth, we suggest public forums to strengthen the dialog between public authorities, private and state-owned investors on infrastructure, the residents, and the local firms. These forums can enable to improve the legitimization of flood risk reduction measures provided by the governmental authorities. For instance, the eviction policy in Jakarta demonstrates that state-led measures are designed without considering the demand of local enterprises. Hence, the flood mitigation policy, including the relocation of inhabitants and enterprises, should consider more thoroughly the interest of the owners of the enterprises. Public forums can facilitate the dialogue between public authorities performing flood mitigation and the relocated residents to ensure that they continue their business at their new location. Moreover, the efficacy of these state-led flood mitigation projects is often limited to a small-scale range because the investment is not synchronized with individual measures of the enterprises. Therefore, it is possible that the scarce financial resources of the governmental and/or corporate budgets are wasted. For instance, the state-led river dredging, and construction of pump houses is a top-down masterplan by the World Bank and the government of DKI Jakarta. Local enterprises and residents in the affected areas are barely included in the design. Hence, bottom-up initiatives and knowledge are neglected to integrate (Bott and Braun 2019). Firms stated that they hardly benefit from the masterplan of flood mitigation because, for instance, the maintenance of the pump house is weak. Even more, they were forced to invest in elevation or one drainage system to prevent their flood exposure.

Conclusions

SMEs—a central pillar for economic development and employment—are heavily exposed to floods and face many business disruptions. To date, state-led flood mitigation policies largely miss considering their susceptibility. Also, flood governance that integrates SMEs is lacking. Therefore, the book chapter had two goals. First, we analyzed how SMEs are affected by flood hazards, and we identified the implemented individual adaptation measures and their deficiencies. Second, we discussed the shortcomings of flood governance. This study was based on a quantitative survey with 120 SMEs and qualitative in-depth interviews with 47 business owners or directors and 21 representatives from public, business associations, industrial park management authorities.

Since research on firms' adaptation strategies is scarce and lacks a detailed understanding of firms' competencies to implement individual adaptation measures, we referred to the concepts from the organizational studies, namely routines and dynamic capabilities. The analysis showed that most enterprises implement individual flood adaptation measures, preferably elevation. The in-depth interviews revealed that individual adaptation measures are not sufficient to reduce flood risk. Furthermore, SMEs commonly miss support or collaboration with public authorities and industrial park management authorities. Due to this shortcoming, we discussed how SMEs could be stronger integrated into flood governance and proposes four policy designs. Overall, the engagement of SMEs should be pooled within joint efforts of flood mitigation in order to achieve a more effective flood risk reduction and a higher resource efficiency.

In sum, the chapter raised a hot topic for further scientific assessment and within policy debates. Considering a tendency of increasing exposures due to Climate Change in coastal cities in Indonesia and globally, there is a high need to utilize and synchronize the potential of all stakeholders, notably the whole private sector (i.e., SMEs) to accomplish an effective flood risk reduction. Integrating the private sector, notably SMEs, into flood governance is still unsatisfactory. However, SMEs should be a pivotal player in achieving flood adaptation because their viability is necessary to sustain and improve livelihoods and economic prosperity. Hence, SMEs also have a strong interest in implementing an integrative flood adaptation policy.

References

Agrawala S, Maëlis C, Kingsmill N et al (2011) Private sector engagement in adaptation to climate change: approaches to managing climate risks. OECD, Paris

Archer D, Almansi F, DiGregorio M et al (2014) Moving towards inclusive urban adaptation. Approaches to integrating community-based adaptation to climate change at city and national scale. Clim Dev 6(4):345–356. https://doi.org/10.1080/17565529.2014.918868

Bahinipati CS, Rajasekar U, Acharya A et al (2017) Flood-induced loss and damage to textile industry in Surat city India. Environ Urban Asia 8(2):170–187. https://doi.org/10.1177%2F0975425317714903

Barnett J, O'Neill S (2010) Maladaptation. Global Environ Chang 20(2):211–213. https://doi.org/10.1016/j.gloenvcha.2009.11.004
Berkes F, Ross H (2013) Community resilience: toward an integrated approach. Soc Natur Resour 26(1):5–20. https://doi.org/10.1080/08941920.2012.736605
Berkhout F, Hertin J, Gann DM (2006) Learning to adapt: organisational adaptation to climate change impacts. Clim Change 78(1):135–156. https://doi.org/10.1007/s10584-006-9089-3
Birkmann J, Garschagen M, Setiadi N (2014) New challenges for adaptive urban governance in highly dynamic environments: revisiting planning systems and tools for adaptive and strategic planning. Urban Climate 7:115–133
Birkmann J, Garschagen M, Kraas F et al (2010) Adaptive urban governance: new challenges for the second generation of urban adaptation strategies to climate change. Sustain Sci 5(2):185–206. https://doi.org/10.1007/s11625-010-0111-3
BNPB (2015) Indonesia: national progress report on the implementation of the hyogo framework for action (2013–2015). BNPB, Jakarta
Bott L-M, Braun B (2019) How do households respond to coastal hazards? A framework for accommodating strategies using the example of Semarang Bay, Indonesia. Int J Disast Risk Re 37. https://doi.org/10.1016/j.ijdrr.2019.101177
Carter JG, Cavan G, Connelly A et al (2015) Climate change and the city. Building capacity for urban adaptation. Prog Plann 95:1–66. https://doi.org/10.1016/j.progress.2013.08.001
Cyert RM, March JG (1963) A behavioral theory of the firm. Blackwell, Cambridge, Mass
Das A, Luthfi A (2017) Disaster risk reduction in post-decentralisation Indonesia: institutional arrangements and changes. In: Djalante R, Garschagen M, Thomalla F et al (eds) Disaster risk reduction in Indonesia. Springer International Publishing, Cham, pp 85–125
Department for Environment, Food & Rural Affairs (2017) Local nature partnerships: map and key contacts. https://www.gov.uk/government/publications/map-of-local-nature-partnerships. Accessed 14 Jun 2018
DiGregorio M, Phong TV, Garschagen M et al (2016) A political economy of urbanisation and climate risk in Vietnam. International Institute for Environment and Development (IIED), London
Disaster Recovery Institute (2016) The international glossary for resiliency. Disaster Recovery Institute, New York
Djalante R, Garschagen M (2017) A review of disaster trend and disaster risk governance in Indonesia: 1900–2015. In: Djalante R, Garschagen M, Thomalla F et al (eds) Disaster risk reduction in Indonesia. Springer International Publishing, Cham, pp 21–56
Djalante R, Holley C, Thomalla F (2011) Adaptive governance and managing resilience to natural hazards. Int J Disast Risk Sc 2:1–14. https://doi.org/10.1007/s13753-011-0015-6
Eisenhardt KM, Martin JA (2000) Dynamic capabilities: what are they? Strateg Manage J 21(10/11):1105–1121
EM-DAT (2018) EM-DAT database. https://www.emdat.be/database Accessed 03 Feb 2019
European Commission (2015) User guide to the SME definition. European Commission, Brussels
Folke C, Hahn T, Olsson P et al (2005) Adaptive governance of social-ecological systems. Annu Rev Env Resour 30:441–473. https://doi.org/10.1146/annurev.energy.30.050504.144511
Garschagen M (2013) Resilience and organisational institutionalism from a cross-cultural perspective—an exploration based on urban climate change adaptation in Vietnam. Nat Hazards 67:25–46
Garschagen M (2015) Risky change? Vietnam's urban flood risk governance between climate dynamics and transformation. Pac Aff 88(3):599–621
Garschagen M (2016) Decentralizing urban disaster risk management in a centralized system? Agendas, actors and contentions in Vietnam. Habitat Int 52:43–49
Garschagen M, Romero-Lankao P (2015) Exploring the relationships between urbanization trends and climate change vulnerability. Clim Change 133(1):37–52
Garschagen M, Surtiari GAK (2018) Hochwasser in Jakarta – zwischen steigendem Risiko und umstrittenen Anpassungsmaßnahmen. [Flooding in Jakarta – between increasing risk and contested adaptation measures]. Geogr Rund 70(4): 10–15

Garschagen M, Surtiari GAK, Harb M (2018) Is Jakarta's new flood risk reduction strategy transformational? Sustain 10(8):2934. https://doi.org/10.3390/su10082934
Gavetti G, Levinthal D (2000) Looking forward and looking backward: cognitive and experiential search. Admin Sci Quart 45(1):113. https://doi.org/10.2307/2666981
Government of Indonesia (2007) Undang-Undang Republik Indonesia Nomor 24 Tahun 2007 Tentang Penanggulangan Bencana Dengan [Law of the Republic of Indonesia Number 24 Year 2007 on Disaster Management]. Government of Indonesia, Jakarta
Grothmann T, Patt A (2005) Adaptive capacity and human cognition: the process of individual adaptation to climate change. Global Environ Chang 15(3):199–213. https://doi.org/10.1016/j.gloenvcha.2005.01.002
Hallegatte S (2009) Strategies to adapt to an uncertain climate change. Global Environ Chang 19(2):240–247. https://doi.org/10.1016/j.gloenvcha.2008.12.003
Hallegatte S (2014) Natural disasters and climate change. An Economic Perspective, Springer, Cham
Indonesia BPS (2015a) Manufacturing industry directory 2015. BPS Indonesia, Jakarta
Indonesia BPS (2015b) Village potential survey 2014. BPS Indonesia, Jakarta
International Labour Organization (2011) Multi-hazard Business Continuity Management. Guide for small and medium enterprises. International Labour Organization, Geneva
InaRisk (2018) ArcGIS REST services directory. http://service1.inarisk.bnpb.go.id:6080/arcgis/rest/services/inaRISK Accessed 12 Mar 2019
IPCC (2018) Global Warming of 1.5 °C. Cambridge University Press, New York
Kato M, Charoenrat T (2018) Business continuity management of small and medium sized enterprises. Evidence from Thailand. Int J Disast Risk Re 27:577–587. https://doi.org/10.1016/j.ijdrr.2017.10.002
Kreibich H, Müller M, Thieken AH et al (2007) Flood precaution of companies and their ability to cope with the flood in August 2002 in Saxony, Germany. Water Resourc Res 43:41
Lassa JA (2013) Disaster policy change in Indonesia 1930–2010. From government to governance? Int J Mass Emer Dis 31(2):130–159
Leitner H, Sheppard E (2017) From Kampungs to Condos? Contested accumulations through displacement in Jakarta. Environ Plann A 50(2):437–456. https://doi.org/10.1177%2F0308518X17709279
Linnenluecke MK, Griffiths A (2015) The climate resilient organization: adaptation and resilience to climate change and weather extremes. Edward Elgar, Cheltenham, UK
Mardiah ANR, Lovett JC, Evanty N (2017) Toward integrated and inclusive disaster risk reduction in Indonesia. Review of regulatory frameworks and institutional networks. In: Djalante R, Garschagen M, Thomalla F et al (eds) Disaster risk reduction in Indonesia. Springer International Publishing, Cham, pp 57–84
Mayring P (2015) Qualitative Inhaltsanalyse. Grundlagen und Techniken, Beltz, Weinheim
McGranahan G, Balk D, Anderson B (2007) The rising tide. Assessing the risks of climate change and human settlements in low elevation coastal zones. Environ Urban 19(1):17–37. https://doi.org/10.1177%2F0956247807076960
de Mel S, McKenzie D, Woodruff C (2012) Enterprise recovery following natural disasters. Econ J 122(559):64–91. https://doi.org/10.1111/j.1468-0297.2011.02475.x
Neise T, Revilla Diez J (2019) Adapt, move or surrender? Manufacturing firms' routines and dynamic capabilities on flood risk reduction in coastal cities of Indonesia. Int J Disast Risk Re 33:332–342. https://doi.org/10.1016/j.ijdrr.2018.10.018
Neise T, Revilla Diez J, Garschagen M et al (2017) Manufacturing firms' adaptations to floods and proposal for integrative adaptive regional development in Jakarta. In: Djalante R, Garschagen M, Thomalla F, Shaw R (eds) Disaster risk reduction in Indonesia. Springer, Cham, pp 281–305
Neise T, Revilla Diez J, Garschagen M (2018) Firms as drivers of integrative adaptive regional development in the context of environmental hazards in developing countries and emerging economies—a conceptual framework. Environ Plann C 36(8):1522–1542. https://doi.org/10.1177%2F2399654418771079

Neise T, Sambodo MT, Revilla Diez J (2019) Are micro-, small- and medium-sized enterprises willing to contribute to collective flood risk reduction? Scenario-based field experiments from Jakarta and Semarang, Indonesia. Organization & Environment
Nelson RR, Winter SG (1982) An evolutionary theory of economic change. Belknap Press of Harvard University Press, Cambridge, Mass
Neumann B, Vafeidis AT, Zimmermann J et al (2015) Future coastal population growth and exposure to sea-level rise and coastal flooding—a global assessment. PLoS ONE 10(3):e0118571. https://doi.org/10.1371/journal.pone.0131375
Pahl-Wostl C, Becker G, Knieper C et al (2013) How multilevel societal learning processes facilitate transformative change. A comparative case study analysis on flood management. Ecol Soc 18(4):58 https://doi.org/10.5751/ES-05779-180458
Pakpahan AK (2013) Micro, small, and medium-sized enterprises in Indonesia. Towards more coordinated politics and more competitive-based SME. In: Heise M, Rucktäschel K (eds) Indonesia's search for democracy. Political, economic, and social developments. Nomos, Baden-Baden, pp 299–318
Pathak S, Ahmad MM (2016) Flood recovery capacities of the manufacturing SMEs from floods: a case study in Pathumthani province, Thailand. Int J Disast Risk Re 18:197–205. https://doi.org/10.1016/j.ijdrr.2016.07.001
Rembeth V (2015) Indonesian Experiences of Private Sector Involvement. Focus on Partnership. In: Izumi T, Shaw R (eds) Disaster management and private sectors. Springer Japan, Tokyo, pp 279–299
Sandee H (2009) Small enterprises and decentralization. some lessons from Java. In: Holtzappel C, Ramstedt M (eds) Decentralization and regional autonomy in Indonesia. Implementation and challenges. Institute of Southeast Asian Studies; International Institute for Asian Studies, Singapore, Leiden, pp 183–195
Solecki W, Pelling M, Garschagen M (2017) Transitions between risk management regimes in cities. Ecol Soc 22(2):38
Sudmeier-Rieux K, Fra Paleo U, Garschagen M, Estrella M, Renaud FG, Jaboyedoff M (2015) Opportunities, incentives and challenges to risk sensitive land use planning: lessons from Nepal, Spain and Vietnam. Int J Disast Risk Re 14(1):205–224
Sydnor S, Niehm L, Lee Y et al (2017) Analysis of post-disaster damage and disruptive impacts on the operating status of small businesses after Hurricane Katrina. Nat Hazards 85(3):1637–1663. https://doi.org/10.1007/s11069-016-2652-y
Teece DJ, Pisano G, Shuen A (1997) Dynamic capabilities and strategic management. Strateg Manage J 18(7):509–533
Thurston N, Finlinson B, Breakspear R et al (2008) Developing the evidence base for flood resistance and resilience. Department for Environment, Food and Rural Affairs Flood Management Division, London
Tseng CP, Penning-Rowsell EC (2012) Micro-political and related barriers to stakeholder engagement in flood risk management. Geogr J 178(3):253–269
UNISDR (2007) Hyogo framework for action 2005–2015. Building the Resilience of Nations and Communities to Disasters. UNISDR, Geneva
Wedawatta G, Ingirige B (2012) Resilience and adaptation of small and medium-sized enterprises to flood risk. Disaster Prev Manag 21(4):474–488. https://doi.org/10.1108/09653561211256170
Yin RK (2014) Case study research. Sage Publication, London, Design and methods
Zollo M, Winter SG (2002) Deliberate learning and the evolution of dynamic capabilities. Organ Sci 13(3):339–351. https://doi.org/10.1287/orsc.13.3.339.2780

Chapter 7
Resilience of Coastal Cities in Facing Climate Change: Reclamation Case of Benoa Bay Bali and North Jakarta Bay

Henny Warsilah

Abstract Climate change in recent times resulted in various natural disasters. Nearly 80% of disaster events in the Asian region are a result of global climate change. In Indonesia, cities located in coastal areas are at high exposure from disasters to climate change. In addition, there are around 60% of Indonesia's population living in coastal areas. Such conditions make coastal areas potentially have problems in basic services, especially in the field of sustainable public policy and livelihoods. In contrast, Indonesia has still lack of international climate policy formulation prior to less research related to the impacts of climate change in Indonesia until now, especially if it is associated with coastal reclamation activities to develop a global tourism economy that requires more new land expansion. The aim of this paper is how does the government lead the effort to make coastal cities a better social resilience in coping with climate change? This paper is extracted from the research that had been conducted for two years (2017–2018) with title "Inclusive Development in Coastal Cities: Cases of the Reclamation of Jakarta and the Bay of Benoa, Bali", under direction of the Indonesian Institute of Sciences (LIPI). This research employs qualitative research approaches by using the concept of Bengen (2001: 56), the key to coastal areas governance are four aspects as follows: (1) ecological integration; (2) Integration of sectors; (3) the integration of scientific disciplines; and (4) the integration of stakeholders. This research also employs ICM Concept by Adrianto and Warsilah (2017–2018). Data collection is obtained from tiered focus group discussions. The objective of this paper is to analyze the social resilience of coastal areas in coping with global climate change impacts, to enrich regional climate studies and climate change policy in Indonesia. The resume of this research is the effort of Indonesian Government in handling climate change disasters but the policy has failed to encourage the formation of social security in coastal areas and their communities.

Keywords Climate change · Coastal and social resilient · Reclamation · Cultural and social adaptation · Bali · North Jakarta

---

H. Warsilah (✉)
Indonesian Institute of Sciences, Jakarta, Indonesia
e-mail: hennywarsilah@gmail.com

R. Djalante et al. (eds.), *Climate Change Research, Policy and Actions in Indonesia*, Springer Climate, https://doi.org/10.1007/978-3-030-55536-8_7

Introduction

Indonesia and other archipelagic nations in the world are currently facing a real and serious threat of climate change. The threats are in the form of sea level rises, uncontrolled coastal abrasion, coral reefs destruction, and coastal residents who have been forced to move from their homes. Geographical facts show that Indonesia is an archipelagic nation consisting of more than 17,000 islands and one of the largest archipelagic nations with a vast marine jurisdiction and a very diverse population. Indonesia itself has lost around 29.000 ha of coastal areas due to sea level rises in the northern region of Java and other regions as well. The Indonesian government has strengthened its awareness and formulated appropriate actions related to climate change. One of the results is the formation of National Action Plan to cope with Climate Change (RAN-PI) which is currently still in legal formation process. The formulation of the RAN-PI certainly shows government's willingness to deal with climate change. Climate change referred to the impact of global warming in the form of significant changes in physical and biological systems. These changes include the increase of tropical storms intensity, changes in precipitation patterns, salinity of seawater, changes in wind patterns, reproduction of animals and plants, distribution of species and population size, frequency of pest attacks and disease outbreaks, and affecting ecosystems in areas with high latitude, high location, and coastal ecosystems (KLH 2007). These changes will gradually result in a real impact, especially for archipelagic nations with long coastlines, in the form of sea level rises and high waves.

Efforts to make coastal cities with a better social security in dealing with climate change require a good adaptation. The concept of adaptation is one that needs to be prioritized and mainstreamed in a long-term planning at the national to individuals level, primarily related to urban space governance. Mainstreaming climate change adaptation is an important process in all development programs in urban areas, small islands, and coastal areas in Indonesia. However, the problems of coastal cities are quite complex, including (1) Ecological problems, for example, the loss of biodiversity, loss of marine habitat, the declining of marine productivity, the contamination and alteration of river flows, and reclamation efforts. (2) Economic problems, for instance, the rising of health costs, loss of work days, rising social services for new infrastructure and rehabilitation. (3) Social Problems, for example, unemployment, loss of income or livelihood, loss of employment opportunities, the decrease of productivity, loss of social protection, and life-supporting infrastructure needs. Based on the above problems, the question of this paper is *How the Government's efforts to make coastal cities have a better social resilience in the face of climate change?*

The research question at the top is how can the government make coastal cities have better social resilience in facing climate change, especially when regulations are applied, of course in accordance with RAN-API, namely the reclamation of marine and coastal areas?

The answers to these questions, based on the results of research and opinions from various sources such as NGO's ForBali, academics and environmental practitioners, they argue that coastal city governance must use a participatory sustainable development approach. So the choice of coastal reclamation efforts should be rejected, because it does not prosper the local population and will increasingly reap flood disasters and climate change disasters. The North Jakarta Bay and Benoa Bay in Bali cases show these facts, the community is getting poorer and the floods are getting more intense.

Contrast

Major cities in the world, historically, are located in coastal areas. Two-thirds of cities in the world with population more than 2.5 million inhabitants, and 75% major cities population in the world are also situated in coastal zones (UNESCO 1993; Edgen 1993; in Kay and Alder 1999). Meanwhile, Indonesian coastal cities are highly concentrated population. Almost 60% population in major cities of Java and areas outside of Java Island inhabited coastal areas such as Jakarta-Special Capital Region, Surabaya-East Java, Semarang-Central Java, Denpasar-Bali and Makassar-South Sulawesi, Padang-West Sumatera, and Lombok-West Nusa Tenggara" (Dahuri et al. 2001). Still, urban development and urban planning in Indonesia are biased to mainland, so high population growth and concentration spread more in coastal areas. In contrast, local government have imposed policies to solve the problem of the limited land availability by conducting coastal reclamation instead.

Problem

The Indonesian government has not yet implemented a climate change disaster management policy to encourage the formation of social security in coastal areas and surrounding communities. However, the policy choice is taken by carrying out reclamation in coastal areas. In fact, reclamation activities in these coastal areas have caused social and physical problems. Increase flood and climate change disasters. In this position, the Government (Central and Regional) has taken the wrong choice of coastal area arrangement policies that do not adhere to the paradigm of sustainable inclusive development and without listening to the aspirations of the community and involving their participation because. Even though the community is the most affected by climate change disasters.

Aim

The aim of this paper is *how does the government lead the effort to make coastal cities a better social resilience in coping with climate change*?

Objective

The objective of this paper is to analyze the social resilience of coastal areas in the face of global climate change impacts, to enrich regional climate studies and climate change policy in Indonesia. The resume of this research is that the Indonesian Government had failed in handling climate change disasters and had not taken a policy to encourage the formation of social resilience in coastal areas and their communities. The policy choice taken instead is to conduct reclamation of coastal areas so that it would result in social and physical disasters when the disaster of climate change hit the coastal areas. Meanwhile, the central and regional governments had taken wrong steps in making policy on structuring coastal areas, so that the people would be affected by climate change disasters.

Methodology

This paper is extracted from the research conducted for 2 years (2017–2018) with the title "Inclusive Development in Coastal Cities of Java and Outside Java" under the Indonesian Institute of Sciences (LIPI). The research team from the Indonesian Institute of Sciences (LIPI) underlined the effects of climate change and disasters that occurred for years in cities located in the coastal region that exposed to various disasters, such as high rainfall but at the same time, the sea space is capitalized through reclamation. The condition of extreme weather changes has resulted in the occurrence of robs, floods, landslides, sedimentation, clean water crises to plastic waste. This article analyzes and provides solutions in the form of modeling to the application of the concept of inclusive development in urban coastal areas.

Arrangement of city space and sea space that considers ecological balance basically adopts the concept of sustainable development and consistently gives satisfaction and quality of life to the community by not damaging the environment and considering the available reserves of resources. The important point of this article is the effort to put forward a new paradigm of market-driven urban development planning (economy), and it also has social and environmental dimensions. The study is located in coastal cities in Java and Outside Java: disaster of coastal areas in North Jakarta Bay and reclamation in Benoa Bay, Bali. As a region Bali often faces disasters, but it is vulnerable to exploitation by capital owners. The study was conducted using the inclusive social development research approach with qualitative research methods.

This research employs qualitative research approaches and using the concept of Bengen (2001: 56), the key to coastal areas of governance are four aspects as follows: (1) ecological integration; (2) Integration of sectors; (3) the integration of scientific disciplines; and (4) the integration of stakeholders and using ICM Concept by Adrianto and warsilah (2017–2018). Data collection is obtained from tiered focus group discussions. The objective of this paper is to analyze the social resilience of coastal areas in the face of global climate change impacts to enrich regional climate studies and climate change policy in Indonesia.

Literature Review

Climate Change in Indonesia and the Government of Indonesia Issued Policy Regarding Climate Change

Indonesia is an archipelagic nation with the fourth-longest coastline in the world, after United States, Canada, and Russia, with a long coastline of 95.181 km and ¾ of the total area of Indonesia consists of coastal zones (5,8 million km^2 from 7,827,087 km^2). Until now, coastal peoples have generated a large amount of marine resources and a great benefit from coastal zones. This zone is often linked with a variety of economic activities and a sharing of various benefits among coastal communities in managing marine resources. Indonesia has faced grave threats due to climate change. Consisted of 17.508 islands, Indonesia is by far the largest archipelagic nation in the world with long coastal zones of jurisdiction and a large population of more than 250 million inhabitants. Indonesia is also one of countries that has frequently coped with the challenges of natural disasters with millions inhabitants living within the coastal zones. It is also supported by the fact that many coastal communities still depend on fisheries although the activity generates an alarming rate of Coast at Risk Index (Beck 2014). Disaster-related statistics from *Centre for Research on the Epidemiology of Disasters* (CRED) also confirms this risk. The effects of climate change in coastal areas in a variety of ways are often worsened by unmanaged developments within the areas; sea level rises, for example (Ward et al. 2012). It is projected that sea level rises will increase 40 cm or ±20 cm in 2050; and it will increase 56 cm or ±32 cm in 2080. If this trend is true, it is estimated that the average of sea level rises will reach ±40 cm this last century (ICCSR 2010 in MOST UNESCO-LIPI: 2018).

Due to sea level rises, until now Indonesia has lost 29.000 Ha of coastal areas, particularly in the North of Java and other regions as well. As a global phenomenon, climate change and global warming pose uncertain wet and dry season, high-intensity rainfall, sea level rises that threaten coastal areas and other disasters due to the impacts of climate change in the form of tsunami waves that frequently hit coastal areas recently in 2019 such as in Tanjung Lesung Banten (Java), Lampung (Sumatra),

Lombok, West Nusa Tenggara (NTB), and Palu (Sulawesi). With a large number of casualties, a devastating tsunami caused many buildings to collapse due to land liquefaction that buried villages with sand, mud, and stone.

Land-use planning seeks to manage land available in an efficient way by land reclamation, such as wet beach, wetland, muddy beach, which are not considered before to economically beneficial. Whereas coastal cities are vulnerable to climate hazards, these threats have been coping with by any archipelagic countries around the globe, without any compromises. Climate change caused more devastating disasters with extreme weather, typhoon, tsunami, underwater earthquake, landslides, floods, and forest fires. Every day we witness disaster happening around the world and each year 250 million people are victims of severe disasters. Since 1992, international organizations have spent $2.7 billion to reduce impacts on typhoon, floods, and drought. Scientists and scholars from DKKV and Global Risk Forum proposed to mitigate actions from hazards and impacts of climate change. There is also an urgency to take further initiatives on disaster management of climate change and its impacts.

Based on the condition above, the government of Indonesia takes initiatives to rising awareness and mitigate further actions to copewith climate change. The government furthermore introduced a new proposal to deal with climate change impacts, called National Action Plan for Climate Adaptation or RAN-PI.[1] Collaboration between Indonesian Institute of Sciences (LIPI) and Program Management of Social Transformation (MOST) UNESCO tackles the study to reduce risks and climate change impacts in coastal areas from the community perspective in 2018. Due to its devastating impacts, this study also points out its impacts of human casualties, ecological risks, and livelihood of people living in the impact zone. The study of MOST UNESCO-LIPI also collaborates with other universities, University of Indonesia (UI), University of Gadjah Mada (UGM), and the program called National Action Plan of Inclusive Policy Design for Climate Adaptation in Coastal Areas from bottom-up (9/1/2019). This research is aimed to address climate change impacts and to give recommendation to National Action Plan for Climate Adaptation (RAN-API) in order to have a more inclusive policy. Hence, all communities will get equal rights and opportunities to increase their capacity building to adapt climate change and its impacts. The outcome of the research underlines strengthening capacity of adaptation and mitigation of coastal communities by rising public awareness with suitable approaches (Hidayati, MOST UNESCO-LIPI: 2018).[2]

To tackle the impacts of climate change, the government of Indonesia issued policy regarding climate change with two concepts: (1) adaptation in managing climate change and (2) mitigation in reducing the risks of climate change. As cited from RAN-PI, adaptation refers to a set of adaptability to socio-ecological system in coping with negative impacts of climate change. However, adapting to climate change will not be effective as the impacts exceed capacities to adapt successfully. Hence, mitigation is a set of actions to reduce the source of hazards, such as to reduce

[1]National Action Plan to Deal with Climate Change (Rencana Aksi Nasional Menghadapi Perubahan Iklim.).

[2]See in https://m.mediaindonesia.com/.../209386-adaptasi-masyarakat (community adaptation).

greenhouse gas emissions. In contrast, climate change gives a different perspective of how the impacts are more vulnerable to people, for instance, sea level rises, high tides, changing patterns in wet and dry seasons that threaten farmers harvest, changing salinity of sea that causes less catch for fishermen. World's populations have suffered these impacts and categorized as devastating disasters.[3]

As a part of adaptation, coastal cities have been urged to have a better social resilience in coping with climate change. As a concept, adaptation should be mainstreamed into a long-term urban planning and channeled from national to individual level. In other words, the mainstreamed adaptation of climate change should be included into urban development, small islands, and coastal zones in Indonesia. However, coastal cities still face a great challenge, for example, (1) Ecological problems such as biodiversity destruction, habitat loss on ocean, contamination and river alterations, (2) Economic problems such as the rising of health care, loss of workdays, the rising of social services for new infrastructures and rehabilitations, (3) Social problems such as unemployment, the loss of income, unexpected job loss, the decreasing productivity, the loss of social protection, the loss of infrastructure that supports daily life, and (4) Sea and coastal reclamation that cause social and ecological threats.

The framework of Indonesia coastal planning had been proposed in Constitution or UU No.24 in 1992, that regulates coastal planning including special zones which coordinate the planning and its outcomes from national, provinces to districts level of coastal zones. However, most cities located in coastal zones do not integrate land-use planning with coastal planning. Therefore, the coastal planning often does not meet its objectives. Many coastal zones were abandoned and designed as a backyard for waste collection. Meanwhile, another model of coastal planning, for example, reclamation, is in contradiction with previous coastal planning which considered coastal areas less important. Based on this problem, the research question of this paper is that *how does the government lead the effort to make coastal cities a better social resilience in coping with climate change, particularly when regulation implemented is incompatible with RAN-API, that is reclamation of sea and coastal areas?*

During the period of government prior to the era of President Joko Widodo, private investors had benefited from marine and coastal resources, but they did not carry out coastal development in an inclusive and sustainable manner. So that the coastal area is not developed in an inclusive and sustainable manner, massive environmental damage has occurred both in Java and outside Java. And the coastal communities become poor and not prosperous. Indonesia experienced a setback in the development sector, this happened during the monetary and economic crisis in 1998. The momentum of the economic and monetary crisis in 1998 was used as a warning to change Indonesia's development vision. Thus, the government of President Joko Widodo began to massively reconstruct Indonesia's coastal areas by implementing sustainable inclusive development through the construction of sea tolls, road and port infrastructure and building environmentally friendly industries. Especially in

[3]See in https://belajarbencana.wordpress.com/2008/06/03/perubahan-iklim-dalam-konteks-kebencanaan-di-indonesia (climate-change-in-context-disaster-in-indonesia).

certain sectors, and at the regional level, because so far many development projects in coastal areas have absorbed a lot of capital. Such as the fish farming industry, tourism resorts, mining industry, sea port development, and coastal reclamation. In this capital-intensive development, private investors have not paid attention to the aspects of sustainability and inclusiveness of marine and coastal ecosystems. Currently, they are required to adopt an inclusive sustainable development approach, if they do not comply with their business permits they are revoked and their companies are prohibited from operating.

The Concept of Inclusive Coastal Management Related to Climate Change (ICRM)

The concept of coastal zone management has been elaborated by scholars, for instance, in Bengen (2001: 56) that points out managing coastal zones refers to managing natural resources and services in coastal zones by making a comprehensive assessment, planning, and organizing each activity that marks potential benefits for further sustainable development. Moreover, Bengen also underlines that the key to success of coastal zone management is the integrated planning and management of coastal zones and beaches that include four aspects as follows: (1) ecological integration, (2) sectorial integration, (3) academic integration disciplinary, and (4) stakeholders integration. According to Martinuzen (1977), the arrangement of coastal spaces certainly must refer to the sustainable development paradigm that pays attention to the balance of ecological, economic, and social functions. From the results of the study, there appears to be a gap, a fairly wide gap between environmental conservation and economic development. In reality, both economic and environmental aspects cannot go hand in hand, what happens is a trade-off so that one aspect must be chosen as a priority. This is in accordance with Law No. 27 of 2007, which states that reclamation is an activity carried out by people in order to increase land resources in terms of environmental and socio-economic aspects by way of extraction, drying, or drainage. Management of coastal spatial planning is also associated with the concept of resilience cities.

Walker (2007) states that "Resilience is one that has developed capacities to help absorb future shocks and stresses to its social, economic, and technical systems and infrastructure so as to still be able to maintain essentially the same functions, structures, systems, and identity". And according to Wildavsky, resilience is the concept that a more resilience to disaster system, not just immune to changes, but also how the system can bounce back, mitigate, and recover from disasters. Wildavsky also underlines the general characteristics of a resilient system as follows, redundancy, diversity, efficiency, autonomy, powers, interdependence, adaptation, and collaboration (Djalante and Thomalla 2011).

The case that occurred in Indonesia is due to the absence of good governance in the coastal areas and the less policy choices to reclaim the coastal areas that caused degradation of coastal areas and the threat of coastal ecology. As a result, coastal areas are not resilient, so people have difficulties in collaborating their objectives to protect the coastal environment and its lack of adaptation process. As investigated by Warsilah (2016), the conditions of the uncontrolled reclamation in coastal areas in North Jakarta Bay leads to a loss of biodiversity in coastal ecosystem that has been formed naturally many years ago. It also causes damage of coastal zones and seashores. Consequently, it will risk the fishermen's livelihood and their activities. Due to the loss of marine biodiversity, phytoplankton that sits at the bottom of the food chain will be diminished. Loss of phytoplankton would mean emigration of fish to other areas. But more worrisome, due to reclamation activities, fishermen would also risk their marine resources and local livelihood. They would force to leave their homes while the area will be replaced by commercial buildings erected by developers who will prevent them from using social facilities and doing fishing activities. Reclamation also causes fewer fish catch to local fishermen. Consequently, it also results in the loss of fishermen's income. Local fishermen will also have difficulties and losing direction of their boats from and to the local ports where the boats usually tie up and sail to the sea as it covered with buildings in reclamation islands. In the near future, reclamation activities should be integrated with urban planning by protecting conservation in coastal areas. The new urban planning should also consider carrying capacity of socio-culture, economy, and ecology of the city development. Reclamation activities in coastal areas cannot be carried out if the functions of ecological conservation are not supported. Carrying capacity of social and environment cannot be maintained continuously to support cities as centers of economic and political activities. Therefore, cities as centers of commerce, services, and industries should separate its functions as centers of administration (Warsilah et al. 2016).

In another study, Warsilah (2014) analyzed that the development of coastal areas that were handed over to developers tended to use the superblock paradigm. It also happened in North Jakarta, the reclamation island was made with a superblock system so that more artificial islands were buried, and it caused reclaimed areas in disaster-prone conditions. Moreover, developers introduced the Superblock Development model in reclamation area, a specific model of Central Business District. This type of development is usually offered to high-end clients who willingly spend their money and enjoy recreational facilities in one place or called "one stop shopping", while the poor in coastal areas will only get a small portion from economic activities, either as consumers or work in a small sector. This superblock on reclamation islands generates less economic profit for the poor living in coastal areas, because they are not direct economic actors. They are usually common laborers or part-time workers. This development model would not also support their prosperity. In most cases, they are expelled socially from their homes because their lands are replaced with condominiums, hotels, malls, marine resorts, restaurants, eco-parks, apartments, that need spacious lands. In addition, the choice of development in areas reclaimed with the superblock system has forced coastal communities to leave their homes and has excluded fishermen socially (Social exclusion). They are marginalized from their settlements and do not have access to reclamation areas (Warsilah 2014) (Fig. 7.1).

Sources: Fokus.tempo.co and Suara.Jakarta.co

Sources: Jurnas.com and Tirto.id

Fig. 7.1 Reclamation Project in North Jakarta Bay and development of Building Superblocks

On the other hand, land reclamation activities in coastal areas may have a significant impact when climate change occurs, for instance, the decline of salinity in Benoa or Jakarta bay due to water-level rises that continuously decrease the amount of seawater filling the bay during the tides. The scale of coastal reclamation, location, and its technical operations will significantly cause impacts on coastal areas and its ecological functions. Hypothetically, ecological impacts on Benoa and Jakarta bay are inundation and flood risk in hinterland because of backwater effects that can cause flooding upstream and the loss of ocean functions. Reclamation will also directly reduce the volume of storage from flooding. Theoretically, if the volume of water flows is equal and the water storage in the bay is reduced significantly due to land reclamation, the volume of water in rivers will be inundating low topography areas and other areas with land subsidence during the tides and when it is raining (Sudiarta et al. 2016). The land reclamation is expected to cause flood because reclamation transforms coastal landscapes and flows of groundwater in coastal areas. These changes will also affect the composition of river sediments, tide patterns, ocean currents along the coastline, and groundwater sources (Hantoro 2002).

It is conceptually said that coastal reclamation and climate change will have a double impact on the bay area and its surrounding while coastal area are rich in biodiversity. Bay area is generally shallow mud habitat, protected, and important habitat for mangrove forests. The forests have an important role related to physical aspect, ecological and economic value. Physically, mangrove forests are coastal protection from abrasion and it has a filter system to protect coral reefs from destruction, absorb pollutants from dumping waste, and wastewater from mainland. In regard to disaster mitigation, the forests can mitigate high waves, storms, even tsunamis due to its tree structures which act as a protective belt slowing down the wind or high waves. The large stretches of mangrove have also an important role in urban environmental planning as a green protective buffer of the city due to its strategic location planted in the middle of the city. From the ecological perspective, mangrove forests are planted to maintain biodiversity and productivity in coastal areas and small islands. Mangrove ecosystem also provides homes for marine resources, shrimps, fish, and crustaceans. The estuarine ecosystem, mangroves are source of carbon and energy for phytoplankton and creating a significant role in food web (Sudiarta et al. 2016). From the socio-ecological aspect, mangrove forests are source of income of many local fishermen where marine resources are easy to find shrimps, crabs, oysters, and small fish as a source of their daily meals. Local fishermen also use mangrove forests as aquaculture that provides breeding and nursery grounds for crabs and prawns with traditional bamboo floating nets (Warsilah 2018a, b).

Reclamation Project: Non-inclusive and Adaptive Governance in Facing Climate Change

A future of Bali tourism still has clouded with problems. Starting from waste dumping, 700 Ha land conversion to new buildings each year and coastal reclamation land, now artificial islands had been built for One Stop Shopping areas on the area reclaimed from the sea. A tourism scholar from Udayana University, I Nyoman Sunarta,[4] said that Bali still copes with challenges in developing tourism industry. Tourism has boosted Bali's economy. On the other hand, there are still problems ahead. According to him, Bali is famous for its soft-sand white beaches, fertile rice paddies, Balinese Hindu culture, and a low-cost tourism. This fame attracts more tourists to Bali. Tourism industry in Bali has progressed ever since and it makes Bali as one of the favorite tourism destinations in the globe. In 2108, for instance, six million tourists visit Bali. However, like a saying "tourism is like a swan that produces not only a golden egg, but tourism also destroys its nest". Tourism industry allows accommodation sectors to grow rapidly, and they consume a huge amount of groundwater. Each room in a hotel consumes 300 L at least per day. He said that in 2025, it is estimated Bali will face groundwater crisis in future. The average of 700 Ha paddies fields have been converted to hotels and luxury accommodations for foreigners.

Meanwhile, there is a reclamation plan in Benoa bay that will affect ecological disasters because it will disturb land and groundwater. Land reclamation is also equipped with roads and highways in order to connect artificial islands, despite tourism industry in Bali remained stagnant. Every day, there are 13.000 m^2 of wastes dumped into the ocean, and only half of them are recycled. Traffic jams are common in roads due to the rapid growth of vehicles on roads that reaches 13% each year. Meanwhile, there is only 2.28% of new roads for motorcycles.[5] In reality, wealthy outsiders have taken Bali's booming tourism industries, while locals only get a small portion of it and many work for an income Rp 50.000 per day.

The concept of social exclusion that underlines citizenship topic is fundamental (Silver 1994; Littlewood; Hera 2016 in Alkatiri 2018). Furthermore, the concept has been elaborated by (Rodger 1995; Kabeer 2002; Okumu and Bonome 2010 in Alkatiri 2018) that reveals a strong relation between social exclusion and limited access to natural resources (land, sea, coast, and forest) of certain problematic groups (fishermen, farmers, labors as vulnerable and marginalized groups). Or in a sociological term they are called marginalized groups due to social exclusion and limitation to have rights as citizenship, for instance, coastal and marine resources. These resources are not only source of their livelihood, but it also an important element in social relation and social integration. To put in context in Benoa and Jakarta bay case, the problem of

---

[4]In-depth interview conducted in June 2018 at Bali.

[5]Tempo, 2019: presented in the opening of Future City Summit (FCS) 2019, in Bali Paragon Resort, Sunday January 13, 2019.

social exclusion in communities is the absence of social integration among communities. It is showed by the lack of participation, social supports, social networks, isolation, or a closed access to natural resources and policy or regulation on coastal and marine management (Jehoel-Gijsbers dan Vrooman 2007 in Alkatiri 2018). Another fact is that these marginalized groups have not been supported by inclusive social development that should give them opportunities and access to participate as citizens so their basic rights are fulfilled (Wirutomo 2015; Warsilah 2016).

The power of exclusion, according to Hall et al. (2011), consists of force of power, force of regulation, force of institution, force of market, legitimation force, and force of social relation. In context of Benoa and Jakarta bay, the force of power is generally exercised in the form of power of exclusion on marine and coastal resources. Power of market that refers to land price is a dominant power in relation with sea access. Force of market is showed by a push–pull dynamic in relationship between private investors, central government, and local government.[6] The price of reclamation land is also a dominant force due to a limited space on mainland and reclamation should be carried out to build man-made islands for the massive project of superblock or One Stop Shopping. Strong economic motives have been exercised by investors in the form of domination over marine and coastal management for the sake of global tourism. In contrast, local people have inhabited the areas for many generations with unique socio-cultural values where they built fishermen *kampong's* settlements. There are 17,000 fishermen living around Jakarta bay and 1000 fishermen in Benoa bay. On the other hand, private investors and local government are using legal obligations in this case to support reclamation activities, including domination over sea access.

Hall et al. (2011) underline that State is the only actor who exercises his power through violence. According to our findings, it reveals that the police as a part of state actors act on behalf of a governmental body and exercise their power to social activists who refuse reclamation activities, including a local representative from Indonesian Democratic Party (PDI-P) in Benoa district and locals from *Adat* villages. Now, they are still arrested by the police. There is also intimidation on many local people who reject reclamation, for instance, a doctor who was always terrifying during the interview and he was obliged to report himself every day at a police station. Hence, violence mechanism is dominant power in reclamation projects, and it soon develops conflicts between local security officers and local people. However, local movements against reclamation have been amplified by NGO's ForBali, Environmental NGO's like Walhi, groups of artists, and local elites. They are still struggling to gather mass for massive demonstration and fight against reclamation. Still, they do not have military equipments to reach their goals, but they only have words and actions to express their political protest.

---

[6]There are four artificial islands built in Jakarta bay by private investors. After the cancelation of reclamation project, reclamation islands are under the authority of Local Government of Jakarta under Anies Baswedan administration (2017–2021), who replaces previous Governor, Basuki Tjahaya Purnama. Now, the management of reclamation islands is taken over by Provincially-Owned Corporations (BUMD) Jakarta. These four artificial islands however are rented for restaurants. This certainly violates the regulations, and there is not a single attempt to conserve the area from reclamation projects.

The research finding conducted by LIPI (2017 and 2018) shows that access to marine resources relates to four power as described by Hall et al. (2011), force of regulation, force of market, force of institution, and force of legitimation, while power of social institution and force of social relation have been exercised in marine exclusion in Benoa and Jakarta bay. Massive protests by social activists against reclamation, as a result, have pushed forward power of social institution, based on solidarity of the community. Each power has its own mechanism, process, and actors. Comparing to five forms of power above, the combination of power between mechanism of rites-cultural and socio-ecological conservation has succeed to suppress reclamation activities and market power. Meanwhile, power of force is contextual, and power of legitimation still has a dominant role supported by force of regulation. The combination of force of regulation and force of market, both usually penetrate into governmental bodies (Ministries of marine and coastal areas) because marine affairs are under direction of central and local government.

A massive Benoa bay reclamation plan conducted by PT TWBI[7] has been rejected by various communities, NGO's, and local elites. TWBI's idea of revitalization of Benoa bay, reclamation activities according their version are as follows:

1. Formation of new islands and to deepen the pit, in order to channel the water efficiently to mangrove forest.
2. Take a distance of water canal to get a better flushing.
   - The distance between mangrove forest and the island is 100 m.
   - The distance between islands is 75–300 m.
3. Reclamation activities in Benoa water will not include mangrove forest.
4. Reclamation is to protect and to fix sea flows.
5. Previously, the size of Pudut Island located in Benoa bay is 5 Ha. Due to abrasion, its size decreases to 1 Ha. By land reclamation, this island will be resized to 8 Ha, and on the island will be built a cultural center.
6. To deepen water canals, minimum 2.5 m on tides.
7. Revitalization and rehabilitation people's plantation which the size previously is 1373 Ha and now is 900 Ha (from various media in Bali, 2016–2018).

On the artificial islands of Benoa bay reclamation plan, investors will establish a Bali cultural center and new exhibitions; center of commerce of Balinese and national products; conservation center and celebrity travel destination; Botanical Garden, Eco park, Eco resorts; public and social infrastructures; new alternatives roads from East to West, international recreational parks, water and waste management areas; green energy infrastructures; iconic buildings; iconic bridges; and center of international tourism. In short, branded by theme "One Stop Shopping", tourists will be entertained with various and luxury things provided in one place. There is no need to visit other places in Bali. Everyone will enjoy shopping, recreation, and sport activities in one

[7]Tommy Winata is one of the most successful conglomerates in Indonesia and the owner of PT TWBI, the investor of Benoa Bay Reclamation, Bali.

place. In the near future, this reclamation plan will replace other tourism destinations in Bali.

On the other hand, the same study conducted by LIPI and other scholars (2018) shows that the reclamation project violates local or *Adat* laws, Balinese cultures, and Hindu rituals. More than seventy sacred places and *Puras* are located in reclamation areas. Therefore, modern buildings will replace these sacred places. Another implication is that Balinese Hindu ceremony will not be taken place easily due to limited access to privatized coastal areas, for instance, the ceremony of "*Ngaben*" or cremation of the dead and the ash of dead body and thrown to nearby seawaters and "*Meulasti*" or purification rituals nearby coastal areas. Hindu cannot perform freely in *Puras*, a place to worship their Gods. Thus, according to the law of cultural improvement No. 5 (2007), there is a violation against human rights to impede Hindus perform rituals or religious activities.

The main problem of coastal areas in Bali and North Jakarta Bay is that reclamation activities have degraded environment and due to the lack of good governance in the management of coastal areas it leads to environmental degradation. Moreover, several beaches in Bali (in Buleleng and northern coast of Bali) are vulnerable area to disasters, especially tsunami threats. Anticipation and mitigation are urgently needed. The progress of development and massive exploitation on nature and environment should be balanced with nature conservation. *Pudut*, one of islands located in Benoa bay, is an island that is considered vulnerable to climate change impacts. However, investors and local government claimed that the reclamation of Benoa bay is to reduce impacts of disaster instead global climate change, and to revitalize destruction on coastal areas. Meanwhile, NGOs argued that the Benoa bay plan is only to gather economic profit in marine tourism industry and making alternative tourism destination by creating tourism competitiveness and a new tourism icon.

Disagreements occur between investors backed up by the Government and NGO's activists: "Investors and local governments claim that the reclamation of Benoa Bay is to reduce the impact of floods, but not to avoid the impacts of global climate change, and the reclamation is deliberately carried out to revitalize the damage to the coastal area". Meanwhile, NGOs argue that "the reclamation plan in Benoa Bay is only to collect economic benefits from the marine tourism industry by making an alternative new tourism destinations without looking at the existing problems". In fact, floods have increased since the beginning of degrading the beaches, and local people as actors in the small tourism industry have been marginalized, making them worse off in their household economy and they are getting poorer. In addition, life their religiosity is disturbed, they can no longer pray freely in the temple which is located in the reclaimed area.

Discussion

In this part of the discussion, the purpose of the exclusive use of the development paradigm in the governance of coastal areas related to pursue economic growth will only result in social exclusion and marginalization. Why is the management of coastal cities in Java and Outside Java necessary to adopt inclusive development? The answer is because the development is oriented to economic benefits only and it will damage the environment and coastal ecosystems. In other words, the success of economic development will ultimately lead to the use of natural resources that exceeds reproductive power and exceed the natural absorption threshold of waste, both solid, liquid, and airborne waste. If natural resource utilization continues to be carried out extractively and with old methods that are less environmentally friendly and do not consider the carrying capacity of the ecosystem, in the future Indonesia will face 3 (three) major crises, namely (i) water crisis, (ii) food crisis, and (iii) energy crisis.

Therefore, this section will discuss the approach of development paradigms in coastal areas by using inclusive or precisely Inclusive Coastal Development as a reference. Inclusive development is a social development approach that broadly analyzes a process of continuous improvement of a society, or an overall social system toward a better or more human life by supporting the sustainability of humanity and ecology. Social inclusion is one of the approaches to human development that develops an increasingly open environment, by involving all people, regardless of differences in socio-economic background, characteristics, abilities, status, physical, ethnic, cultural, and environmental conditions (Lenoir 1974).

An inclusive coastal development requires community involvement and participation in building coastal areas. In addition, this inclusive development paradigm also opens the community's accessibility to marine and coastal resources as well as economic resources and government policies. So that it becomes the most rational choice to the development of land and coastal communities. The inclusive development paradigm in the coastal region or region can be implemented with two development concepts, namely the concept of city management based on green development and the concept of ecological city management (Warsilah 2018a, b), as shown in the Fig. 7.2.

Figure 7.2 shows that for the realization of sustainable development, there must be a link between environmental, economic and social aspects. These three aspects must not defeat each other, for example if the economic aspect is the foundation, the goal is economic growth. While the social aspect in the form of social welfare is

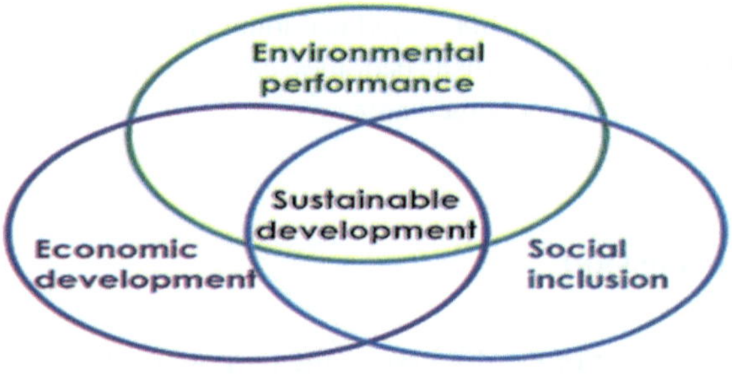

Fig. 7.2 Pillars of sustainability *Source* Bappenas, 2016

neglected, environmental damage will also become more prevalent. So sustainable development becomes an important offer to build cities, especially coastal cities.

The implementation of inclusive and sustainable development in Indonesia is proclaimed globally through the declaration of Sustainable Development Goals (SDG's) that is also integrated with economic development, environmental development, and the application of social inclusion. This inclusive development will be carried out until 2030. SDG's are a continuation of the Millennium Development Goals (MDG's) which have been implemented for 15 years (2000–2015). Among the objectives of the SDG's are as follows:

- The 8th Goal is aimed at "Enhancing Inclusive and Sustainable Economic Growth, Productive and Overall Employment Opportunities, and Decent Work for All";
- The 9th Goal is "Building Resilient Infrastructure, Enhancing Inclusive and Sustainable Industries, and Encouraging Innovation";
- Goal 10th is "Reducing Gaps In Country and Interstate";
- Goal 11th namely "Making Cities and Settlements Inclusive, Safe, Resilient, Sustainable;
- Goal 16th, namely "Strengthening an Inclusive and Peaceful Society for Sustainable Development, Providing Access to Justice for All, and Building Effective,
- Accountable and Inclusive Institutions in all Levels" (Warsilah 2018a, b and the National Development Planning Agency/Bappenas—SDG's: 2017).

Based on the explanation of Bappenas above (2017), Indonesia should be in the form of a maritime and archipelagic country, in all municipalities or regencies that have large coastal areas, should seriously start implementing national planning that has been outlined by the National Development planning Agency. Denial of the master plan for sustainable inclusive development is a violation of state law.

Reclamation in Benoa Bali and North Jakarta bay, according to our findings, is against with principles of United Nations, based on working reports of Informal Consultation Process on Sea and Law of the Sea in the seventh summit: UNGA, A/61/56 from July 17, 2006. However, ecosystem approach should refer as follows: (a) imposing conservation of ecosystem structure, its functions, key process in maintaining ecosystem of good and services; (b) it will be applied on specific geographical area based on ecological criteria; (c) underlining interaction between human activities and ecosystem and in between ecosystem and inter-ecosystem; (d) considering various factors beyond management area that affect marine ecosystem in management area; (e) to balance community goals; (f) to create inclusive collaboration with various stakeholders participation and local community in planning, implementation and management; (g) referring to best approaches available, including traditional information, original, scientific that enable to adapt with new knowledge and experiences; (h) to value risks and implementing prudent approaches; (i) to use integrated decision process and management related to various activities and sectors; (j) to seek solution of marine ecosystem degradation if possible with revitalization; (k) to evaluate cumulative impacts and various human activities on marine ecosystem; (l) to consider ecological perspective, social, cultural, economic, law and technics; (m) to seek accurate balance between integration, conservation, and sustainable use of marine biodiversity; (n) to minimize devastating impacts on human activities to marine ecosystem and biodiversity, especially rare and fragile marine ecosystem (Maes 2008).

Furthermore, United Nation Assembly in 2006 also addresses that (1) environmental degradation happened in many parts of the world and the demand of competitiveness needs an urgent response to manage main priorities and to intervene conservation of integrated ecosystem; (2) ecosystem approaches in marine management should be focalized to human activities in order to protect and if possible to revitalize ecosystem health prior to maintain environmental goods and services; to give social and economic benefit for enhancing food resilience; to defend livelihood of community to support international development objectives. Another point written in United Nations of Millennium Declaration is to conserve marine biodiversity (Look Resolution UNGA 61/222, paragraph 119).

United Nations General Assembly proposed that the implementation of ecosystem approaches would be succeed if: (a) embedded in policy and national plan; (b) to support marine scientific research, through intensive research, the impacts of climate change to marine ecosystem health, the impacts of sea noise in marine ecosystem and to consider the impacts to the ecosystem inward and outward area of national jurisdiction, according to international law; (c) if needed, enhancing regional fisheries management organization, adapting their mandates and modernize their operations according to international law; (d) strengthening and optimizing coordination and internal collaboration based on international law between nations, intergovernmental organizations, regional scientific research, advisory organization, and management bodies; (e) implementing a full and effective mandate from existing multilateral organization, including one organization established under UNCLOS; (f) implementing Rio Principles and utilization of management tools in regards to conservation and sustainable utilization of marine biodiversity, including management tools based on integrated and specific region case per case, based on available scientific research suggestion and implementing prudent approaches and make it consistent with international law; (g) to identify and to involve stakeholders in advanced collaboration; (h) sectorial approaches, integrated management and planning on various levels, based on international law; (i) effective and integrated management inter-sector; (j) the progress Plan Implementation in the World Summit on sustainable development, including harvesting and farming fish stocks in a sustainable way without destructing ecosystem, and to set marine conservation based on international law and scientific data and; (k) to evaluate, according to national legislation and international law, related to marine activities, the possibility or significant impacts on environment.[8] Hence, article 123 based on United Nations Convention on the Law of the Sea (UNCLOS) can be implemented as legal foundation to bring MSP's initiatives together with nations of sea borders enclosed and semi-enclosed sea. These nations should cooperate and coordinate to (a) management, conservation, exploration, and exploitation of marine resources; (b) implementing rights and their obligations related

[8]Report on the work of the United Nations Open Informal Consultation Process on the Ocean and the Law of the Sea at the seventh meeting: UNGA, A/61/56 dated July 17, 2006.

to protection and conservation of sea environment; (c) policy implementation of their scientific research. This collaboration can be implemented through reliable regional organization (Maes 2008).

Figure 7.3 shows the plan of building reclamation in Benoa Bay, Bali built by a developer who was rejected by the community.

Figures 7.4 and 7.5 show the massive social Movement Rejects for Reclamation in Benoa Bay, Bali initiated by NGO's ForBali and Balinese Communities.

Gambar Maket Rencana Pembanguan di Kawasan reklamasi Teluk Benoa

800 × 482Images may be subject to copyright.

Sumber: http/twbi.co.id/index.php?pages=news3.

Fig. 7.3 Model of the reclamation development plan at Benoa Bay, Bali

Fig. 7.4 Social movement rejection for Benoa Bay Reclamation (*Source* NGOs ForBali 2019)

Fig. 7.5 Social Movement rejection on Beno Bay Reclamation (*Source* NGOs ForBali 2019)

A constant struggle of communities to fight against Benoa bay reclamation initiate NGO's ForBali and Walhi Bali for five years is finally fruitful. This is a victory for Balinese along with the faulty reports of environmental impact analysis (Amdal) by National Commission of Human Rights (KOMNAS HAM) and KONTRAS. In Jakarta, the movement against Jakarta bay reclamation by various NGOs, scholars, and many elements of community leads to a complete stop of reclamation activities, and four islands built on reclamation area will be used for conservation and public use.

Principles on Inclusivity Model on Sea and Coastal Development (ICM)

Since 1990s, coastal development in Bali and Jakarta had been conducted exclusively. The findings of the research show that coastal areas in Benoa Bay, Bali, and North Jakarta are managed with an exclusive approach, for the sake of economic interests, not to develop areas inclusively. An exclusive approach with the only aim of economic interests has led to social exclusion which resulted in the marginalization of coastal communities and fishermen from the coastal environment. They no longer have access to marine and coastal resources Using social exclusion approach, it nurtures social conflicts related to domination of marine and coastal resources by capital investors. From the fisherman's perspective, this exclusion model has triggered coastal and ocean grabbing, and it leads further to marginalization process of coastal and fishermen communities. Managing and decision process of marine and coastal resources should be conducted inclusively, that is, transparent, accessible, participative and supported by social, cultural backgrounds, and scientific knowledge. A further development of marine and coastal areas in Indonesia should implement inclusive sustainable development and referring to local wisdom, rights of ownership, customary communal land rights, rights of acquisition, and establishing local institutions.

An inclusive sustainable development in Indonesia is often called Sustainable Development of Integrated Coastal Management (SDICM). As stated by Bengen (2013: 13), it needs to organize a sustainable SDICM program in Indonesia prior to the above reasons; there are still degradation and depletion of marine resources in Indonesia. However, coastal communities, especially fishermen, are still poor. Meanwhile, private investors still derive benefits from marine and coastal resources that are unsustainable and exclusive. Indonesia still gains momentum when monetary and economic crises occurred in 1998. Massive reconstruction in coastal areas had been conducted exclusively in certain sectors, regional and high involvement of private investors and local people. As a consequence, in coastal areas there are many development projects that absorb intense capital, for instance, fish farming industries, tourism resorts, mining sites, seaports, and coastal reclamation. Private investors still do not consider sustainability aspect and inclusiveness of marine and coastal ecosystems.

Community participation and accessibility are the most important variables in inclusive development paradigm. Community participation or public involvement in marine and coastal development processes is one of the important inclusive developments. As stated by Pollnac et al. (2003) in two different research sites, in Cilacap Central Java (Coastal Resources Management/CRM) and Segara Anakan Conservation and Development Project, and in Bunaken National Park in South Sulawesi (Natural Resource Management Project/NRMP), he found that stakeholders have significant roles in participation, planning, and execution of the projects, individually or collaboratively. It leads to main factors in sustainability of SDICM program. Participation does not stand alone, but it comes from various reasons. Stakeholders are willing to participate prior to these reasons: (a) benefits they will perceive, (b) the even distribution among stakeholders, (c) sustainability of benefits after the projects. Due to stakeholders' participate in project plan, they feel that they have roles in creating the project. They also collaborate with community surrounding. As a result, the community will have sense of belonging to the project. This participation process will tie a mutual relationship from both parties and the project will be in accordance with community's willing. Community participation, planning, and implementation of the project will result in a further community's empowerment.

As stated also by Christie (2003), he points out that stakeholders' support in coastal areas is an important factor to sustainability of the programs. Conflict of interest, or even only conflict of perception, among the constituents (fishermen, marine tourism operators, scholars, government officials, NGO's, and conservationist) will nurture dissatisfaction among them if necessary steps are not taken proactively. Dissatisfaction of one or more constituents would threaten the sustainability of SDICM program because they violate rules and previous agreements. The project outputs of SDICM from planning and decision making process are in accordance with six parameters of sustainability as stated by Bengen (2003), as follows: (1) accordance with local policies, formal and informal policy; (2) accordance with social and cultural conditions of local community; (3) supported by available human resources and institutions; (4) involvement of active stakeholders; (5) having a clear plan and program; (6) having benefits to environment, including socio-culture and economy of local community;

(7) these six factors will be completed with an addition of scientific information supports.

Sievanen (2003) in the study of Mobility of Community and Sustainability Implication points out that marine tourism as a mechanism to introduce international capital and state power often results in negative impacts, that is marginalization of coastal users. Moreover, he argued that marine tourism inherently widens possible marginalization of marine and coastal users, for instance, fishermen. Thus, Sievanen recommends two things, related to marine and coastal management:

1. First, it needs to have a clear definition of term "community", the notion being targeted and empowered actors by SDICM program, because there is often unclear information about who takes benefit of the SDICM program.
2. Second, collaboration between public and private sectors that should have pushed forward to reduce poverty and environmental development. It will soon create coastal and marine-dependent communities, because they are often targeted community marginalization, for instance, local farmers and fishermen.

Adrianto and Warsilah (2018a, b, 2019) in their study of Inclusive Development in Coastal Areas (ICM) in coastal management positions have included the Inclusive Coastal Management or often abbreviated as ICM. It requires integration between social development and ecosystem-based management and an adaptive management. The ultimate goal of the ICM is to produce a healthy and prosperous coastal ecosystem with stakeholder involvement from various sectors, so that a standard regulation is obtained related to coastal ecosystems. The coastal ecological system consists of the following factors:

(a) It has a focus on interactions within ecosystems
(b) It recognizes human as essential components of the ecosystems
(c) It is a participatory approach
(d) It is a comprehensive system of components which also takes into account the external drivers
(e) Best available knowledge in decision-making.

This finding is in accordance with the research conducted by Warsilah (2015) in coastal areas located in Kemijen, Semarang, Kenjeran, and Surabaya which reveals that the key community prosperity in coastal areas is to conduct inclusive development, or involving community participation in the area. Moreover, inclusive development is an approach that includes groups and untouched areas of development or socially exclusion and they will be involved in developmental process or social inclusion. Community participation in development and rehabilitation of coastal areas is urgently needed. If we manage to build marine and coastal areas, including its natural resources productively, efficiently, inclusively, and environmentally friendly, we will succeed to tackle the main nation's problems. Government of Indonesia has a huge concern with marine issues. Referring to President Joko Widodo's vision on maritime sector, there are five groups of policy and main programs to be executed (a) protecting sovereignty of the Unitary State of Republic of Indonesia (NKRI) by securing sea borders, (b) ending illegal fishing and other illegal activities, (c) the development of marine economic sector, (d) to preserve marine resources, (e) to build capacity development of knowledge and technology and enhancing national maritime culture (Warsilah 2015).

Referring to the study conducted by Adrianto (IPB 2017) and Warsilah (LIPI 2018) the model of marine and coastal developments should at least refer to Inclusive Coastal Management (ICM) concept that stands on three principles as follows:

(1) System Integration, from spatial, temporal dimension, and coastal resources system that leads to problems in ecology, social, and economy
(2) Functional Integration, internal horizontal consistency, or inter-sectors and between management of actions, collaboration, and partnership among related parties
(3) Policy Integration, local government policy or national-hierarchical and local government policy and consistency in each economic development, as a complimentary between projects and programs (see Fig. 7.6).

Moreover, according to Adrianto (2018) and Warsilah (2018a, b), ICM principles first and foremost should implement integrative approaches (economy, social, and environment) and system approach. Secondly, the outcomes should rehabilitate the ecosystem health, productivity, and create coastal ecosystem more resilient. Third, these principles should channel the connectivity between ecological system and social system. Fourth, based on these principles, ecology, economy, and social values should not be included with corporate values. Fifth, to use participatory approach or public involvement. Sixth, the principles should be adaptive (Fig. 7.7).

A case of Benoa and North Jakarta bay reclamation protest shows a high degree of representativeness from various aspects, social, cultural, economic, and religious aspect. It also reveals socio-cultural response, social conflicts, economy, culture,

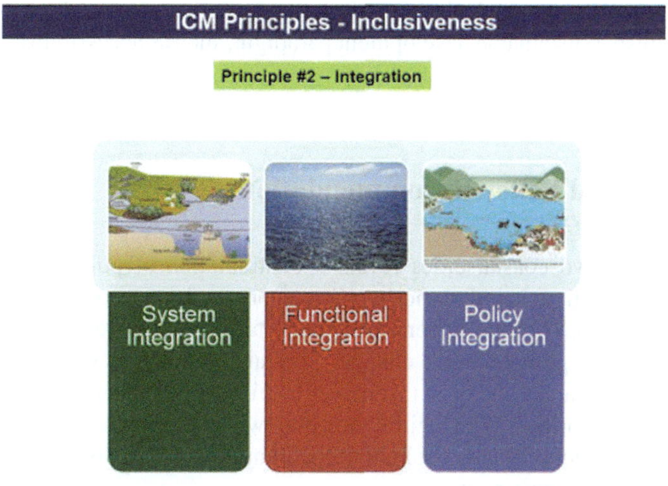

Fig. 7.6 ICM principles (*Source* Adrianto and Warsilah 2018)

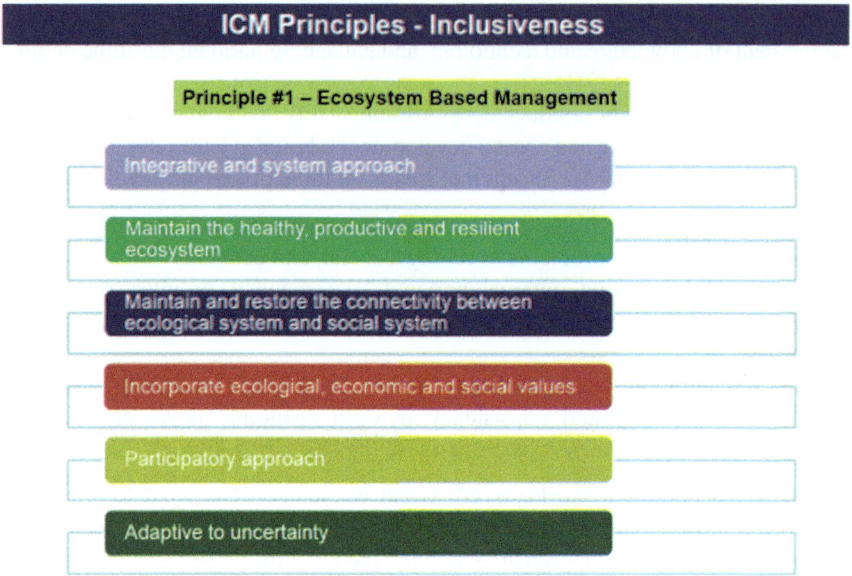

Fig. 7.7 ICM Principles—inclusiveness (*Sources* Adrianto et al. 2018)

and spatial dimension. Social approaches and social capital among communities against marine and coastal reclamation are also high, prior to its impacts to social environment, economy, local culture, and local religion (Hindu). Reclamation project conducted by PT TWBI is not in accordance with social, cultural, and environmental dimensions. Moreover, environmental risks are also a great concern due to its impacts that threaten environmental quality of sea and endanger coastal ecosystem. These factors also contribute to inclusiveness of matrix formation of coastal areas that still relate to inclusive sustainable development paradigm, such as accessibility, freedom, and participation.

Conclusion and Policy Recommendation

The findings in two coastal cities reveal a contradiction of policy implemented with RAN-API policy. Local authority (Governor) prefers to conduct marine and coastal reclamation. This may underrate the importance of social resilience of coastal community against global climate change. Coastal communities still feel that their marine and coastal areas had been taken over by high profile investors who collaborate with local authorities. With legitimation of power, they conduct coastal reclamation and build artificial islands for luxury condominium, one stop shopping, malls, apartments, marine tourism, etc.

The study of climate change impacts is important to enrich regional climate studies and climate change policy, particularly to Indonesia where its sea is a two-third of its mainland. Social resilience will be achieved if communities' adaptation capacity is growing stronger in coping with climate change challenges. Community involvement is a key to resolve negative impacts of coastal development and marginalized process of the poor in coastal areas. In the last two years, communities living in coastal areas in Indonesia reached 60% of total population. This condition makes them more vulnerable to disaster threats, for instance, flooding and tsunami waves. The key to people's prosperity in coastal areas is implementing inclusive development which integrates development and its process with communities' participation in surrounding areas, because inclusive development is an approach that attempts to include marginalized groups and inaccessible areas with development process or socially exclusion.

Communities living in coastal areas should implement certain strategy, notably in managing sea and coastal resources, as follows:

1. Adaptation strategy should be integrated with coastal cities development, particularly in spatial planning
2. Mainstreaming a long-term adaptation in all programs of urban and coastal planning and everyone is involved in development process or called social inclusion.

Moreover, it shows that community participation in development and rehabilitation of marine and coastal areas is urgently needed. If we enable to develop marine and coastal areas along with its natural resources productively, efficiently, inclusively, and eco-friendly, we may solve the main nation's problems. By advocating marine issues, the government of Indonesia has a huge concern, referring to Joko Widodo's vision on maritime sector. There are five groups of policy and main programs as follows: (a) protecting sovereignty of the Unitary State of Republic of Indonesia (NKRI) by securing sea borders, (b) ending illegal fishing and other illegal marine activities, (c) the development of marine economic sector, (d) to preserve marine resources, (e) to build capacity development of knowledge and technology and enhancing national maritime culture.

Indonesia as one archipelagic nation has a three-fourths of its territories covered with sea, enriched with natural resources, a huge environmental services, and its marine economy and all can be transformed into a competitive superiority and a new economic growth. Based on these reasons, it is an excellent vision made by President Joko Widodo to prioritize marine economy development and Indonesia's vision as global maritime axis. Moreover, maritime axis policy is an approach to make Indonesia more advanced, prosperous, and sovereign nation based on marine economy, defense or security affairs, and marine cultures.

In order to rehabilitate two places being studied of sea and coastal areas, there are three points that should be implemented by activists who are against reclamation and Governors in two regions:

1. Governors should make immediate correspondence with President Joko Widodo to cancel (or to revise at least) Presidential Regulation No. 51 in 2014, that refers

to revitalization of North Jakarta bay and Benoa bay as conservation area (a key to save coastal areas in coping with global climate change).
2. To have correspondence with Ministry of Marine Affairs and Fisheries of Indonesia, Susi Pujiastuti to address the application of location permit without consideration from Governor of Bali (mainly, it is a formal protest to permit application and ask Minister Susi Pujiastuti to withdraw the permit).
3. To push and ensure all local regulations related to North Jakarta bay and Benoa bay as conservation area (for instance, local regulation or RZWP3K/Regional Regulation Plan for Regional Regulations concerning Zoning Plans for Coastal Areas, Islands and Small Islands (Ranperda RZWP3K) would include two bays as marine conservation due to the preservation of local cultures and religious rites).

References

Adrianto L (2016) Catatan Diskusi Rancang Penetian Tentang Pembangunan Inklusisf di Daerah Pesisir Jawa, Koordinator HennyWarsilah, PMB-LIPI (Note of the Design of Research Inclusive Development in the Java Coastal City, PMB- LIPI)

Adrianto L, Warsilah H (2018) Wetting of research design "Inclusive development in coastal cities in Java and outside Java", PMB-LIPI. Jakarta

Alkatiri F (2018) *Eksklusi Sosial dan Inklusi Sosial Masyarakat Perbatasan di Nusa Tenggara Timur*. Disertasi Ph.D. Departement Sociology Universitas Indonesia ["Social Exclusion and Social Inclusion The People at The Border Areas, East Nusa Tenggara (NTT)]" of in Disertation of Sosiology Department, Universitas Indonesia

Bengen DG (2003) "The Evolution and Urgency of Management of Coastal and Marine Resources" delivered in the close-out symposium of the East Kalimantan Coastal Project in 2003. IPB, Bogor, p 13

Blog: https: //www.slideshare.net/kamakamaruzzaman/304187011-reklamasipantaimakalah

Christie P, Makapedua D, Lalamentik LTX (2003) Bio-physical impacts and links to integrated coastal management sustainability in Bunaken National Park, Indonesia. Indonesian J Coast Mar Resources, Special Edition, No. 1, 2003

Dahuri R et al (2001) Pengelolaan Sumber Daya Pesisir dan Kelautan secara teratur (Regular Resources Management of Coastal and Oceanic). PT.Prad Paramitra, Jakarta

Djalante R, Thomalla F (2011) Community resilient to natural hazards and climate change impacts: a review of definitions and operational frameworks. Downloaded from: 26 Nov 2011

Hall D, Hirsch P, Li TM (2011) Powers of exclusion: land dilemmas in Southeast Asia. Penerbit: National University of Singapore (NUS) Press, Singapore

Hantoro WS (2002) The influence of marine and coastal characteristics on coastal city development. LIPI National Priority Research Report

https://www.academia.edu/4432623/Reklamasi_Pantai. Beach reclamation as an alternative for regional development. Access date 10 Jan 2014

https://bulletin.penataanruang.net/index.asp?mod=_fullart&idart=267. Reclamation and revitalization. Access date 10 Jan 2014

https://www.slideshare.net/YogieVianto/reklamasi-dan-revitisasi

Impact of coastal reclamation on economic-social conditions of fishermen in Lampung Bay, www.blog.unila.ac.id

IPB (2017) IPB coastal and ocean resource center: marine as a source of new economic growth, published by IPB: Bogor

Kay R, Alder J (1999) Coastal planning and management. E & FN Spon. London

KLH, KLH, Ministry of Environment (2007) Draft national action plan in dealing with climate change

Law No. 24 of 2007. Disaster management. Ministry of Home Affairs, 2007

Law No. 26 of 2007. Spatial planning. Ministry of Home Affairs, 2007

Law No. 27 of 2007. Management of Coastal areas and small islands

Lenoir R (1974) Un Francais Sur Dix, Social Exclusion, published Les exclus

Martinuzen J (1997) State, society and market: a guide to competition theories of development. Zed Books Ltd., London

Pollnac R, Pomeroy R, Bunce L (2003) Factors influencing the sustainability of integrated coastal management project in central Java and North Sulawesi, Indonesia. J Coastal Marine Resour (Special Edition, no 1, 200)

Sievanen L (2003) "Shifting Communities and Sustainability Implications" dalam Indonesian Jounal of Coastal and Marine Resources, Special Edition, no 1, 2003

Warsilah H (2014) Social transformation of the people of the city of Jakarta from the new order period to the reform Era. LIPI Research Professor's Oration Manuscript, published by LIPI Press, Jakarta

Warsilah H (2015) Coastal areas become a priority for inclusive development. http://lipi.go.id/article/coastal-so-development-inclusive-priorities/10478

Warsilah H (2018) Research design "Inclusive development in Coastal Cities in Java and Outside Java", PMB-LIPI. Jakarta

Warsilah H (2018a) Social resilience in resilient cities disasters. Publisher, Yayasan Obor, Jakarta

Warsilah H (2018b) Inclusive development in the outer Java coastal area: the case of Benoa Bay reclamation, Bali (Inclusive development coastal area in outside Java). Publish by LIPI Press, Jakarta

Warsilah H, Jan S, Hatoro W, Alan K, Kusumawijaya M et al (2016) Policy paper: save Jakarta (From the reclamation disaster), published by Rujak Center For Urban Studies, Jakarta

Wirutomo P (2015) Finding the meaning of social development: a case study of the informal sector in Solo city. J Sociol Soc 18(1):101–120

www.darx.com: Blog by Darius Arkwright, Son of Loloda. Accessed 9 Aug 2018

www.tempointeraktif.com

Chapter 8
Measuring Vulnerability of Coastal Ecosystem and Identifying Adaptation Options of Indonesia's Coastal Communities to Climate Change: Case Study of Southeast Sulawesi, Indonesia

Ma'ruf Kasim

Abstract General Impact of climate change on coastal regions always affected societies and ecosystems. There are three ecosytems vulnarable to climate change, for example, coral reefs, mangrove forests, and seagrass habitats. Indonesia is one of the countries identified to be highly at risk to climate change. Majority of Indonesia's population reside in low-lying coastal areas. Recently, most coral reefs in the eastern part of Indonesia were bleached. Bleaching has significantly reduced fish populations associated with coral reefs and demersal fish. The main factors causing coral bleaching are the increase and extreme temperature fluctuations. This chapter aims to explore the level of vulnerability of several coastal ecosystems such as coral reefs, mangroves, and seagrasses and Identifying the best choice of livelihoods in order to adapt to climate change in Eastern Indonesia's coastal areas. Methods to explore some impact of climate change on coastal community were explored by Focus Group Discussion (FGD). Response of each ecosystem to the changes that affect it by providing quantitative assessments. This research results show that the vulnerability of coral reefs occupies a significant level of decline compared to mangrove forests and seagrass habitats. One of the factors responsible for this vulnerability sensitivity is the physiology of zooxanthellae as the main organism that composes coral reefs. Temperature increases in general will have an impact on coral reef turning white, or known as coral bleaching. Different from coral reefs, the impacts of climate change on mangrove forests are indirect because naturally the fluctuating and increasing temperatures affect the activity of mangroves in absorbing CO_2. Temperature fluctuations generally can trigger the reproductive s'ystem of the main organisms inhabiting mangroves. This is believed to cause complexity in utilizing economic value resources from mangrove forests. In seagrass ecosystems, climate change can spur an increase in the abundance of filamentous algae which is a competitor of seagrass in absorbing nutrients, increasing the disruption of the seagrass reproduction process. Climate change also will affect the spread and resilience of

M. Kasim (✉)
Faculty of Fishery and Marine Science, Halu Oleo University, Kampus Bumi Tridarma, Universitas Halu Oleo, Jl. Mokodompit, Andounohu. Kendari, Indonesia
e-mail: marufkasim@uho.ac.id

R. Djalante et al. (eds.), *Climate Change Research, Policy and Actions in Indonesia*, Springer Climate, https://doi.org/10.1007/978-3-030-55536-8_8

seagrasses. The low support of coastal ecosystems due to the effects of climate change will encourage people to find an alternative livelihoods as a strategy to adapt to climate change. Furthermore, changes in the ecosystem will minimize options for the coastal community to utilize available resources in the coastal habitat. Some people however maintain/cultivate seaweed (*Kappaphycus alvarezii, Eucheuma denticulatum*, and *Gracilaria* sp.) as an alternative source of livelihood.

Keywords Coral reefs · Mangroves · Seaweed · Climate change · Coastal areas

Introduction

Coastal areas could be affected by climate change in a variety of ways. There is strong evidence explaining that climate change causes significant changes across the world's oceans such as increased ocean temperatures, ocean acidification, and sea level rise during the twentieth century (McClanahan 2015; IPCC 2018), and this condition is expected to continue (IPCC 2018). Climate change is expected to affect ecosystems and species in the sea through various direct and indirect mechanisms (Brander 2007; Munday et al. 2008; Cheung et al. 2010; Ding et al. 2017). Coastal areas are sensitive to human activity, environmental change such as sea level rise, increases in precipitation frequency and intensity of storms from oceans, and increases in ocean temperatures. Marine and coastal areas are among the world's most diverse ecosystem which can attract optimum utilization of resources. This abundance of resources makes humans exploit almost every place in the sea to get resources to support their lives. Barange et al. (2014) reported that the growing human population and changes in food preferences increase world demand for fish, and other marine resources, adding to the pressure on fisheries sustainability concerns. Among countries that show high dependence on fisheries resources, those that have a strong influence from climate change are in West Africa, South Asia, and Southeast Asia. This condition causes coastal ecosystems to be vulnerable to human activity and climate change. FAO explained that the vulnerability of an ecosystem is a condition where the population or ecosystem experiences large changes due to short-term and long-term disturbances, and this ecosystem will be able to recover within a certain period of time.

Indonesia is a country in Southeast Asia that is vulnerable to climate change (Aldrian et al 2011; Taylor 2015; Djalante and Thomalla 2012). Physical and biological pressures occur not only by population growth and its relation to dependence on resources (Solecki et al. 2015; Takagi et al. 2014; McGranahan et al. 2007). Moreover, the effects of climate change have changed the pattern of distribution and availability of marine resources (Handayani et al. 2017). Eastern Indonesia is experiencing significant pressure from climate change. This pressure really has an impact on the amount of resources that can be utilized by small fishermen and coastal communities. Observed and anticipated changes in the global climate present significant opportunities and challenges for society and the economy. This vulnerability is caused by the combined effects of predicted warming, the relative importance of fisheries for the economy and national diet, and limited community capacity to adapt to potential impacts and opportunities (Allison et al. 2009).

Eastern Indonesia is an area with a high level of fisheries production and other resources. 80% of the Eastern Indonesian communities depend on marine resources, especially in shallow marine fisheries, pelagic fisheries, coral reef resources, seagrass resources, and mangrove forests, especially for Southeast Sulawesi which is in the middle of the mega biodiversity area. Coastal resources are very vulnerable to climate change.

The Southeast Sulawesi Climate Vulnerability and Risk Assessment Report confirms that climate change has a direct impact on the number of small pelagic fisheries production and traditional fisheries. Some ecosystems that experience physical and biological changes are seen in areas close to people's access to resource exploitation. Changes in rising sea levels occur in several districts in Southeast Sulawesi. Extreme environmental changes caused by climate change will have an impact on the existence of resources on coral reefs, mangroves, and seagrasses. The level of vulnerability will depend on the sensitivity of the ecosystem to physical contact or the level of environmental changes that occur. The most vulnerable ecosystems are habitats that can be easily disturbed and also very slow to recover or may not be able to recover.

Some cases of damage that occur can be assessed from the extent of damage caused by these activities or the extent of damage due to environmental changes. In general, the level of damage in coastal areas is caused by two main factors. The first factor is direct and indirect physical activity carried out by humans that causes environmental changes that occur in the short or long term, and the second is changes in the local and global environment as the impact of world climate change.

Methods

Data Collection

This research was conducted during May–December 2018 at several areal in Southeast Sulawesi. Some impact of climate change on coastal community was explored by Focus Group Discussion (FGD) on several areal in Southeast Sulawesi, Indonesia. The population targeted by FGDs are fishermen, seaweed farmers, women fishermen, government, and communities who have access to marine resources.

Data Analysis

The data collected by the FGD is then analyzed using the appropriate approach. Vulnerability assessment uses an approach made by Helper et al. (2007) and an illustration built by Polsky et al. (2007), including consideration of the judgments put forward by Turner et al. (2003). We tried to consider the local conditions found in Southeast Sulawesi. Some approaches to local valuation are the area of each ecosystem that exists, including coral reefs, mangroves, and seagrasses. Other changes include: a decline in ecosystem conditions in the past 10 years, high community access to ecosystems (coral reef, mangroves, seagrass), change in the type and capacity of fishing gear, high human activities and impact on the environment (including mining activities).

Response of each ecosystem will be different to enviromental changes and this can be explored with quantitative assessments. This can be done in order to make it easier to provide a good assessment of the impacts and best efforts that might be made to overcome changes and possible environmental damage. The value of each possible threat will clarify any differences in response and vulnerability of each ecosystem. There are several ways to provide an assessment of vulnerability to ecosystems or species in relation to environmental changes. This is done to facilitate the prioritization of habitat or species conservation (Bryant et al. 1998; Zacharias and Gregr 2005; Kappel 2005), some focus on the assessment of a single ecosystem in relation to a number of threats (Bryant et al. 1998), and improvement of approaches to species (Keppel 2005). Halpern et al. (2007) have a separate scoring system based on several aspects, including responses to ecosystem-based approaches to the marine environment by including species in the whole ecosystem, secondly incorporating a complete set of marine ecosystem types and potential threats, and third is calculating certainty in threat rankings. Additional approaches taken include input from experts regarding vulnerabilities and impacts of environmental change.

Results and Discussion

Climate change is more addressed in increase of earth's temperature and melting of ice in the polar regions. The increase in temperature is closely related to increasing the amount of carbon dioxide in our atmosphere (CO^2, CH^4, N^2O, CF^4, C^2F^{6+}). Both of these conditions have a direct impact on coastal ecosystems, especially coral reefs, seagrasses, and mangroves.

Climate Change Impact on Coral Reef

The impact of climate change on coral reef has occurred in these decades is the increase in sea temperatures, causing coral reefs to experience bleaching and this condition occurs in most of the world's coral reefs such as Australia, Indonesia, Thailand, Sri Lanka, Malaysia, Singapore, Burma, India, Maldives, Caribbean, Jamaica, Brazil, Colombia, Venezuela, and in various parts of East Africa. Coral bleaching is an interesting phenomenon related to how zooxanthella as main organism of coral reefs is affected by the temperature changes. Zooxanthella belongs to dinoflagellates groups that can live and develop as a coral reef builder. Zooxanthella is very susceptible to changes in temperature, both temperature increases and decreases. Rapid increase in seawater temperature causes zooxanthellae to leave coral reef tissue with strong indications of reduced number of zooxanthellae in host reefs (Hoegh-Guldberg dan Smith 1989a; Glynn dan DíCroz 1990; Lesser et al. 1990; Edward 1995; Gattuso et al. 2014; CRITC COREMAP–LIPI 2016b).

Some experts report that decrease in the population of zooxanthella in coral reef in particular *Acropora* sp. occurs when changes in temperature are at least 1 °C from the initial temperature (Hasanan et al. 2018; IPCC 2018). Other studies confirm that the bleaching process occurs for Acropora corals due to changes in temperature with very small variations either due to an increase or decrease in temperature (Williams and Bunkley-Williams 1990). Marshall and Baird (2000) also noted the same thing in observing sensitivity for coral bleaching for *Acropora* spp. on Great Barrier Reef which is vulnerable and influences at temperatures of 30–31 °C. in fact, some species such as *Acropora palifera, A. hyacinthus*, and *A. cytherea* have the highest rates of death and only leave about <8% of the population that remains alive. However, several other species of corals have a strong resistance to relatively narrow temperature changes such as *Montastrea annularis, M. cavernosa, Agaricia lamarcki, A agaricites*, and *Siderastrea radians* which tend to survive with temperatures < 30 °C (Fitt and Warner 1995; Warner et al. 1996). Longest stress period for *Isopora palifera* and *A. hyacinthus* affects severe stress on the physiological functions of zooxanthella (Hasanah et al. 2018). Weis (2008) and Higuchi et al. (2010) revealed that long-term changes would damage the formation of photosynthetic membranes and mitochondria of coral reefs and could cause oxidative stress.

Coral reefs in Indonesia are at constant risk of degradation. Long-term monitoring from 1993 to 2012, covering more than 1000 stations in 78 areas, indicated that, by the end of 2012, only 5.3% of Indonesia's coral reefs were in very good condition, 27.2% were in good condition, 37.3% in reasonable condition, and 30.5% in bad condition (Susanto et al. 2015). The results of research on 15 coral reefs in three regions of Indonesia showed a relative decline in the diversity of habitat species which resulted in the degradation of coral reef species. Reefs that experience land-based pollution (waste, sedimentation, and/or industrial pollution) show 30–50% diversity decreases at 3 m, and 40–60% diversity decreases at a depth of 10 m. Reefs that are bombed or damaged by anchors are approximately 50%. The reefs in the Java Sea are around 20% less diverse than in Ambon, Maluku. The Spermonde

Islands found a 25% reduction in diversity. The reduced diversity of coral reefs affected by land pollution implies a dramatic, rapid decline in Indonesia's coral-based fisheries resources (Edinger et al. 1998). Indonesia provided around 41% of coral exports worldwide from 1995 to 2001 (Spalding et al. 2001). Many people also value coral reefs because of their unique aesthetic and cultural values. Coral reefs are an invaluable resource for local communities throughout the world, functioning as a source of food, work, and livelihoods, and as a protection for the coast.

Bleaching event usually coincides with El Nino in Indonesia, which was first recorded in 1982–1983, which caused massive coral mortality in the Southern China Sea, Java Sea, and Sunda Strait. Coral deaths on the Thousand Islands and Karimunjawa Islands reach 90%. In 1997–1998, El Nino occurred again and resulted in mass coral death in Luat South China, Java Sea to reach 60–70%. At the same time in the western part of Southern Sumatera, there was also a massive coral death but the cause was not due to rising temperatures but a decrease to below 26 °C. Coral mortality in the western part of Sumatra reaches 90%. The El Nino incident at that time coincided with the incident of IOD (Indian Ocean Dipole mode). In 2010, there was coral bleaching in the Natuna Islands, Riau Islands, and around Aceh, especially on Pulau Weh. But this warm mass of water does not continue so that coral reefs in Bangka Belitung and Pulau Seribu are safe. In 2016, bleaching process was also reported to have occurred in several Indonesian waters. This coral reefs bleaching also occurs in the western part of Sumatra, South Java, Bali, Lombok to the south of NTT and NTB. The period of warm water appears in March, and mass mortality of corals varies greatly. Coral deaths this year also occur in Australia's Great Barrier Reef, and coral mortality is reported to have reached 90%. The current bleaching event does not seem to be continuing because the anomalous warming period in the south of the Indian Ocean has moved back to normal. El Nino and bleaching events are believed to be the result of global climate change (CRITC COREMAP–LIPI 2016b).

Coral reefs in Southeast Sulawesi until 2018, of a total of 53,153 ha, still show dominance of coral cover with moderate conditions reaching 13–60%. The distribution of many coral reefs is found in shallow areas with up to 10 m deep. The highest dominance of coral reefs is in the subtidal area with an average depth of 7 m. The dominant coral species are branching coral, encrusting coral, and massive coral. In addition, there is also a fairly varied distribution of soft corals. The total number of coral species in Southeast Sulawesi reaches 500 species. Most people use coral reefs as an area to catch demersal fish. However, some people also take coral as building material.

In 2009, the seawaters of Southeast Sulawesi experienced extensive coral bleaching reaching 80%. Bleaching also occurs in several locations in Indonesia. This event of bleaching is actually a coral death that can reach 80–90% (COREMAP II 2009). In 2016, the bleaching process in Southeast Sulawesi was repeated again and fulfilled a location that was not too large when compared to 2009. Some coral reefs were seen by algae. Some locations seem to recover well. The process of recovering coral reefs in several locations, especially in the waters of Buton Regency, Southeast Sulawesi Province, looks very varied. The bleaching process that occurred in Southeast Sulawesi occurred at several points, among others, in the waters of Buton Regency, Wakatobi River, and South Buton waters (CRITC COREMAP–LIPI 2016a).

Climate Change Impact on Seagrasses

Seagrass beds are one of the shallow marine ecosystems that have the highest productivity. Seagrass meadows can be found on the coast that is spread throughout the world, and are generally found on small islands (Short et al. 2007; Duarte 2000; De Iongh et al. 2007). The ecological function of this habitat is very large. Seagrass beds are used as a place for enlargement of various types of marine organisms, as a place to find food and as a place of reproduction for millions of marine organisms. This ecological function that places seagrass as one of the ecosystems is very important in shallow sea areas.

Seagrass beds are increasingly threatened with several stress triggers including temperature. However, the combined effects of various factors that trigger climate-related stress still need more careful observation. Usually aquatic plants such as seagrasses will be very disturbed by one of the stressors and will be more severe with a combination of other stressors. The temperature is one of the stressors that is very possible for marine plants such as seagrass. Temperature is considered as one of the most important limiting factors, and warming sea temperature is a very big threat to seagrass populations on earth (Diaz-Almela et al. 2007; Moore et al. 2012; Jueterbock et al. 2013; Araújo et al. 2016; Assis et al. 2017; Repolho et al. 2017).

Some reports state that currently the seagrass population continues to decline every year (Cullen-Unsworth and Unsworth 2013; Waycott et al. 2009; Coles et al. 2011). Short et al. (2016) further emphasizes that, besides human physical activity, the loss of population seagrass that occurs is also due to climate change and global warming. From other studies, it was reported that at present there has been a loss of carbon from seagrass after a period of significant increase in temperature in Australia (Arias-Ortiz et al. 2018). While in the Pacific region, loss of seagrass habitat is estimated at around 35% (Waycott et al. 2011).

It was further reported that climate change is expected to exacerbate the anthropogenic impact on seagrasses in the Pacific intertidal region. This seagrass loss is caused by an increase in temperature, increased sedimentation and turbidity, rising sea levels that carry high turbidity, the location and distribution of seagrass, and the physical effects of changes in water movement patterns. Rising sea levels have been reported to have an impact on turbidity and will affect seagrass habitats. In general, increasing ocean temperatures as a result of climate change will be one of the most important environmental stressors in seagrass habitats. This is evidenced by the recent loss of massive seagrass carbon storage that occurred after very high seawater temperatures in Australia (Arias-Ortiz et al. 2018). Although it is realized that each seagrass will have a different tolerance to temperature changes, temperature stress will also trigger stress for other environmental factors. Some seagrass species in the tropics have a different tolerance for increasing seawater temperatures (Campbell et al. 2006).

In general, seagrass species in the tropics have a higher optimal growth rate than temperate species. This can be seen with the response of seagrass which can continue to survive with temperatures as high as 32 °C, while others begin to show leaves death rates above 26 °C (Lee et al. 2007). In most tropical species, reduced growth and increased respiration rates are seen at higher temperatures (Short et al. 2016), and extreme temperatures are known to reduce seagrass growth and cause death (Collier and Waycott 2014).

In Indonesia, various studies have identified more than 200 species of fish, 85 crustaceans, and other marine species in seagrass beds. At low tide, many coastal villagers collect shells, shrimp, and crabs in seagrass beds. In Banten Bay, West Java, a stretch of 330 ha of *Enhalus acoroides* and *Thalassia hemprichii* (ha) provides economic opportunities for 50 fishermen (Hutomo 1997). The record of seagrass rehabilitation in Indonesia is very limited, with information available only from two institutions. Oceanographic Research Center–Indonesian Institute of Sciences (RCO-LIPI) began rehabilitation in 1988 at the Seribu Islands, Jakarta, and then continued in 2000 in Banten. The Seribu Islands National Park initiated rehabilitation in 2006. Both institutions use seagrass transplants (Asian Development Bank 2014).

Seagrass beds in Southeast Sulawesi up to 2018 reached an area of 10,762 ha with nine seagrass species and were dominated by four other main species *Enhalus acoroides*, *Thalassia hemprichii*, *Cymodocea rotundata*, and *Halodule uninervis*. The distribution pattern of seagrass is spread in tidal areas up to subtidal. Most coastal communities make use of seagrasses by extracting large amounts of resources in the seagrasses including *Portunus pelagicus*, *Holothuria* sp., *Tripneustes gratilla*, and various other commodities. There is no record on climate change impact on seagrass in Southeast Sulawesi.

Climate Change Impact on Mangrove Forest

Mangrove ecosystems are also affected by global climate change, although the effects of damage will not be as large as coral reefs. Some of the effects of climate change on mangrove forests are an increase in sea levels, rainstorms, changes in temperature, dryness of the land, moisture for a long time, and an increase in temperature for a long time. All changes that are driven by climate change including the anthropogenic effects of land use over and including water resources in watersheds can directly and indirectly affect the distribution and performance of mangroves. However, several other environmental factors will also greatly affect the physiological and ecological performance of mangroves.

Several studies revealed that there has been loss of mangrove forests in recent years including in the Caribbean region. FAO explained that the mangrove forest felled was around 3660 ha and partially suffered significant damage in the period 1980–2005 (FAO 2007a, b). Other studies revealed that the world's mangrove forests had disappeared by more than 25% (Spalding et al. 2010; UNEP 2014). Especially for the Caribbean Islands which have fairly dense mangrove forests, they have lost around 24% in the last quarter century (Polidoro et al. 2010). However, researchers still doubt whether the loss is a direct influence on climate change or other factors. The possible emphasis is on mangrove loss which is addressed in human destruction activities and environmental changes that will continue to occur in the future (Nurse et al. 2014), although the possibility of indirect influence from climate change may be influential. Some studies offer predictive analysis of what might occur in mangroves if temperatures increase, sea levels continue to rise, and changes in the composition of gases in the atmosphere and changes in ocean conditions occur (Alongi 2007; Gilman et al. 2008; Record et al. 2013; Lugo et al. 2014; Alongi 2015; Short et al. 2016). Climate change affects mangrove forests including in the Caribbean region especially in changes to air and ocean temperature variations, ocean chemical composition, high rainfall, increased wind strength and direction in coastal areas, sea level rise, wave changes such as extreme storms and droughts (Nurse et al. 2014).

Mangrove forests in Indonesia are found in the estuary areas of large islands, such as on the east coast of Sumatra Island, Kalimantan, several beaches of Sulawesi Island and Java, and along the coast of Irian Jaya. On Irian Island, Kalimantan and Sumatra Island have many large and long rivers with various types of deltas, as a result of the river currents that carry material to the estuary and tide from the sea (Pramudji 2000). Mangrove forests in Indonesia have a high diversity. About 15 plant families, with 18 genera and 41 species, and 116 related species are found in the mangrove ecosystem in Indonesia. Mangrove forests are also home to hundreds of animal species, including mammals, reptiles, birds, and marine fauna (Santoso 2000). The ecological and economic functions of mangroves are very high and directly affect the lives of coastal communities. Besides the mangrove tree line, it can protect coastal communities from various wind pressure and extreme conditions in coastal areas. At present, most of the mangrove land is finally used as residential land or pond cultivation at the expense of a large area. In several regions in Indonesia, traditional use of mangrove products, usually on a small amount (Kusmana and Sukristijiono 2016).

Measuring Vulnerability Values on Coastal Ecosystems

Several studies have revealed the results of evaluating the level of vulnerability of ecosystems and their relationship to human activities (Halpern et al. 2007). In this discussion, we try to make several approaches, especially for Eastern Indonesia by looking at the potential specific threats to an ecosystem by taking into account the recovery and sensitivity of each species in specific ecosystems, especially coral reefs, which address the vulnerability of zooxanthella, seagrass habitats which are addressed in susceptibility to increasing temperatures and rising sea levels that have an impact on sea turbidity, mangrove forests specifically on sea level rise in the structure and potential for successful reproduction of mangroves. The specific approach

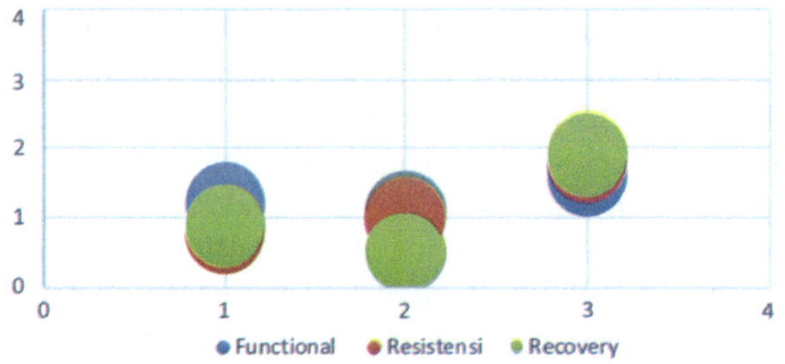

Fig. 8.1 Level of vulnerability of coastal ecosystems 1. Seagrass, 2. Mangrove, and 3. Coral reef in Southeast Sulawesi, Eastern Indonesia, for specifications at functional, resistance, and recovery levels, with each vulnerability level values ranging 0.0–1.0: Low vulnerability, 1.1–2.0: Moderate vulnerability, 2.0–3.0: High vulnerability, and 3.0–4.0: Very High Vulnerability

taken is to look at the potential vulnerability of each habitat and its relation to possible recovery power. The species approach is the most likely given that the vulnerability of each species will be very different. A reasonable consideration is that each species has a different adaptation (Fig. 8.1).

The level of vulnerability of coral reefs occupies a relatively higher position compared to mangroves and seagrass beds. The level of vulnerability in terms of impact on ecosystem services occupies a relatively high level of 2.0 which is categorized as high vulnerability, while for mangroves and seagrasses it occupies a moderate level of vulnerability. This condition can change and is not absolute because it is very dependent on changes in the area and role of each ecosystem due to exposure (physical and non-physical) that affects its function as a service ecosystem for organisms and their support to the surrounding community, existing recovery levels, and resistance levels.

Climate Change Impact on Coastal Community

There are three main ecosystems that are very important in shallow marine areas, among others, coral reefs, mangrove forests, and seagrass beds. Community interest is not solely on the existence of the three ecosystems, but rather on the many resources that exist in these three ecosystems. Mangroves, seagrasses, and coral reefs are known to protect land from speed current, waves, and storms, reducing damage from natural disasters to coastal communities. They also play an important role in purifying heavy metal wastes and toxic substances and removing harmful air and water pollutants that can cause serious illness or even death (Cook et al. 2005; UNEP 2006). Access to resources is very large and intense, causing a decline in population and ecosystem function. This condition is exacerbated by environmental factors that directly and indirectly affect the quality of the environment and coastal and marine resources. Exploitation of mangroves has increased, and there is a tendency to destroy or disrupt fallen trees. The impact is natural resources and organisms living in the area will be destroyed. Large-scale conversion will cut the ecological food chain and reduce the production of ecosystems around mangrove forests. More research has reported regional scale changes in coral calcification and mortality that are consistent with the scale and impact of ocean warming and acidification when compared to local factors such as decreasing water quality and overfishing (Hoegh-Guldberg et al. 2007).

Changes in ecosystems can affect the provision of these services and directly affect human well-being. The main drivers of changes in marine and coastal ecosystems that affect the flow of benefits for humans ranging from global problems to regional problems. Anthropogenic climate change has been a global concern for decades and many believe that climate change will cause uncontrolled large-scale changes in ocean temperatures and ecosystems that support marine life (Gjerde and Breide 2003). Ocean acidification is related to the phenomenon of global climate change and increased levels of atmospheric carbon dioxide (CO_2). The world's oceans absorb most of the anthropogenic CO_2 released into the atmosphere, which causes a decrease in pH and ocean levels to become increasingly acidic, with critical implications for marine ecosystems and biodiversity (Bates and Peters 2007; Worm et al. 2006).

Coastal communities are generally categorized as small-scale fisheries which include fishing activities, especially in developing countries that have an important role in small-scale economies, especially rural economies (Bene 2006). The livelihoods and welfare of coastal communities are very dependent on the resources and services provided by coastal ecosystems. From basic needs, such as food, water, and shelter, to services, such as tourism and recreation, coastal ecosystems meet many human needs. Marine and coastal resources and ecosystems have a direct and indirect impact on the lives of coastal communities. The community is more in the habit of taking fish and non-fish resources in the area around their village. Most coastal communities are classified as small-scale fisheries that depend heavily on the surrounding natural resources.

Coastal communities are very vulnerable to changes that arise from both natural and anthropogenic symptoms (Ferrol-Schulte et al. 2015). Most coastal residents work as fishermen as the main source of income with very limited alternative livelihood resources. If there is a disturbance from climate change and other natural hazards or climate variability, there will be economic shocks. One of the supporting factors for this is that their isolation makes them more vulnerable to natural changes and will fall into the trap of poverty (Allison and Ellis 2001; Sievanen 2014). This situation occurs to the people who live in Siompu Islands, Buton Regency, Southeast Sulawesi. Located in the southwest of Buton Island, in Southeast Sulawesi. This region has a comparative advantage over marine resources. Siompu Island should be developed economically based on marine resources such as fisheries and tourism. However, most people still live in poverty. This shows that the island's natural resources have not been utilized optimally. Characterized by small islands, natural hazards, climate variability, and isolation may be associated with poverty and not optimal use of island resources. Fishing as the main source of employment for the population is vulnerable to climate variability and natural hazards. This will have a significant impact on the livelihoods of the community as a whole.

At present, the influence of climate change has truly become one thing that is very prominent in several coastal areas of Indonesia, especially in Eastern Indonesia. Climate change can affect coastal areas in various ways. Beaches are sensitive to sea level rise, changes in frequency and intensity of storms, increased rainfall, and warmer ocean temperatures. In addition, the increased concentration of carbon dioxide (CO_2) in the atmosphere causes the ocean to absorb more gas and become more acidic. This increase in acidity can have a significant impact on coastal and marine ecosystems. There is evidence of an increase in the intensity or frequency of many extreme events such as floods, soil erosion, heat waves, tropical cyclones, high rainfall, tornadoes, drought, storm surges, salinity intrusion, etc., which causes the loss of cultivation areas, damage to the land where fishermen live, increases the scarcity of sources of fish and shrimp seedlings, the destruction of several settlements, decreases production of fisheries and cultivation, increases the management costs of fisheries management and the emergence of diseases, etc., (USAID 2018). The impact of climate change tends to exacerbate the problems that have been faced in coastal areas. Facing existing challenges that affect man-made infrastructure and coastal ecosystems, such as shoreline erosion, coastal flooding, and water pollution, has become a concern in many areas. Overcoming additional pressures from climate change may require new approaches to managing land, water, waste, and ecosystems.

The results of our research in Southeast Sulawesi indicate that there are several main impacts of this climate change that not only have an impact on reducing fisheries production, but also have an impact on the social life and demographics of the community. Some of the real impacts that have occurred include drought, extreme rain, seasonal pattern changes, rising sea levels, rising temperatures, and strong winds. Some events that occur as a result of climate change can be seen in several locations in Southeast Sulawesi as given in the Table 8.1.

The conditions described in the table above, although not evenly distributed, have strong evidence in several regions in Southeast Sulawesi. It is possible to see that the availability of resources has not been fully affected in general. The clear indicator can be seen from the socioeconomic conditions of the coastal communities of Southeast Sulawesi, most of which continue to operate as usual, even though in some regions there are specifications for environmental changes that have an impact on reducing seaweed production or coral bleaching.

The three main ecosystems in Southeast Sulawesi that include coral reefs, mangrove forests, and seagrass habitats are allegedly affected by climate change. In 2009–2010, there were 85% coral reefs in Buton district, especially in the three West Coast waters, South Coast waters, and Wakatobi waters which were bleached due to the increase in temperature which occurred evenly at depths of up to 5 m. This condition then recovers gradually, even though some locations provide a picture that cannot recover with a tightly covered macroalga. There is no strong evidence in the Southeast Sulawesi region that climate change really has a bad influence on the level of people's economic life. Except for a number of indicators, there is a decrease in catch due to the more distant current fishing area (COREMAP II 2009; CRITC COREMAP–LIPI 2016a).

Table 8.1 An overview of the influence of climate change in Southeast Sulawesi (Results of FGD)

Main impact	Chain impact	Conditions that have occurred in Southeast Sulawesi, Eastern Indonesia
Extreme temperature and drought	Increased sea surface temperature	Coral bleaching Decrease in fishing productivity Seaweed disease Decrease in aquaculture production
	Coastal forest fires	Economic losses Safety of the population is threatened Health coastal communities are disrupted
	Dehydration	Low productivity Declining health
Extreme rain	Coastal floods	Disease attack Material loss Ecosystem loss
	Failure of cultivation activities	Ecosystem loss Economic losses Loss of livelihood
	Malaria, dengue fever, diarrhea	Productivity decreases Loss of life Need for medical expenses Welfare has dropped dramatically
Changes in seasonal patterns at sea	Uncertain fishing season	Less catch Catching productivity decreases Loss/Loss of profits Lack of food sources
	The difficulty of determining the fishing ground	Catching productivity decreases Loss/Loss of profits Lack of food sources Prolonged rain, floods, landslides Ecosystem loss
	Prolonged rain	Floods, landslides Ecosystem loss
	Strong winds	Loss of houses and public facilities Ecosystem loss Potential for loss of life
	The strength of the tidal wave	Loss of houses and public facilities Ecosystem loss Potential for loss of life

(continued)

Table 8.1 (continued)

Main impact	Chain impact	Conditions that have occurred in Southeast Sulawesi, Eastern Indonesia
Sea level rise	Occurrence of flood tides	Floods, landslides Ecosystem loss Potential diseases and poor sanitation systems Loss of advice and infrastructure
	The existence of permanent pools	Loss of houses and public facilities Ecosystem loss Potential for loss of life
	Damage to the pond	Losing cultivation commodities Loss of business capital, economic loss
	Widespread abrasion	Loss of facilities and infrastructure Ecosystem loss Economic losses

Several categories of fish are targeted for capture because economic value is decreasing in the area near community housing. The results of our study explain that coral fish populations in Buton District did not significantly increase from 2009–2016, especially in marine protected areas in Buton District. This condition is thought to be closely related to the high number of dead corals since 2007–2009. In 2014, the population of reef fish belonging to the target fish did not experience significant additions. The percentage of dead coral cover is around 40% and live corals are only about 30%, and the rest are other organisms and algae and abiotic components. Environmental conditions like this show that the target fish is only around 25%, indicator fish is around 17%, and the rest are major fish with a dominance value of 58%. In 2016, at the same location, the average live coral was only about 28.89% and dead coral and algae and sand components covered 8.5–52.9%. And the target fish is only around 30% of the total fish population in the coral reef area (CRITC COREMAP–LIPI 2016a). In 2018 the distribution of live coral cover in the Southeast Sulawesi region, Eastern Indonesia, reached 53,153 ha. These conditions include moderate on 42% and good conditions 16%, and damaged conditions ranging from 42%. The main damage that occurs is not entirely a physical change by human activities but also from changes in the physical conditions of the environment as an effect of global warming. Climate and non-climate triggers affect corals and coral reefs, and negative impacts have been observed. Bleaching involves damage and loss of endosymbiotic algae, which live in coral tissue and play a key role in supplying energy to coral hosts. Mass coral bleaching and death, triggered by positive temperature anomalies (high confidence), are the most widespread and striking effects of climate change.

For example, the level of heat stress in most of the 47 reef locations where bleaching occurred during 1997–1998 was unmatched in the period 1903–1999 (Lough 2000). Ocean acidification reduces the biodiversity and rate of coral calcification while at the same time increasing the dissolution rate of coral skeletons through stimulation of biological erosion and chemical dissolution. Overall, these changes will balance the calcium carbonate content of coral reefs which have an impact on the surrounding organims.

Community Adaptation Strategies to Climate Change

At present, climate change and variability are global challenges. The impact of climate change will continue to occur and be more prominent in developing countries, especially Southeast Asian countries. It is projected that climate change will have a significant impact on the livelihoods and living conditions of poor people in developing countries due to lower adaptive capacity. This condition has implications for threatening many sectors including those related to food security, water resources, marine resource productivity, sea level rise, desertification, and the spread of diseases. Climate change is more about changing the hydrological cycle, weather patterns, rising sea levels, and increasing the intensity and frequency of extreme weather conditions which all have important impacts on livelihoods and living conditions of poor people in developing countries, especially in rural communities (FAO 2007a, b; Carter 2001). The impact of climate change in coastal areas of Indonesia occurs during the transition period of the season. Some areas experience delays at the beginning of the rainy season and the beginning of the dry season. Changes at the beginning of the season are not consistent in all regions of Indonesia. Some regions even experience an early season change. Current rainfall intensity is also different from that of the previous years. At present, there is a tendency for a greater difference between the intensity of rain during the rainy season and the intensity of rain during the dry season, and sometimes the intensity difference is very extreme. Flood, drought, and landslide incidents are evidence of the phenomenon of climate change caused by changes in the intensity of rainfall in Indonesia. This incident certainly has an impact on the development process in the affected area. Another major threat as a result of the impact of climate change in relation to Indonesia's geographical conditions is sea level rise. The threat of rising sea levels and high sea wave incidents can occur throughout the coastal area, especially during the transition season (BAPPENAS 2011). Climate change continues to have a major impact on coastal communities. Changes in rainfall patterns and high sea level rise forced fishermen to face unpredictable weather and high tides. Climate change has also disrupted livelihoods on many islands, as fishermen in Maluku claim that they can no longer predict the right time and location to catch fish because of changing climate patterns. Of the 2, 7 million fishermen in Indonesia, around 80% of them are small-scale and traditional fishermen who are vulnerable to extreme climate variations in the sea (BAPPENAS 2011).

The most important thing that coastal communities can do to survive and continue to carry out their activities with the threat of climate change is adaptation strategies (Aldrian et al 2011). Long-term strategies can be developed by coastal communities in response to the effects of climate change. This is due to the fact that sustainable livelihoods can only be achieved if the long-term strategy is fully implemented (FAO 2007a, b). To deal with and minimize the effects of climate change, Indonesia can take more specific steps, including increasing knowledge about hazards and the effects of climate change on ecosystems in coastal communities, integrating climate change issues in coastal area spatial planning, formulating strategies and activities for climate change adaptation and to recognize, support, and facilitate smart adaptation efforts carried out by the community, developing and using technology to support adaptation efforts and to improve the sustainability of coastal communities in the face of climate change impacts, and improving the application of climate information, policy formulation, and protection tools for the most vulnerable groups who are unable to anticipate and adapt to climate change (BAPPENAS 2011).

Overcoming the impact of climate change in the future needs to have an appropriate mitigation strategy to counteract the impact. However, mitigation strategies differ from one location to another depending on the socioeconomic characteristics of the community trying to overcome the impacts of climate change that are expected to occur in the future (Hizza 2013). Adaptive capacity depends on economic welfare, ecological well-being, the degree of dependence on natural resources, infrastructure (man-made or natural), effectiveness of institutions and governance systems, insurance, guaranteed land tenure and mediation measures, and information and communication systems. Communities with adaptive capacity are likely to be more resistant to impacts or able to recover from pressing events and conditions (USAID 2009). Implementing adaptation strategies to climate change means carrying out adaptation measures by reducing the impact and/or increasing the resilience of communities and ecosystems in dealing with climate change and variability while carrying out various daily activities. Adaptation to the impacts of climate change must be mainstreamed into economic development planning and implementation processes. Adaptation and mitigation strategies will increase benefits for people who are most vulnerable to climate change (Murdiyarso and Kauffman 2011). A strong relationship between government organizations, non-governmental, and international organizations holds the key to the success of the strategy in facing the effects of global warming. Besides that, the strategy is to build the strength of the community itself (Djalante 2013).

In the event of changes and climate variations that affects fishing patterns and fishing areas, the community will experience a decrease in catch production. This has been felt by some fishermen in Southeast Sulawesi. The high cost of fuel oil for fishing boats forces fishermen to carry out fish catches near their area, and some even use explosives and fish poisons. This will further aggravate the sustainability of the fish population. The adaptation strategy that can be developed is to use a variety of simple efforts with marine cultivation. Aquaculture developed can be in the form of fish farming in ponds, large-scale seaweed cultivation, maintaining mangrove forests, maintaining seagrass, and not using tools that can damage the environment. Even though this action is a local action that has a small impact, it is important to continue to carry it out as an effort to defend against climate change. In Table 8.2, we are trying to explore some of the strategic steps of local community adaptation on climate change.

Table 8.2 Adaptation strategy of specific changes in local area due to climate change

Local impact/Environmental change	Impact	Specific changes in impact	Adaptation strategy
Increasing sea temperature	+	Increase in amount of catch for certain types of fish (Pelagic) (*El Nino effect*)	Processing of fishery products is durable Diversification of processed fishery products
	–	Coral bleaching Seaweed disease Coastal forest fires Economic losses Dehydration	Eco-friendly fishing Reducing the sedimentation process Using better seaweed seeds Good seaweed cultivation method Minimizing the impact of forest degradation Developing alternative livelihoods Availability of clean water that is suitable for consumption
Extreme rain	–	Coastal floods Failure of cultivation activities Malaria, dengue fever, diarrhea	Arrangement of flood-prone areas Good and strong management of cultivation systems for climate change Improving the system of health services and nutritional resilience and being aware of the coastal environment
Changes in seasonal patterns at sea	–	Uncertain fishing season The difficulty of determining the fishing ground Prolonged rain Strong winds The strength of the tidal wave	Alternative aquaculture (Seaweed, fish, and non-fish) Increase artificial reef and marine protected areas Regional arrangement that is resilient to hard rain and strong wind Maintaining mangrove forests Management of the layout of coastal village development Resilient housing and settlement

(continued)

Table 8.2 (continued)

Local impact/Environmental change	Impact	Specific changes in impact	Adaptation strategy
Sea level rise	–	Occurrence of flood tides Damage to the pond Widespread abrasion	Resilient rural area arrangement Construction of settlements and public facilities toward the highlands Resilient management and management of coastal cultivation Construction of beach abrasion protection

Coastal communities, in Southeast Sulawesi, Eastern Indonesia, have developed a number of adaptation steps to the impacts of climate change including the development of aquaculture (fish, non-fish, and seaweed), developing alternative livelihoods with gardening, farming, and livestock, developing processed products, long-term fisheries, use of environmentally friendly fishing gear, use of modern fishing technology, and developing marine community base tourism. Apart from the adaptation strategies developed by the community, it is possible to continue to experience more impacts on their daily livelihoods because some strategies are not environmentally friendly, for example, fishing technology that might damage the environment. The percentage of each adaptation strategy carried out can be seen in the following table (Fig. 8.2).

One strategy that can be developed by coastal communities, especially those living in remote areas and small islands, is to develop seaweed farming. Seaweed is a commodity that is very easy to strengthen the economy of coastal communities. The main indicators for being able to develop people's lives with seaweed are very easy cultivation methods and easy access to funds to support the daily lives of coastal communities. Recommendations for improving the way seaweed is cultivated are important. A good cultivation method will increase community seaweed yields. The level of income of people with seaweed cultivation is quite significant. The method that can be used to further increase production is by applying protected seaweed cultivation. Protected seaweed cultivation is a seaweed cultivation strategy that can increase growth and overall production value. The use of nets in seaweed cultivation has a specific growth rate that is much higher than the longline method. From the results of the research we did, it was found that seaweed SGR with longline (current method) can reach 3–4%/day, while AlgaNet (projected cultivation equipment with a net system) has SGR of up to 10%/day. The production of seaweed growth in one particular area with AlgaNet can exceed 200–300% of total seaweed production (Kasim et al. 2016; Kasim and Mustafa 2017; Kasim et. al 2017a; Kasim et. al 2018) (Fig. 8.3).

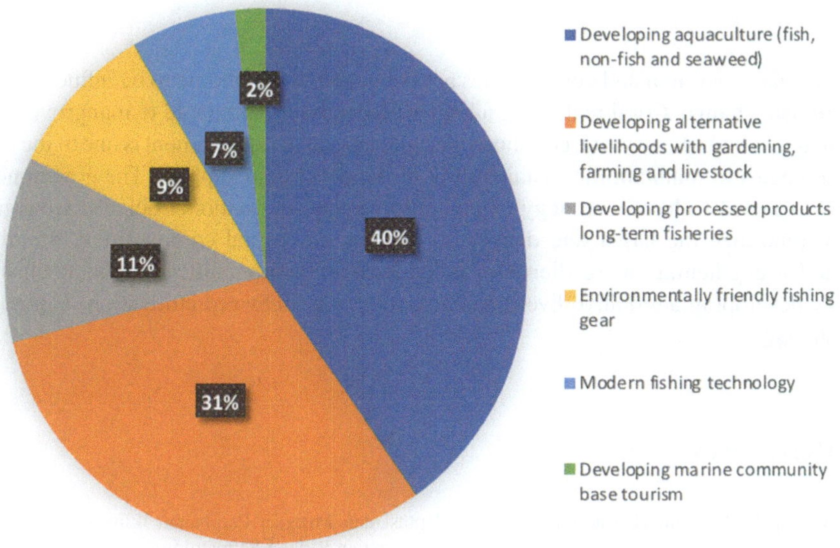

Fig. 8.2 Percentage of adaptation strategies for Southeast Sulawesi coastal community (For example, in Buton District, Konawe Selatan District, Muna Regency, Wakatobi, Kolaka, Baubau City, and Kendari City)

Fig. 8.3 Seaweed cultivation is a good livelihood to build the strength of coastal communities in the face of climate change (Kasim et al. 2017b)

Conclusion

Coastal ecosystem and communities are being and will continue to be influenced by climate change. Coral reefs have a higher vulnerability compared to mangroves and seagrass beds. Structured, collaborative, and integrated management is one of the best strategies to maintain the sustainability of the ecosystem support. The government is expected to have a strategy of law enforcement and national defense strategies to minimize the larger and dangerous impacts on coastal communities. Building and strengthening the resilience of individuals and communities to climate change by developing alternative livelihood activities (e.g., seaweed cultivation) is a good strategy.

References

Aldrian E, Karmini E, Budiman (2011) Adaptasi dan Mitigasi Perubahan Iklim di Indonesia, Pusat Perubahan Iklim dan Kualitas Udara Kedeputian Bidang Klimatologi Badan Meteorologi, Klimatologi, dan Geofisika, p 174

Allison EH, Ellis F (2001) The livelihoods approach and management of small-scale fisheries. Mar Policy 25(5):377–388

Allison EH, Perry AL, Badjeck MC, Adger WN, Brown K, Conway D, Halls AS, Pilling GM, Reynolds JD, Andrew NL, Dulvy NK (2009) Vulnerability of national economies to the impacts of climate change on fisheries. Fish Fish 10:173–196

Alongi DM (2007) Mangrove forests: Resilience, protection from tsunamis, and responses to global climate change. Estuar Coast Shelf Sci 76:1–13

Alongi DM (2015) The impact of climate change on mangrove forests. Current Clim Change Rep 1(1):30–39

Araújo RM, Assis J, Aguillar R, Airoldi L, Bárbara I, Bartsch I (2016) Status, trends and drivers of kelp forests in Europe: an expert assessment. Biodivers Conserv 25:1319–1348. https://doi.org/10.1007/s10531-016-1141-7

Arias-Ortiz A, Serrano O, Masqué P, Lavery PS, Mueller U, Kendrick GA, Rozaimi M, Esteban A, Fourqurean JW, Marbà N, Mateo MA, Murray K, Rule MJ, Duarte CM (2018) A marine heatwave drives massive losses from the world's largest seagrass carbon stocks. Nat Clim Change. https://doi.org/10.1038/s41558-018-0096-y

Asian Development Bank (2014) State of the coral triangle: Indonesia Mandaluyong City Philippines, p 69

Assis J, Araújo MB, Serrão EA (2017) Projected climate changes threaten ancient refugia of kelp forests in the North Atlantic. Glob Chang Biol 24:e55–e66. https://doi.org/10.1111/gcb.13818

BAPPENAS (2011) Indonesia adaptation strategy : improving capacity to adapt, ministry of national development planning/national development planning agency (BAPPENAS). https://www.bappenas.go.id/files/6414/1171/7069/INDONESIA_ADAPTATION_STRATEGY_-_Improving_Capacity_to_Adapt.pdf. Access 17 Apr 2019, p 39

Bates N, Peters A (2007) The contribution of atmospheric acid deposition to ocean acidification in the subtropical north atlantic Ocean. Mar Chem 107:547–558

Bene C (2006) Small-scale fisheries: assessing their contribution to rural livelihoods in developing countries. Food and Agriculture Organization of the United Nations, Rome, Italy

Brander KM (2007) Global fish production and climate change. Proc Natl Acad Sci 104(50):19709–19714. https://doi.org/10.1073/pnas.0702059104

Bryant D, Burke L, McManus J, Spalding M (1998) Reefs at risk: a map-based indicator of threats to the world's coral reefs. World Resources Institute Washington D.C.
Campbell SJ, McKenzie LJ, Kerville SP (2006) Photosynthetic responses of seven tropical seagrasses to elevated seawater temperature. J Exp Mar Biol Ecol 330(2):455–468
Carter TR (2001) Climate change 2001: the scientific basis, Cambridge University Press Cambridge United Kingdom, IPCC
Cheung WWL, Lam VWY, Sarmiento JL, Kearney K, Watson R, Zeller D, Pauly D (2010) Large-scale redistribution of maximum fisheries catch potential in the global ocean under climate change. Glob Chang Biol 16:24–35. https://doi.org/10.1111/j.1365-2486.2009.01995.x
Christianen MJA, Govers LL, Bouma TJ, Kiswara W, Roelofs JGM, Lamers LPM, van Katwijk MM (2011) Marine megaherbivore grazing may increase seagrass tolerance to high nutrient loads. J Ecol 100:546–560
Coles R, Grech A, Rasheed M, McKenzie L, Unsworth R, Short F (2011) Seagrass ecology and threats in the tropical Indo-Pacific bioregion. Nova Science Publishers, pp 225–239
Collier CJ, Waycott M (2014) Temperature extremes reduce seagrass growth and induce mortality. Mar Pollut Bull 83(2):483–490
Cook A, Weinstein P, Centeno JA (2005) Health effects of natural dust: the role of trace elements and compounds. Biol Trace Elem Res 103(1):1–15
COREMAP II (2009) Report on Implementation of Implementation Activities of the Local Research of COREMAP II of Buton Regency in 2009 (Indonesian)
CRITC COREMAP—LIPI (2016a) Baseline study for monitoring the health of coral reefs in the Buton archipelago. Access 12 Apr 2019. https://www.coremap.or.id/berita/1173
CRITC COREMAP–LIPI (2016b) Bleaching coral and bleaching events in 2016. Access 12 Apr 2019
Cullen-Unsworth L, Unsworth R (2013) Seagrass meadows, ecosystem services and sustainability. Environ Sci Policy Sustain Dev 55:14–28
De Iongh HH, Kiswara W, Kustiawan W, Loth (2007) A review of research on the interactions between dugongs (Dugong dugon Muller 1776) and intertidal seagrass beds in Indonesia 73–83. https://doi.org/10.1007/s10750-007-0785-4
Diaz-Almela E, Marbà N, Duarte CM (2007) Consequences of Mediterranean warming events in seagrass (*Posidonia oceanica*) flowering records. Glob Chang Biol 13:224–235. https://doi.org/10.1111/j.1365-2486.2006.01260.x
Ding Q, Chen X, Hilborn R, Chen Y (2017) Vulnerability to impacts of climate change on marine fisheries and food security. Mar Policy 83:55–61
Djalante R (2013) Identifying drivers, barriers and opportunities for integrating disaster risk reduction and climate change adaptation in Indonesia: an analysis based on the earth system governance framework. In: Leal Filho W (eds) Climate change and disaster risk management. Climate change management. Springer, Berlin. https://doi.org/10.1007/978-3-642-31110-9_9
Djalante R, Thomalla F (2012) Disaster risk reduction and climate change adaptation in Indonesia. Int J Disaster Resilience Built Environ 3(2):166–180
Duarte C (2000) Marine biodiversity and ecosystem services: an elusive link. J Exp Mar Biol Ecol 250(1–2):117–131
Edinger EN, Jompa J, Limmon GL, Widjatmok W, Risk MJ (1998) Reef degradation and coral biodiversity in Indonesia: effects of land-based pollution, destructive fishing practices and changes over time. Mar Pollut Bull 36(8):617–630
Edwards AJ (1995) Impacts of climate change on coral reefs, mangroves and tropical seagrass ecosystems. In: Eisma D (ed) Climate change: impact on coastal habitation. Lewis Publishers, Boca Raton, pp 209–234
FAO (2007a) Adaptation to climate change in agriculture, forestry and fisheries: perspective framework and priorities. Interdepartmental. Working group on climate change. Food and Agriculture Organization of the United Nations Rome, p 32
FAO (2007b) Mangroves of Asia 1980–2005: country reports. Forest Resources Assessment Working Paper No. 136. Rome, www.fao.org/forestry/site/mangrove/statistics

Ferrol-Schulte D, Gorris P, Baitoningsih W, Adhuri DS, Ferse SC (2015) Coastal livelihood vulnerability to marine resource degradation : a review of Indonesia national coastal and marine policy framework. Mar Policy 52:163–171. https://doi.org/10.1016/Soo25-326X(98)00047-2

Fitt WK, Warner ME (1995) Bleaching patterns of four species of caribbean reef corals. Biol Bull 189(3):298–307

Folke C, Carpenter S, Walker B, Scheffer M, Elmqvist T (2004) Regime shifts, resilience, and biodiversity in ecosystem management. Annu Rev Ecol Evol Syst 35:557–581. https://doi.org/10.1146/annurev.ecolsys.35.021103.105711

Gattuso JP, Hoegh-Guldberg O, Pörtner HO (2014) Cross-chapter boxes. Climate change impacts, adaptation, and vulnerability 97–166. https://doi.org/10.1017/cbo9781107415379.005

Gilman EL, Ellison J, Duke NC, Field C (2008) Threats to mangroves from climate change and adaptation options: a review. Aquat Bot 89:237–250

Gjerde KM, Breide C (2003) Towards a strategy for high seas marine protected areas: proceedings of the IUCN, WCPA and WWF experts workshop on high seas marine protected areas, 15–17 January 2003, Malaga, Spain. IUCN, Gland, Switzerland and Cambridge, UK

Glynn PW, D'Croz L (1990) Experimental evidence for high temperature stress as the cause of El Nino-coincident coral mortality. Coral Reefs 8:181–191

Halpern BS, Kimberly A, Selkoe KA, Fiorenza MF, Kappel FC (2007) Evaluating and ranking the vulnerability of global marine ecosystems to anthropogenic threats. Conserv Biol 21(5):1301–1315. https://doi.org/10.1111/j.1523-1739.2007.00752.x

Handayani W, Rudiarto I, Setyono JS, Chigbu UE, Sukmawati AM (2017) Vulnerability assessment: A comparison of three different city sizes in the coastal area of Central Java, Indonesia. Adv Clim Change Res. https://doi.org/10.1016/j.accre.2017.11.002

Hasanah AN, Rukminasari N, Yunus B, Yanuarita D, Jompa J, Suharto, Inaku DF (2018) The effect of temperature on zooxanthellae of isopora palifera and acropora hyacinthus from karanrang island, Indonesia, SPERMONDE 4(1):7–11

Higuchi T, Fujimura H, Hitomi Y, Arakaki T, Oomori T, Suzuki Y (2010) Photochemichal formation of hydroxyl radicals in tissue extracts of the coral Galaxea fascicularis. Photochem Photobiol 86:1421–1426

Hizza EL (2013) Coping and adaptation strategies to climate change impacts: the case of rural community in Vuga village-Lushoto district, MUCCoBS Working Paper Series, Moshi University College of Cooperative and Business Studies

Hoegh-Guldberg O (2011) Coral reef ecosystems and anthropogenic climate change. Reg Environ Change 11:215–227

Hoegh-Guldberg O, Smith GJ (1989) The effect of sudden changes in temperature, light and salinity on the population density and export of zooxanthellae from the reef corals Stylophora pistillata Esper and Seriatopora hystrix Dana. J Exp Mar Biol Ecol 129:279–303

Hutomo M (1997) Struktur komunitas padang lamun perairan Indonesia. In: Inventarisasi dan evaluasi potensi laut-pesisir II geologi, kimia, biologi, dan ekologi. Prosiding Kongres Biologi Indonesia XV. Universitas Indonesia. Jakarta. (*In Indonrsian*), pp 54–61

IPCC (2018) Global warming of 1.5 °C: An IPCC special report on the impacts of global warming of 1.5 °C above pre-industrial levels and related global greenhouse gas emission pathways, in the context of strengthening the global response to the threat of climate change, sustainable development, and efforts to eradicate poverty. IPCC, Switzerland, p 26

Jueterbock A, Tyberghein L, Verbruggen H, Coyer JA, Olsen JL, Hoarau G (2013) Climate change impact on seaweed meadow distribution in the North Atlantic rocky intertidal. Ecol Evol 3:1356–1373. https://doi.org/10.1002/ece3.541

Kappel CV (2005) Losing pieces of the puzzle: threats to marine, estuarine, and diadromous species. Front Ecol Environ 3:275–282

Kasim M, Mustafa A (2017) Comparison growth of *Kappaphycus alvarezii* (Rhodophyta, Solieriaceae) cultivation in floating cage and longline in Indonesia. Aquaculture Rep 6:49–55. https://doi.org/10.1016/j.aqrep.2017.03.004

Kasim M, Mustafa A, Munier T (2016) The growth rate of seaweed (Eucheuma denticulatum) cultivated in longline and floating cage. AACL Bioflux 9(2):291–299

Kasim M, Jamil MR, Irawati N (2017a) Occurrence of macro-epiphyte on Eucheuma spinosum cultivated on floating cages. AACL Bioflux 10(3):633–639

Kasim M, Mustafa A, Male I, Muzuni JW (2017b) New methods on cultivation of Eucheuma denticulatum and Kappahycus alvarezii in Indonesia. J Fish Aquat Sci 12:207–217

Kasim M, Asjan, Efendy, IJ, Ishak E (2018) Influence of initial weight of seeds in variation of growth and carrageenan content of Eucheuma spinosum. AACL Bioflux 11(4):1155–1163

Kendrick G, Waycott M, Carruthers TJB, Cambridge ML, Hovey R, Krauss SL, Lavery PS, Les DH, Lowe RL, Vidal O, Ooi JLS, Orth RJ, Rivers DO, Ruiz- Montoya L, Sinclair EA, Statton J, Van Dijk JK, Verduin JJ (2012) The central role of dispersal in the maintenance and persistence of seagrass populations. Bioscience 62:56–65

Kilminster K, McMahon K, Waycott M, Kendrick GA, Scanes P, McKenzie L, O'Brien KR, Lyons M, Ferguson A, Maxwell P, Glasby T, Udy J (2015) Unravelling complexity in seagrass systems for management: Australia as a microcosm. Sci Total Environ https://dx.doi.org/10.1016/j.scitotenv.2015.04.061

Kusmana C, Sukristijiono (2016) Mangrove resource uses by local community In Indonesia. J Pengelolaan Sumberdaya Alam dan Lingkungan 6(2):217–224

Lee KS, Park SR Kim YK (2007) Effects of irradiance, temperature, and nutrients on growth dynamics of seagrasses: a review. J Exp Biol Ecol 350:144–175

Lesser MP, Stochaj WR, Tapley WR, Schick JM (1990) Bleaching in coral reef anthozoans: effects of irradiance, ultraviolet radiation, and temperature on the activities of protective enzymes against active oxygen. Coral Reefs 8:225–232

Lough JM (2000) 1997–98: Unprecedented thermal stress to coral reefs? Geophys Res Lett 27:3901–3904. https://doi.org/10.1029/2000GL011715

Lugo AE, Medina E, McGinley K (2014) Issues and challenges of mangrove conservation in the Anthropoceno. Madera Bosques 20:11–38

Marshall PA, Baird AH (2000) Bleaching of corals on the great barrier reef: differential susceptibilities among taxa. Coral Reefs 19:155–163

McClanahan T (2015) EH Allison JE, Cinner Managing fisheries for human and food security. Fish Fish 16:78–103

McGranahan G, Balk D, Anderson B (2007) The rising tide: assessing the risks of climate change to human settlements in low-elevation coastal zones. Environ Urbanization 19(1):17–37

Moore KA, Shields EC, Parrish DB, Orth RJ (2012) Eelgrass survival in two contrasting systems: role of turbidity and summer water temperatures 44:247–258. https://doi.org/10.3354/meps09578

Munday PL, GP Jones GP, Pratchett MS, Williams AJ (2008) Climate change and the future for coral reef fishes. Fish Fish 9:261–285

Nurse LA, McLean RF, Agard J, Briguglio LP, Duvat-Magnan V, Pelesikoti N, Tompkins E, Webb A (2014) Small Islands. In: Climate change 2014: impacts, adaptation, and vulnerability. Part B: regional aspects. contribution of working Group II to the fifth assessment report of the intergovernmental panel on climate change

Polidoro BA, Carpenter KE, Collins L, Norman C, Duke A, Ellison M, Joanna CE, Elizabeth JeanFarnsworth E (2010) The loss of species: Mangrove extinctionrisk and geographic areas global concern. PLoS ONE 5(4):e10095

Polsky C, Neff R, Yarnal B (2007) Building comparable global change vulnerability assessment: the vulnerability scoping diagram. Glob Environ Chane 17:472–485

Pramudji (2000) Hutan Mangrove Di Indonesia: Peranan Permasalahan Dan Pengelolaannya, Oseana XXV (1):13–20

Record S, Charney ND, Zakaria RM, Ellison AM (2013) Projecting global mangrove species and community distributions under climate change. Ecosphere 4(3):1–23

Repolho T, Duarte B, Dionísio G, Paula JR, Lopes AR, Rosa IC, Grilo TF, cacador I, Calado R, Rosa R (2017) Seagrass ecophysiological performance under ocean warming and acidification. Sci Rep 7:41443. https://doi.org/10.1038/srep41443

Salm RV (1993) Coral reefs of the sultanate of Oman. Atoll Res Bull 380:1–85
Salm RV, Coles SL (2001) Coral bleaching and marine protecte areas. In: Proceedings o the workshop on mitigating coral bleaching impact through MPA design. Asia Pacific Coastal Marine Program report 0102. The Nature Conservancy, Honolulu. Also available from https://www.conserveonline.org. Accessed 18 May 2003
Santoso N (2000) Pola Pengawasan Ekosistem Mangrove. Makalah disampaikan pada Lokakarya Nasional Pengembangan Sistem Pengawasan Ekosistem Laut Tahun. Jakarta, (In Indonesian)
Short FT, Carruthers T, Dennison W, Waycott M (2007) Global seagrass distribution and diversity : a bioregional model 350(3):20 https://doi.org/10.1016/j.jembe.2007.06.012
Short FT, Kosten S, Morgan PA, Malone S, Moore GE (2016) Impacts of climate change on submerged and emergent wetland plants. Aquat Bot 135:3–17
Sievanen L (2014) How do small-scale fishers adapt to environmental variability? Lessons
Solecki W, Rosenzweig C, Blake R, de Sherbinin A, Matte T, Moshary F, Rosenzweig B, Arend M, Gaffin S, Bou-Zeif E, Rule K, Sweeny G, Dessy W (2015) New York City panel on climate change 2015 report: indicators and monitoring. Ann New York Acad Sci 1336:89–106. https://doi.org/10.1111/nyas.12587
Spalding M, Kainuma M, Collins L (2010) World atlas of mangroves (version 3.0). A collaborative project of ITTO, ISME, FAO, UNEP-WCMC, UNESCO-MAB, UNU-INWEH and TNC. Earthscan, London, London, p 319. https://www.routledge.com/books/details/9781844076574, https://data.unep-wcmc.org/datasets/5. Accessed 18 may 2019
Susanto HA, Suraji, Suraji TM 2015 Management of coral reef ecosystems in Indonesia: past, present, and the future. Coast Ecosyst 2:21–41
Takagi K, Saeki Y, Yashiroda H, Yagi H, Kaiho A, Murata S, Yamane T, Tanaka K, Mizushima T, Kato K (2014) Pba_3-Pba_4 heterodimer acts as a molecular matchmaker in proteasome α-ring formation. Biochem Biophys Res Commun 450(2):1110–1114
Taylor MA (2015) Why climate demands change. The grace kennedy foundation lecture, p 97 www.gracekennedy.com/images/lecture/GKL2015-Climate.pdf
Turner BL, Kasperson RE, Matsone PA, McCarthy JJ, Corell RW, Christensen L, Eckley N, Kasperson JX, Luers A, Martello ML, Polsky C, Pulsipher A, Schiller A (2003) A framework for vulnerability analysis in sustainability science. Proc Nat Acad Sci 100(14):8074–8079
UNEP (2014) The Importance of Mangroves to People: A Call to Action. In: van Bochove J, Sullivan E, Nakamura T (eds) United Nations Environment programme world conservation monitoring, Centre Cambridge p 128
USAID (2009) Adapting to coastal climate change : a guidebook for development planners, Coastal Resources centre–University of Rhode island (CRC–URI) and international Resources group (IRG) p 144
USIAD (2018) Report on Vulnerability and Climate Risk Studies in Southeast Sulawesi Province. USAID Climate Change Adaptation and Resilience (APIK). Southeast Sulawesi, Indonesia. p 86
Waycott M, Duarte CM, Carruthers TJB, Orth RJ, Dennison WC, Olyarnik S, Calladine A, Fourqurean JW, Heck KL Jr, Hughes AR, Kendrick GA, Kenworthy WJ, Short FT, Williams SL (2009) Accelerating loss of seagrasses across the globe threatens coastal ecosystems. Proc National Acad Sci (PNAS) 106:12377–12381
Waycott M, McKenzie LJ, Mellors JE, Ellison JC, Sheaves MT, Collier C, Schwarz A, Webb A, Johnson JE, Payri CE (2011) Vulnerability of mangroves, seagrasses and intertidal flats in the tropical Pacific to climate change. Chapter 6, pp 297–368. In Bell JD, Johnson JE, Hobday (eds) Vulnerability of tropical pacific fisheries and aquaculture to climate change. Secretariat of the Pacific Community. p 925
Weis VM (2008) Commentary cellular mechanisms of cnidarian bleaching: stress causes the collapse of symbiosis. J Exp Biol 21:3059–3066. https://doi.org/10.1242/jeb.009597
Williams Jr EH, Bunkley-Williams L (1990) The world-wide coral reef bleaching cycle and related sources of coral mortality. Atoll Res Bull 355:1–72
Zacharias MA, Gregr EJ (2005) Sensitivity and vulnerability in marine environments: an approach to identifying vulnerable marine areas. Conserv Biol 19:86–97

Chapter 9
Protection of Rights of Community of *Bajo* Tribe from the Impacts of Climate Change

Muhammad Sabaruddin Sinapoy and Susanti Djalante

Abstract Global warming is causing rise in sea surface temperatures and atmospheric instability in the lower layers close to the surface of the earth, which then leads to the expansion of seawater. The process of global climate change stages begins with rising temperatures, unstable rainfall, and seasonal shifts. This has a direct impact on the ecological conditions of coastal communities such as rising sea levels, seawater intrusion, changes in catch and fish season, storms, and declining biodiversity. As a result, the disruption of fishing activities Bajo tribal fishermen (dependence on SDLP). This condition also has an impact on socioeconomic life and income. The purpose of this paper is to develop adaptation strategies that can be applied to the fishing communities of the Bajo tribe as an effort to tackle ecological changes as well as adaptation efforts in the form of responsive actions to minimize and anticipate impacts by way of technology adoption, infrastructure development, altering arrestment area, network and mobilization of family members, and diversification of livelihoods. Adaptation strategies include physical adaptation strategies, socioeconomic adaptation, and adaptation of human resources both proactively and reactively in order to save the fishing economy of the Bajo tribe.

Keywords Protection of human rights · Bajo tribe · Impact of climate change

Introduction

Indonesia is a very rich country and is the largest archipelago state in the world with both biotic and abiotic biodiversity. Indonesia is also known as a country

---

The original version of this chapter was revised: The omitted reference section have been inserted. The correction to this chapter is available at https://doi.org/10.1007/978-3-030-55536-8_16

M. S. Sinapoy (✉)
Faculty of Law, University of Halu Oleo, Kendari, South East Sulawesi, Indonesia
e-mail: sabaruddinsinapoy@yahoo.com

S. Djalante
Faculty of Engineering, University of Halu Oleo, Kendari, South East Sulawesi, Indonesia
e-mail: sdjalante@yahoo.com

R. Djalante et al. (eds.), *Climate Change Research, Policy and Actions in Indonesia*, Springer Climate, https://doi.org/10.1007/978-3-030-55536-8_9

rich in natural resources (SDA) and has various customs, although it continues to be displaced by advances in technology/information globally. Until then, the environmental carrying capacity is exceeded and threatens the sustainability of natural resources (environmental damage and pollution). Geographically, it is located around the equator between 94° 45′ BT-141° 01′ BT and 06° 08′ LU-11° 05′ LS and flanked by two continents (Asia and Australia) and two oceans (Pacific and Indian). Spatially, the territorial area stretches from West to East along 5110 km and from North to South 1888 km, and 65% is covered in the sea. The total area of seawater reaches 5.8 million km^2, consisting of 0.3 million km^2 of territorial waters, and 2.8 million km^2 of archipelago waters, coupled with an area of EEZ (Indonesian Exclusive Economic Zone) of 2.7 million km^2 (UNCLOS 1982). The number of islands reaches 17,507, and has a coastline of 81,000 km, which is the second longest coastline in the world after Canada. In addition, 40 inhabitants inhabit their coastal areas, to a degree which has affected the existence and status of aquatic ecosystems. This means that coastal communities with the simplicity of knowledge and technology are still able to maintain and preserve the surrounding marine environment. With the potential of Marine Resources (SDL), Indonesia is a country with the most complete biodiversity in the world.

Coastal and oceanic ecosystems have characteristics that must be understood so that management meets the principles of sustainability. These characteristics include (1) biological and non-biological components in coastal areas form a complex ecosystem resulting from a variety of biophysical (ecological) processes from terrestrial and marine ecosystems, including wind, waves, tides, temperatures, and salinity. Coastal ecosystems can be very resilient or otherwise very susceptible to environmental disturbances caused by both humans and natural disasters; (2) In coastal areas, due to the variety of ecological components and advantages of location factors, there are usually various uses for development interests, such as ponds, capture fisheries, tourism, mining, industry, and settlements; (3) In a coastal area, in general, there are more than one community group having different skills/expertise and pleasure in working, such as farmers, fishermen, pond farmers, and others; and (4) Ecologically and economically, the use of a coastal area in monoculture (single use) is very vulnerable to internal and external changes that lead to business failure.

Not only has abundant natural resources, Indonesia is also unique in terms of ethnic diversity. In 2016, there were 1340 ethnic groups in Indonesia. With this number, Indonesia ranks first as the country with the largest number of ethnic groups in the world. Each of these tribes has a variety of life, traditions, or different cultures. Some of them even have quite different ways of life with some tribes in Indonesia. One of these tribes is the Bajo tribe, dubbed the "boat man."

Bajo tribe, which was once part of the Sriwijaya Royal Navy, has the toughness and skill in navigating the ocean. This tribe is also famous as a tribe that lives nomadically. This tribe also reportedly lives freely around the sea between Indonesia and the Philippines. However, as time went on, some of them decided to start living in a certain place, and one of these places is in the Southeast Sulawesi Province. Southeast Sulawesi is one of the provinces in Indonesia with a topography consisting of land and sea complete with several small islands. The Southeast Sulawesi community consists of several ethnic/ethnic cultures and religions, including people living in coastal/marine areas along Southeast Sulawesi known as the Bajo tribe (Bajo people).

The existence of the Bajo people in coastal/marine areas, when associated with the state of the phenomenon of climate change, shows us all that the conditions that are directed will be felt directly and firstly by the coastal communities and the Bajo people. Although knowledge of local wisdom can be used as an effort to act, the occurrence of climate change from the consequences of global warming can no longer be predicted using only the knowledge below since a long time ago. Local wisdom has limited reach; so, to know accurately the effects of extreme climate change, they must use knowledge and technology. Therefore, it has become an absolute necessity for the state to provide assurance of the protection of the existence of the coastal communities and Bajo tribes on the effects of extreme climate change as one of the human rights guarantees in the field of environment that needs serious treatment.

Because, in the context of state development and community empowerment, all activities cannot exclude environmental existence at certain points and limits. Therefore, if development and empowerment do not give serious attention to the environment and the existence of coastal communities and Bajo people, the state can be said to have denied its loyal promise to its citizens as contained in the fourth paragraph of the 1945 Constitution of NRI which states "… to protect all of the Indonesian people and all of Indonesia's bloodshed." The relevance of the fourth paragraph is in the view of Mattias Finger who said that the global environmental crisis as at present is at least caused by various things, namely wrong and failed policies; inefficient technology even tends to damage; the low level of political commitment, ideas, and ideology that ultimately harms the environment; and actions and behavior deviate from state actors. Mattias Finger further said that the road taken to overcome environmental problems would be through better policy making; new and different technologies; by strengthening political and public commitments; by creating new ideas and ideologies that are pro-environment (green thinking); and by changing the pattern of culture, behavior, and awareness of each individual.

Research Methods

The type of research/approach used by the author is empirical normative legal research. Normative legal research, according to Peter Mahmud Marzuki, namely legal research, is carried out by a legal approach (statute approach), case approach (case approach), historical approach (historical approach), comparative approach (comparative approach), and conceptual approach (conceptual approach). Whereas empirical normative juridical research is done by examining library material which is secondary data and also called library research. Empirical legal research is done by examining the field which is the primary data.

Analysis and Discussion

Climate Change and Its Impact

The issue of climate change has recently become an interesting issue discussed both at global and local levels. Climate change is an implication of global warming, which is caused by increases in greenhouse gases, especially carbon dioxide (CO_2) and methane (CH_4), which are mostly produced by industrialization, transportation, forest fires, and deforestation. Thus, climate is one of the basic elements for understanding natural phenomena and the development of civilization throughout history. Climate change determines most of the modification of human nature and culture because human beings must adapt to changing conditions which sometimes become important elements that can increase or threaten their existence.

This climate change affects all its main parts, namely the atmosphere, hydrosphere, geosphere, and biosphere. There are two main things that occur in the lower atmosphere, namely high rainfall fluctuations and sea level rise. This is due to the geographical difference in solar radiation, which is much higher in the Equator than in the Pole, resulting in a movement of air. The period of heated air in the Equator area causes severe rainstorms when hot air rises. Climate variation is a major factor in the diversity and emergence of new species.

As an archipelagic country, Indonesia is most vulnerable to sea level rise. Sea level rise projections for the Indonesian region have been carried out, until the year 2100; it is estimated that sea level rise of up to 1.1 m has resulted in the loss of 90,260 km^2 of coastal areas and small islands. Indeed, if we consider that there are a number of small islands in Indonesia that have sunk, conditions like this need to be dealt with because if not long after the small islands there will sink. Emil Salim also said that climate change would eliminate 3000 islands and reduce the region's borders.

So, Indonesia has a high vulnerability to climate change. Some climate studies and projections from domestic and foreign institutions show that the climate in Indonesia has undergone changes; it appears that climate change has an impact on multisector. In terms of climate projections based on the Global Climate Model (GCM), there are at least three sources of uncertainty that must be taken into account, namely the scenario of greenhouse gas emissions, global climate sensitivity to greenhouse gas emissions (selection of GCM models), and regional climate system response to global warming (downscaling model). Saad in Ari Wibowo stated that this was related to the existence of Indonesia as the largest archipelago in the world with a total of 17,504 islands. About 12.38% or around 2342 islands are inhabited. The remaining 87.62% or as many as 15,337 islands are uninhabited. Indonesia's vulnerability is also caused by human activity factors that are less concerned with environmental sustainability aspects because of the large-scale conversion of forests without regard to sustainability, the use of fossil fuels, the opening of mangroves in coastal areas, and the destruction of coral reefs that are still being carried out.

Farmers and fishermen are a group of people who have a high vulnerability due to climate change because their type of work is directly related to nature. Almost all the activities of farmers and fishermen are very dependent on climate, its variability, and changes. Although the contribution of farmers and fishermen to the occurrence of climate change is small, they are the leading victims of the various changes that have taken place. On the other hand, Indonesia is a region that is very vulnerable to climate change; in the agricultural sector, climate change is felt with the shift of seasons and changes in rainfall patterns and increases in earth's temperature as well as changes in global air temperature which also affect the temperature in the sea, resulting in changes in ocean currents and also ocean waves. In several places in Indonesia, there has been a greater tide in one to five years. Likewise, sea currents and sea winds are felt to be erratic, which is marked by the West season which is felt to be longer.

Whereas, the impact on coastal areas has experienced an increase in the frequency of storms, an increase in the frequency of high waves, rob whose periods have become unclear, and changes between the uncertain West–East season. This reduces their production and threatens the lives of farmers and coastal communities. Therefore, it is important to determine strategic steps in efforts to adapt and mitigate in response to the occurrence of climate change. Breakthrough efforts are needed in socialization for those who are very interested because they are direct victims of climate change, namely farmers and fishermen who are the biggest component of the Indonesian community and become the backbone of national food supply.

With this fact, farmers and fishermen must adapt. One of the factors that supports the success of adaptation is the understanding of climate change. Knowledge of climate change information is important to be known by the general public, especially farmers and fishermen to anticipate extreme weather conditions so as to reduce the risk of the impact. Deny Hidayati stated that understanding and knowledge of the community related to climate change and environmental change is one of the factors determining the response of the community to the effects of climate change that they experience.

Regarding the impact of climate change, it is necessary to formulate a climate change adaptation action program that aims to guarantee or secure the achievement of the main targets of development and improve community resilience, both physically and economically, socially and environmentally to impact climate change. National development with an adaptation agenda to the impact of climate change has the ultimate goal of creating an adaptive or climate change development system that is currently happening. This is necessary because climate change will affect all aspects of the development of each sector, including the Southeast Sulawesi Province.

Southeast Sulawesi Province is one of the areas with a high risk of experiencing hydrometeorological disasters. The Southeast Sulawesi Province geographically, hydrologically, and climatologically allows various threats or disasters to occur. Such natural conditions pose a high disaster risk. Most of the area is in disaster-prone areas. The diversity of disaster threats really requires systematic and integrated disaster management so as to reduce disaster risk. Disaster history in Southeast Sulawesi Province shows a significant impact in the form of loss, damage, and loss of assets of life and livelihoods of both the community and the government. This is inseparable from the geographical location and the nervous conditions of Southeast Sulawesi Province which are very vulnerable to various natural disasters. The livelihoods of the majority still depend on natural resource management from the agriculture and fisheries sectors, increasing the level of risk from the threat of climate change impacts. Noting these matters, it can be said that Southeast Sulawesi Province is one of the most vulnerable regions to the negative impacts of climate change, so it is natural that Southeast Sulawesi Province seeks to overcome and face the challenges of climate change and global warming that affect human life, especially for the Bajo people.

Bajo Tribe

Indonesia is a country rich in customs and diverse cultures, and has several elements of culture as indicators that can apply to all ethnic groups in Indonesia. As an example, the Bajo people are commonly called sea nomads or boat people because since ancient times they were true sea adventurers who lived entirely on simple boats. The Bajo have their own ethical and cultural systems. The system becomes their guideline (read Bajo) and directs their daily lives from generation to generation. Their culture includes various rules that result from their experience and spirituality with their lives dependent entirely on nature. Because their whole life is faced with the sea, the characteristics of their social, cultural, and economic life were greatly influenced by their views on the forces of nature which surrounded everyday life. The existence of the Bajo community in a coastal area has an emotional closeness to SDA which gives birth to real behavior by considering ecology in their daily lives. The closeness of the Bajo people with the sea and coast allows them to have a variety of local knowledge about the symptoms of nature. There are natural symptoms and signs of the atmosphere that are still used by the Bajo people when they sail. This ethical, custom, and cultural system has proven to be able to make the Bajo community to survive and live in harmony with the coastal and oceanic areas to become their livelihood until now. Wisdom and local knowledge are the result of a long process from generation to generation.

The Bajo people also have a philosophy "Papu Manak Ita Lino Bake Isi-Isina, Kitanaja Manusia Mamikira Bhatingga Kolekna Mangelolana," meaning God has given this world all its contents, we as humans think about how to obtain and use it so that the sea and its results are a place to pursue life and defend themselves while continuing to pass on the ancestral culture of the Bajo. The Bajo people call themselves the Same Tribe, and call the people outside their tribe the Bagai Tribe. The Bajo people are a tribe that originally lived on a boat always in the ocean. The boat is called leppa or Soppe or bidoa.

The Bajo are a tribe often referred to as the sea tribe because they depend on the sea for the fulfillment of their daily lives. Even in the past, almost all of their life was spent on a boat constantly navigating the ocean. The Bajo are widely spread. Like other cultures, the Bajo also have their own ethical and cultural systems. The system is their guide. The Bajo are widely spread across the archipelago, both Indonesia, the Philippines, and Malaysia and Brunei Darussalam. They used to spend their lives on boats and sail from one place to another. Now, the Bajo have settled in the coastal areas, but their homes must remain on the side of the sea, not on the land side. Although the Bajo are geographically separated, there are similarities in the rules of custom, culture, and value systems that apply. So that the Bajo can be classified as a separate ethnic group with the same ancestry. Their culture encompasses various rules of life from their experience and spirituality by relying entirely on nature. This ethical, custom, and cultural system is interesting to study because it is proven to be able to make the Bajo community survive and live in harmony with the coastal and oceanic areas into their livelihoods. The Bajo people have wisdom in maintaining and utilizing coastal areas in a sustainable manner.

Bajo people move around in groups to different places according to the choice of fishing location. Sea is used as a source of life (panamamie ma di lao). They (read Bajo tribe) have the principle that pinde kulitang kadare, bone pinde and kadare which means moving the Bajo to the land as well as moving turtles to land or in other words taking away his life. Even their heads will get dizzy if they do not hear the sound of the waves. Bajo is a concept given by other ethnic groups. The Bajo are born and live in the sea. They have the toughness to navigate the ocean as part of their history and identity. It absorbs and is inherent in their daily lives about traditions and local wisdom to manage marine ecosystems in any part of the archipelago, including in Southeast Sulawesi Province.

The Bajo (Bajo people) generally live in the coastal areas of Southeast Sulawesi spread in almost all coastal areas of Southeast Sulawesi, including in Konawe Kepulauan Regency (Wawonii), on the coast of North Konawe Regency (Lemo Bajo Village), Wakatobi Islands Regency, Muna Regency, Buton Regency North, and several other coastal areas in the district/city of Southeast Sulawesi. They (the Bajo tribe) dwell, gather, move and interact with the marine environment, up to several generations of their offspring. Their nuances of life are very familiar with the coastal/marine realms. Local culture/local wisdom obtained from generation to generation is inherited from one generation to the next, being the provision of their knowledge in utilizing natural resources as well as in the management of coastal/marine areas so that sustainability is maintained.

The dependence of the Bajo people in various places regarding the use of resources on the coast has almost the same characteristics. In general, they live as their ancestors, namely living from the sea. In addition to the head of the family, teenagers are obliged to help parents who work as fishermen. While the role of housewives in addition to helping husbands sell their marine catches to the market is guiding and educating their children in teaching about local wisdom, honesty, manners, courtesy, planting religious values and responsibilities, especially to children still under five. Examples of the ways in which resources, customs, and community organizations are used will be explained in a number of different cases in the Bajo community below:

(a) Bajo tribe in Bone Bay: there are still many examples of traditional Bajo tribe fishermen groups in Bone Bay who depend on fishing activities, namely fishing. The entire area of Bajo village covers an area of 5.58 km^2. The livelihood of most residents is as fishermen. As fishermen, they recognize three fishing locations, namely in deep waters, in coral clusters, and on the coast. They use a variety of traditional equipment made according to the location of the arrest. For deep water fishermen, the equipment used to catch fish is in the form of arrows, spears, and fishing rods. For fishermen who are looking for fish in coral clusters, other than using arrows and spears, they also use a tool called a bunre.
(b) The pattern of settling settlements: the general phenomenon of the Bajo community in South Sulawesi. Based on the results of research revealed by Munsi Lampe that in South Sulawesi, Bajoe Bone Village, Sembilan Sinjai Island, and several places in Selayar (Appa Tana, Jampea, Kayuadi, Bonerate, Taka Bonerate Area, etc.), Bajo communities generally have villages or villages permanent master. Nevertheless, there are still a number of Bajo Villages that make seasonal migrations and live a few months in the East season on several coral islands such as Small Tarupa Island, Tinabo Bungin Belle Island, and West Pasitallu to approach fishing grounds.
(c) Marine cultivation in Batanglampe, Pulau Sembilan: some Bajo fishermen family groups are now not only extracting or catching fish, Bajo fishermen on Pulau Sembilan (Sinjai) have also tried seaweed cultivation, enlargement of lobster and shellfish seeds (japing, pearls) in cages for several years since 1995. The cultivation business on Pulau Sembilan is actually sought as an alternative fisherman business from the main activity of catching fish, which results from day-to-day decline due to overfishing and damage to coral reef habitat. Although it was recently tried by some fishing communities, marine aquaculture practices have been shown to show signs of positive impacts both in terms of the economy and the recovery of damage to coral reef ecosystems caused by the behavior of fishermen using toxic explosives and anesthetics.

The social interaction of the Bajo people with the surrounding community is very intense. The Bajo are known to adapt easily to their social environment, and will always try to speak the local language. Therefore, the Bajo are very fluent in Bugis, Buton, Kaledupa, and so on. In the trade network, the Bajo develop economic relations with the people around where they live. The term Bajo is better known in the eastern part of Indonesia, which is a social unit or community group that can be found in the islands and seawaters of South Sulawesi (Makassar strait, Bone bay, Bajoe), Northern Sulawesi (Kimabajo, Nain island, Torosiaje), Central and East Sulawesi (Tomini bay, Togian and Banggai Islands), Southeast Sulawesi (such as, Lasolo, Tinanggea, Saponda Island, Tiworo Islands (Tikep), Wakatobi (Wanci, Kaledupa, Tomia), Pasarwajo, Mawasangka, and Kolaka, Pomala, Watubangga, Wolo, Lasusua, Pakue) regions; East Kalimantan (around Bontang, including Sabah Timur and the Sulu islands); East Nusa Tenggara (Alor, Kupang and Flores and surrounding areas); West Nusa Tenggara (Bungin); and North Maluku (Bacan). In the western part of Indonesia (such as Riau), community groups such as the Bajo are called Laut Laut,

Orang Laut, or Suku Laut. While in Sabah Malaysia, Brunei Darussalam, and the Philippines, as well as the Portuguese and researchers who inhabit small islands, living in boats and/or seminomaden in the deep sea of the Southeast Asian region, many use the name Bajau, Indigenous Tribes, Sama-Bajau, Sama Difaut, Bajau Laut, Samai people, or Samai Bajau Laut.

Local Wisdom of the Bajo Tribe

Management of Coastal Resources (SDP) must be done wisely and sustainably because coastal and marine areas have a variety of biodiversity that have their respective roles and functions in maintaining the balance of the ecosystem. This biodiversity is a potential resource that can support the lives of coastal communities in improving the quality of life better. Therefore, in order to achieve sustainable and effective management, identification and direction of utilization of the potential resources is needed. As is known, a coastal region is in which there is one or more environmental systems (ecosystem) and SDP. Coastal ecosystems can be either natural or artificial (man-made). Natural ecosystems in coastal areas include coral reefs, mangrove forests, seagrass, sandy beaches, pes-caprae formations, baringtonia, estuary, lagoon, and delta formations. Artificial ecosystems include tidal rice field, tourism area, industrial area, agro-industrial area, and residential area. SDP is one of the natural resources that is widely used by the community, but the utilization of these resources to date has not paid attention to its sustainability; as a result, there has been a decline in function, quality, and biodiversity.

Rokhimin Dahuri states that the SDP potential is generally divided into four groups, among others, as follows: (1) Resources that can be recovered (renewable resources); (2) Resources that cannot be recovered (non-renewable resources); (3) Marine energy; and (4) Environmental services.

Traditional knowledge (traditional knowledge) or local wisdom in managing natural resources plays an important role in conservation and sustainable management. According to the Indonesian dictionary, local wisdom comes from two wise words, namely wise and local is happening somewhere or unevenly.

Understanding of local wisdom is seen from an English–Indonesian dictionary, consisting of two words, namely wisdom and local. Local means local and wisdom equal wisdom. In other words, local wisdom can be understood as ideas, values, local views that are wise, full of wisdom, good value, embedded and followed by members of the community. In anthropological disciplines, it is known as local genius. This local genius is the first term introduced by Quaritch Wales. Anthropologists discuss at length the notion of local genius, including Haryati Soebadio who said that local genius is also a cultural identity, identity/personality of the national culture that causes the nation to be able to absorb and cultivate foreign cultures according to their own character and abilities.

While Moendardjito said that the element of regional culture is potential as a local genius because it has proven its ability to survive until now. Its characteristics are

(a) Able to defend against external culture; (b) Having the ability to accommodate outside cultural elements: (c) Has the ability to integrate external cultural elements into indigenous cultures; (d) Having the ability to control; and (e) Able to give direction to cultural development.

Traditional knowledge is built, maintained and preserved widely known and this greatly helps scientists develop a bottom-up strategy in the management of Coastal and Marine Resources (SDPK) and the government. So, the form of concern of the Ministry of Maritime Affairs and Fisheries (KKP) on this matter is the adoption of local people's knowledge and wisdom in preparing Law No. 27 of 2007 concerning Management of Coastal Areas and Small Islands, as one of the traditional knowledge and wisdom that must be known. Through the laws and regulations, the use of local wisdom as the foundation of resource management is a relatively new approach in Indonesia. This kind of management is usually termed participatory management, collaborative management or sometimes also called community-based management.

Related to the management of SDPK, it basically has a goal to improve the welfare of the whole community social well-being in a sustainable manner, especially the local community living in the coastal zone (wilayah pesisir). Therefore, in the utilization of SDP, the ecological aspects in terms of the sustainability of resources and ecosystem functions must be maintained as the main foundation for achieving this welfare. The use of SDP is expected not to cause damage to fishing ground, spawning ground, or nursery ground fish. In addition, it also does not damage the function of mangrove (bakau) ecosystems, coral reefs (terumbu karang), and seagrass beds that have ecological relevance to the sustainability of resources in coastal areas.

The local wisdom of the community regarding the environment comes from two words, namely wisdom and local. In general, local wisdom can be understood as a local idea that is wise, full of wisdom, good value, embedded and followed by members of the community. According to the formulation by the Ministry of Social Affairs (now the Ministry of Social Affairs), local wisdom is defined as a view of life and knowledge as well as various life strategies in the form of activities carried out by local communities in answering various problems in meeting their needs. In anthropological concepts, local wisdom is also known as local knowledge (indigenous or local knowledge), or local intelligence (local genius), which forms the basis of cultural identity. Local wisdom is one of the heritages of the ancestors; the inheritance can be in the form of values of life that are united in the form of religion, culture, or customs. The existence of local wisdom is not without function.

Local wisdom has many functions; as written by Sartini, the function of local wisdom is as follows: (1) Serves for conservation of natural resources; (2) Serves for the development of human resources; (3) Serves for the development of culture and science; (4) Serves as advice, trust, literature, and taboos; (5) Social meaning such as communal integration ceremonies/relatives; (6) Social meaning; (7) Meaningful ethics and morals; and (8) Meaningful politics, local wisdom related to culture, has an important meaning to maintain the sustainability of culture, as well as to maintain its sustainability.

Preservation of local wisdom by itself will help themto preserve the village environment they live in. Because the values of local wisdom will make the community have a strong character in accordance with the culture and norms applicable in their environment. In addition, villages directly adjacent to the conservation area will preserve conservation areas because usually the area is closely related to the values of local wisdom adopted. Conservation activities in an area should come from community awareness in the region concerned. Awareness of the importance of maintaining biodiversity is needed not only for the interests of the Indonesian people but also for the benefit of the world community as a whole and directed toward long-term interests.

Related to the local wisdom of the Bajo people in Southeast Sulawesi in maintaining the sustainability of the coast and the sea, it can be seen from the surrounding environment of the Bajo community where mangrove ecosystems, seagrass beds, and coral reefs are still well maintained. This condition does not appear in other coastal communities. The Bajo community that inhabits the coastal area in Southeast Sulawesi has an emotional closeness to the SDA, which gives birth to real ecological considerations. Bajo people generally make a living as fishermen. His daily activities are very dependent on the coast and the sea. It can be said that 99.99% of the natural resources they use for everyday life are those from the coast and the sea around their homes.

The Bajo people have a myth that the Gods made oceans for the Bajo people. There is a concept similar to the one in laok (sea belonging to the Bajo people), which means that the land environment is for land people. On the progress of the times and economic needs, now most of the Bajo in various regions have settled. Nevertheless, they still settled on the coast at the top of the sea, not on land.

The Bajo community has local wisdom in the form of traditions, rules, or hereditary taboos that are practiced, nurtured, and adhered to by the Bajo community. Overcoming the ecological crisis is not merely a technical matter, but it needs to be traced to the intricacies of human spirituality, the outlook on life, and awareness of nature and ecological behavior. For this reason, we need human ecological intelligence, in the form of understanding and translating human relations with all elements and other living things. Ecologically intelligent human beings, he said, put themselves as environmental controls as outlined in real attitudes and behaviors when treating nature. The universe is not only a source of exploitation but a house of shared life that continues to be protected, cared for, laid out, and not destroyed.

Based on the description above, the local wisdom of the Bajo people can be grouped in several ways, including.

1. ***Local Wisdom Meaning Ethics and Morals***;
   In relation to maintaining ancestral traditions, the emotional closeness of the Bajo community with Marine Resources (SDL) gave rise to the mamia kadialo tradition. The mamia kadialo tradition is in the form of grouping people when they go to sea for a certain period of time and the boat used. There are three groups of this tradition: palilibu, bapongka, and sasakai.

Palilibu is the custom of fishing using a rowing soppe boat. Going to sea in just one or two days and returning to the settlement sells the catch and some is enjoyed with the family. Bapongka (babangi) is the activity of fishing for several weeks and even monthly using a large boat measuring approximately 4×2 m called leppa or sopek. This activity often includes families, such as wives and children, and some even give birth on a boat. Then, sasakai, which is the habit of fishing using several boats for several months with a range between islands. As long as the group is undergoing mamia kadialo (fishing), there are restrictions for families left behind and those who go to sea.

The prohibition includes throwing into seawaters such as sea cucumber washing water, wood charcoal or kitchen ash, cigarette butts and ash, chili water, ginger and orange juice, and washing of cooking utensils in seawaters. Both washing water and materials should be collected and disposed of on land. Another obstacle that turns out to also have the value of environmental sustainability is taboo eating turtle meat. Sea turtles are believed to help many people who experience calamities, therefore these animals should not be killed. Another norm that applies is respect for the sea. The Bajo people really respect the sea because it is the source of their livelihood. Bajo people will not say inappropriate words or swear at the sea. This shows how the Bajo people really value the sea. The Bajo are very afraid of breaking the pamali. They think it will bring havoc and calamity if they break the pamali. Such calamities can come in various forms such as not getting fish, storms, big waves, disturbances of evil spirits and not even getting results at sea.

2. ***Local Wisdom Which Serves for Conservation and Conservation Of Natural Resources***;

   Bajo people, especially the older generation, still believe in certain coral groups as a place to inhabit the spirits of their ancestors. Parents forbid family members from catching fish and other biota around coral clusters, unless first performing certain rituals by preparing offerings for ancestors. Various restrictions contain the value of preserving marine and coastal aquatic ecosystems.

   The public's knowledge of natural phenomena, he said, has ecological value. Coral reefs, among others, as a barrier to currents and waves, and around the coastal area also provide a calm. When the night comes, there will be a sparkle of moonlight due to the calm reflection of the water surface. The eagles approach the surface of the sea because, when the tide is low, more marine biota appears to fall prey to eagles.

3. ***Local Wisdom Which Works for Human Resource Development (SDM)***;

   Seeing the local wisdom of the Bajo people is not just a reference for someone's behavior, but further, namely being able to realize a life full of civilization. Local wisdom that has occurred has brought changes to always be lived, practiced, taught, and passed down from one generation to another which at the same time form patterns of empowerment of the Bajo people themselves, in terms of knowledge and customs about humans, nature, and how the relationship between all residents of ecological community empower communities who are consciously not in management and sustainable development.

Human resource development (SDM) based on local wisdom is guided by local wisdom values as a criterion that determines a quality for action. As a criterion, local wisdom can also be a foothold for the development of learning for people with more character. Local wisdom can be expressed as science or technology that takes place as human efforts to face life's challenges. Local wisdom owned by the Bajo people in various coastal areas of Indonesia can be the basis and direction in achieving a goal related to the improvement of Bajo Human Resources, especially in the economic sector to increase income or family income. The activity must be carried out, considering that the phenomenon of natural events in the form of climate change cannot be predicted anymore by relying on the knowledge of local wisdom obtained. As a result, the tradition of going to sea as a day-to-day job in fulfilling the daily necessities of living in the home also ended in a decline in income, not even enough.

Therefore, innovative and creative economic developments are needed so that the life needs can be fulfilled. One such effort is to improve human resources based on local wisdom, especially in the sector of fisheries production or other sectors based on the geographical conditions of the residential area. Local wisdom is a historical fact that can be developed to empower coastal communities through human resources, natural resources, and marine resources in improving processing, production, fisheries development, and marine tourism. Therefore, the same understanding is needed by coastal communities to understand local wisdom and maritime culture in order to increase marine and fisheries production. The involvement of elements of knowledge and technology really requires the ability of Human Resources to develop old traditions into something that has more economic value. At present, with the phenomenon of weather uncertainty due to climate change, the Bajo people are no longer just collecting catches from the sea. However, you have to think about the attributes of arrest, alternative livelihoods, ways to collect capital, labor, and market distribution systems. This can be done if the potential of the Human Resources owned increases and becomes a new force in maintaining and increasing the family economy in the midst of the effects of climate change.

Thus, the elements of knowledge and technology in question include the meaning of technology, knowledge of production tools, and techniques of making and utilizing technology. The meaning of technology that guarantees survival for himself and his grandchildren is technology that is not destructive to the environment; therefore, the marine environment is a shared responsibility not to be damaged. With the understanding of technology that does not damage the environment, every capture technology made by the community must incorporate these non-destructive noble values. From this point of view, it is felt in the field that the people use their fishing gear in the form of their technology that is environmentally friendly and maintained together. The fishing gear used in the form of boats, fishing rods, traps, and nets is passive in size and still small in scale. The technology of fishing gear used by coastal communities is the technological inheritance of previous generations of fishermen, and the technology has been very compatible with the marine environment where their lives are. So, the meaning of technology for the community as a tool of life is able to balance the natural needs of the sea, users of equipment, and the belief that the technology used provides comfort, security, and safety because it does not conflict with applicable natural law.

4. ***Local Wisdom Which Serves for Culture and Science Development***;
   In terms of fishing, the Bajo people are very adaptive to the environment, such as protecting coral reefs as a place for fish to live, lay eggs, and eat fish. As well, there is a prohibition to catch small-sized fish and eat them. They may only consume fish that have large sizes or are worthy of harvest. The Bajo people also have quite good conservation awareness, such as the taboo ban, which is the prohibition of catching sea cucumbers that stand because they are believed to be the king of sea cucumbers; after the sea cucumbers have fallen, they are allowed to catch the sea cucumbers around them. Scientifically, the sea cucumber stands in a laying state so that the Bajo people are unconsciously maintaining the sustainability of sea cucumber resources. The wisdom of the Bajo people in managing Marine Resources is also seen in reef fishing activities only during the East wind season.
5. ***Local Wisdom Which Serves as Advice, Trust, Literature, and Taboos***;
   As a nomadic sea tribe (sea nomads), the Bajo are an ethnic group that has successfully spread across the archipelago. The concept of their local wisdom in the form of a ban is actually not limiting their lives, but rather guiding their lives to remain sustainable. The preservation of nature where they seek livelihood will certainly have a good impact on their lives. The concept of pamali plays an important role in preventing damage to the oceans by Bajo tribes because the Bajo people must maintain the condition and sustainability of the sea. If the sea is always in good condition, of course it will guarantee their lives. Conversely, if the condition of the ocean is damaged, the source of their livelihoods is in danger. Without realizing it, by obeying the rules or pamali, the preservation of marine and coastal ecosystems will be maintained. Thus, indirectly, the Bajo tribe has made efforts to preserve the marine and coastal ecosystems. Local wisdom is obtained from generation to generation; in the view of the Bajo people having a philosophy of living in harmony with nature, they see nature as a part of their lives. Therefore, the Bajo have a conservation ethic that treats nature wisely, without over-exploitation. Pamali is an ethic of the Bajo people in building relationships with their natural environment. In the pamali, there is an environmental ethic that must not be violated because, if it is violated, the culprit will get a katula. Katula is an ancestral punishment for those who break the pamali, in the form of illness, disaster, or get a bad future. Some Bajo pamali related to conservation ethics, for example pamali cut down mangrove trees carelessly, and pamali throws into the forest area carelessly. The two pamali contain messages related to environmental preservation. Pamali felling trees carelessly is a message that prohibits damaging forests, but that does not mean that they should not take advantage of the necessities of life. Likewise the prohibition of throwing into the forest carelessly containing messages is prohibited from damaging the ecosystem in the forest. This contains the message that the forest area not only consists of

a collection of trees in the form of mangroves, but also there are other ecosystems in the form of birds, fish, and others. Throwing into the forest carelessly is feared to damage other ecosystems. Moreover, the concept of the rules is part of their cosmology as a way of life, values, and spiritual beliefs. The implications of environmental conservation values have so far been residing in the logical consequences of the Bajo tribal guards. While the non-logical consequences of pamali must be appreciated as spiritual values from the Bajo. Therefore, there needs to be an effort to provide an understanding that the concept of Pamali has educational values for the preservation of marine and coastal ecosystems, which in turn will affect the catch.

6. ***Meaningful Local Social Wisdom (Communal Integration Ceremony/Relatives)***;
Since ancient times, the tradition has never faded in the Bajo people, namely the duata tradition. According to the beliefs of local residents, the duata were gods who descended from the sky and became human. The duata tradition is the culmination of all efforts of traditional Bajo tribal medicine. This habit will be carried out if one of them experiences severe pain and can no longer be cured by other methods including medical treatment. While undergoing a two-treatment procession, a number of traditional elders usually gather in a room and concoct a type of ritual complement, such as colorful rice, incense, betel leaf, coconut, and banana. Meanwhile, the people to be treated were taken to the sea accompanied by songs from the Bajo community, namely lilligo and ngigal dance. Finished from the sea, sick people and traditional elders meet in the original place and treatment takes place. Not limited to treatment, the duata tradition can also be done in thanksgiving and celebration events. Duata is also held when welcoming guests.

As for some ceremonies that still exist in the life of the Bajo people, such as ceremonies fresh nampo, refusing reinforcements, thanksgiving, arriving apart, threatening, and arriving at the raft. The ceremony is still ongoing and is believed that the ceremony can provide a sense of security and comfort to its followers. The beach area according to them is a sacred place, and before carrying out their daily activities, they always hold rituals for their safety in carrying out their daily activities. Some of the ceremonial rituals that take place aim so that what is done can run well without any significant obstacles. In addition, the implementation of a religious ritual as mentioned above is interpreted as a form of gratitude to the Almighty God. The form of ritual ceremonies starts from the life cycle ritual rites such as the ritual ceremonies of birth to the ritual of death. Besides that, ceremonies related to the existence of ancestral spirits were also carried out. With this belief, every movement and pattern of community behavior is strongly influenced by the superstructure or ideology espoused primarily in the concepts of local knowledge in dealing with its natural conditions. Every step in the life of the Bajo tribe through traditional ceremonies is always sacred; this indicates that there is a connection between religious elements that are mutually integrated in an integral unit so that there is harmony in the community.

Furthermore, the ancak ceremony is a communal ritual. This activity is carried out jointly using various equipments. The ancak ceremony is often called "bebalian"; bebalian contains offerings with simple sequences. The offerings are made from a series of young coconut leaves containing a series of flowers and bananas, yellow rice, white rice, incense, and others. All of them are supporting equipment in the series of ancak ceremonies, beginning with the making of ancak places. All equipment and offerings that will be offered are put on wood placed on the border of the village or the entrance gate. This activity has become a tradition of conducting prayer together led by a sandro who plays a role in providing service to the community, especially in the rituals of salvation. The purpose of this arrived ancak ceremony is to provide protection to the public from epidemics and various negative effects caused by nature or humans.

And the ceremony to arrive at separation is actually also a ritual that requires for a person or group of people to be peaceful and safe both in work and in other forms of activity. This ceremony is also very related to the activities of fishing communities that are thick with sea salvation ceremonies. This ceremony is carried out if family members or the fishing community will carry out fishing activities which are of course up to the night; according to the local community's trust, it is necessary to carry out a ritual of salvation which in the local language is often called a separate ceremony (tiba pisah).

Then, local wisdom that is socially meaningful is marked at the Teacher's Pillar ceremony. This ceremony is done when it will make a home for those who have a family. In making their house do it by mutual cooperation and the shape of the house they made looked rectangular viewed from front to back. This teacher's pole is usually placed in the number two position seen from the left on the second pole line from the front. The installation of the teacher's pole is carried out by certain people only, namely those who have spiritual deepening and also know the occult. The function of the teacher's pole is to support the house and magically is also a tool to keep the house from negative things. When carelessly using mangrove wood, the wood is easily brittle. People believe it will bring disaster to the occupants of the house. The occupants of the house will also suffer greatly from diseases that are unrecognized; in other cases, there can also be consecutive accidents. The teacher's pole (tiang guru) also has a central function in every traditional or religious ceremony. The entire ritual procession in the house is a symbol that the ancestors of the Bajo have come down and attended the ceremony. Even though the science is progressing, local knowledge about the natural phenomena possessed by the Bajo community is still a reference for them in living life at sea.

Strategic Community of the Bajo Tribe in Facing Climate Change

The recent increase in extreme weather/climate events and climate-related disasters (such as floods, landslides, tornadoes) has forced us to try to understand what is really happening to the earth and our environment. Related to this matter through the United Nations Convention (1990) on climate change which states, "Climate change means a change of climate which is attributed directly or indirectly to human activities that alters the composition of the global atmosphere and which is in addition to natural climate variability observed over comparable time periods." Diposaptono mentioned climate change resulted in changes in the physical environment in coastal areas and small islands, among others in the form of seawater intrusion to land, tidal waves, floods, droughts, inundation on low land, and coastal erosion. These physical changes certainly affected all sectors of life and livelihood in coastal areas and small islands. These physical changes have an impact on coastal morphology, natural ecosystems, settlements, water resources, fisheries, and socioeconomic and cultural conditions.

Seeing the impact of climate change in Indonesia, the most vulnerable category of society is rural communities, especially coastal communities and small islands not to mention the existence of the Bajo people in which almost all coastal areas in Indonesia are inhabited by them (read Suku Bajo). Basically, the Bajo tribe which has an average fishermen's profession in daily activities is very dependent on nature. Because, from birth, the Bajo tribe has been introduced to life on the surface of the water. Because, the sea is the main source of livelihood for the Bajo people (they live from fishermen). If the prediction of the impact of climate change really occurs between 2050–2100, the Bajo people can be said to be the people who are most prepared to face change or vice versa as the first community to face such a disaster or impact.

The Bajo have a range of local scholars who support in activities utilizing SDL, including (1) knowledge of the wind season; (2) the number of catches; (3) knowledge in terms of marine navigation; (4) knowledge of the characteristics of gathered fish; and (5) and natural signs that show time. These things are very useful for facilitating the fishing process that they have known for generations. Likewise, the system of local knowledge related to climate used in the use, preservation and management of marine resources by coastal communities, among others, is

1. Knowledge related to the wind, in the West season (namberek), they are generally experienced from January to August, East season from September to December. However, in the West season, there is a difference in wind blows, namely the West season with East wind blows occurring from June to October, whereas West wind blows occur from November to August.
2. Regarding rain, which is usually the rainy season, they usually experience from October to April, while the dry season is May to September.
3. Relating to currents and winds in the West season, the currents are stronger and the wind blows faster than in the East season.

With regard to climate change, what people understand and feel is the change in the rainy season and the dry season. Drastic changes in nature and those that are difficult to anticipate will further add to the panic about natural shocks that are ready to eliminate plants or their catches. Fishing communities live in uncertainty about the results of their livelihoods because they depend on nature (season and weather). With the impact of climate change that has a direct effect on the environment, this uncertainty is increasing in the aspects of fishermen's livelihoods. Climate change is a global issue and has become an important issue in the world. Climate change caused by global warming has an impact on various changes to the earth. To reduce the rate of impact of climate change, the Government of Indonesia has issued Presidential

Regulation No. 61 of 2011 concerning the National Action Plan for Reducing Greenhouse Gas Emissions as a basic reference for the implementation of Climate Change Adaptation and Mitigation by all regional governments in Indonesia. Therefore, it is affirmed and also as a matter of government attention, related to Adaptation and Mitigation to climate change regulated in the Minister of Home Affairs Regulation Number 30 of 2010 concerning Guidelines for Management of Resources in the Sea Region fourth part Article 18 affirming Regional Governments in preparing plans management of resources in the sea area, must include material containing efforts to adapt and mitigate the impact of climate change. Furthermore, it is referred to in Article 19 that Climate Change Adaptation and Mitigation is carried out by involving the Government, Regional Government, and/or the community. Article 20 solidifies again with the explanation of the implementation of Climate Change Adaptation and Mitigation as referred to in Article 18 carried out by taking into account aspects such as social, economic, and cultural community; environmental sustainability life; benefits and effectiveness; as well as a wide area.

The series of laws and regulations mentioned above are also inseparable from the spirit that have been published and stated clearly in the constitution/1945 Constitution of the Republic of Indonesia, especially in Article 28H paragraph (1) and Article 33 paragraph (4) of the 1945 Constitution regulated norms regarding the environment in the constitution. In succession, the two Articles read as follows:

Article 28H paragraph (1): "Every person has the right to live in physical and spiritual prosperity, to live in, and to get a good and healthy environment and the right to receive health services."

Article 33 paragraph (4): "The national economy is held based on economic democracy with the principle of togetherness, efficiency with justice, sustainability, environmental insight, independence, and by maintaining the balance of progress and unity of the national economy."

Based on the two Articles above, it is clear that the 1945 Constitution has also accommodated constitutional protection for its citizens to obtain an adequate environment and guarantee the preservation of a sustainable environment for the negative effects of national economic activities. The impacts of climate change are very directly felt by the Bajo people, including.

1. Season Pattern Shift. Changes or shifts in seasonal patterns occur. The shift in this season's pattern has an effect on the difficulty of fishermen predicting high waves and wind. Forms of uncertainty or seasonal shifts in principle also cannot be used as a benchmark because when fishermen determine when West Season and when the east season they are very difficult to determine. Within a period of one year according to the differences in the arrival of the West and East seasons. The season is classified according to wind direction. Changes or shifts in this season's pattern according to fishermen have an impact on aspects of fishermen's knowledge of nature. If, in the past, they as tribes who always interacted with the sea had knowledge about nature, now they feel that they cannot predict nature. Changes in the knowledge of Bajo fishermen about nature, especially about seasons, affect their fishing activities because they are very dependent on nature. At present, they

must strive to determine when to go to sea and where the fishing location is. When at the conditions of bright clouds and calm winds, Bajo fishermen decide to go to sea; but often they suddenly have to cancel the departure because the clouds darken and the wind blows hard so that the big ocean waves.

2. Increased Temperature. Climate change in the form of an increase in temperature is also felt by Bajo fishermen, where the current temperature exceeds the average temperature in the previous period; even according to fishermen the temperature can reach 360 °C. The impact of increasing temperature is on health aspects. Many fishermen fall ill because they are not strong with the heat they feel.
3. Increased Tide of Seawater or Rob. Climate change in the environment includes the rise of tide (rob). The phenomenon of sea tides currently felt by fishermen is higher than in previous years. Increased sea tide reaches 30 cm. Sea tides occur every bright month and are higher in the West Season, compounded by large waves that bring seawater up to the mainland, suddenly crashing into the residential area. The increase in sea tides is felt by fishermen which is very disturbing to their lives, especially in the aspect of housing, especially for those who have non-stage homes. The resting place of fishermen under the house becomes unusable. Sea activities are also sometimes disturbed because fishermen have to evacuate furniture and items under their homes. The rob impact felt by fishermen is not only on housing aspects but also on other aspects such as transportation, public services, infrastructure, and health. Educational activities are also disrupted because the tide goes to schools located around their homes.
4. Increased Storm Intensity. One other climate-related hazard in Indonesia is the movement of tropical cyclones in the South East region of the Indian Ocean and East of the Pacific Ocean. This in some parts of Indonesia can cause one of the strong winds. Strong winds also often occur during the turnover of Munson (rainy season) from the Northeast to Southwest. Extreme events such as strong winds known as tornadoes are also felt by Bajo fishermen. These strong winds affected the collapse of fishermen's houses, especially houses that jutted into the sea.
5. Impact on livelihoods. Climate change is felt to have an impact on livelihoods. Bajo people who generally work as fishermen have felt the effects of climate change on their livelihoods. Climate change affects the certainty of the time of the sea, changes in the location of fishing, and the reduction in the number of catches of fishermen. This has an effect on the level of welfare of fishermen. Climate change is perceived by Bajo fishermen. Changes and shifts in seasonal patterns have had an impact on the certainty of sea time and fishing locations. If in the past Bajo fishermen could count the number of days at sea and did not sail for sure, now they could no longer confirm it. Likewise, with fishing locations, currently there is no longer a definite location for fishing fishermen. In general, the location of fishing fisheries is currently shifting further than before. Changes in water temperature that damage coral reefs have exacerbated fisheries productivity that has been bad due to massive pollution and fishing, thereby reducing fish populations.
6. Search for New Arrest Locations. According to fishermen, the fishing locations where there were originally many fish were no longer showing that, so they had

to find new fishing locations that were generally farther away. In general, fishing locations shift further from the coast to 3 nautical miles. The more distant fishing locations increase the expenditure costs, both for fuel (fuel oil) and for mining such as food, beverages, and others. As well, the reduced number of catches due to climate change exacerbates ecological degradation caused by the use of destructive fishing equipment such as bombs and potassium cyanide.

Coastal communities are the biggest parties to receive the effects of global climate change. The impact of climate change is in the form of sea level rise, tidal waves, increased rainfall, and extreme weather events which will all disrupt the livelihoods of coastal communities. In the past ten years, the central and regional governments have collaborated with various parties to strive to improve the resilience and adaptability of coastal communities/Bajo tribes to the effects of climate change. The coastal area inhabited by the Bajo people is a potential area that will be very severe to face extreme climates. Changes and strategies that will be built and conditioned based on the type and environment are influenced by the level of risk, vulnerability of the community, and its environment.

So, based on the impact of climate change, it can be said that climate change, in addition to influencing the uncertainty of the time of the sea, the more distant fishing locations, and the reduced number of catches, also affects the shifting patterns of fish season. In general, in the face of climate change, fishermen still cannot determine the adaptation strategies they can do. Currently, fishermen do more trials and errors. The form of adaptation strategies in dealing with the impacts of climate change identified by Bajo fishermen can be categorized into two aspects, namely knowledge and technology. The forms of adaptation strategies they have identified are more focused on dealing with the impacts of climate change on their livelihoods. The resilience strategy built by the Bajo people in climate change adaptation and mitigation includes.

(1) Knowledge and Education Aspects. This change in natural conditions forces Bajo fishermen to reconstruct their knowledge. Climate change that occurs requires them to adapt to aspects of knowledge. Bajo fishermen must rebuild the construction of knowledge about the West, East, and Transition seasons, in order to adjust fishing gear. They also have to build construction knowledge about fishing locations. The construction of fishermen's knowledge of the season and the location of fishing through knowledge of local wisdom has long been a guideline. This is because the condition of the season is very difficult to predict. Because, the pattern of the season one year ago cannot be used as a guideline for the following year. Therefore, changes in fishing locations are caused by constantly changing wind directions so that ocean currents also change. The fish that were originally in a certain location shifted following the direction of the ocean currents.
(2) Technology Aspects. The adaptation strategies of Bajo fishermen in the form of new knowledge construction are implemented in the form of fishing gear technology. Uncertainty in the fishing season and reduced catches make Bajo fishermen try to develop technology that can adapt to changes in existing natural

conditions. The technology they developed is by switching to a fishing gear or what they call lempara. Lempara is a fishing gear that is relied on by Bajo fishermen. Lempara is a fishing tool that can be relied upon in natural conditions that experience season uncertainty. Lempara can be used throughout the year because it can capture various types of fish, so it does not depend on a particular season.

(3) Food Security Aspect. If there is extreme climate change in the region, such as increasing temperatures and increasing rainfall, the Bajo people who inhabit the coastal area are very vulnerable from the social, economic, and biophysical aspects that exist. Life that generally relies on the fisheries sector as a main source of livelihood will also be affected. The Bajo community, for food security in the face of extreme climate change, uses a strategy of adaptation (adapted) to natural conditions. In such a situation, the Bajo people usually choose to depend on the yard of their home for domestic animals such as goats, ducks or other poultry, and plant tubers, bananas, vegetables (as a source of carbohydrates), and coconuts, using yard of the house.

(4) Health Aspects. Factual climate change has occurred at the local, regional, and global levels. Variability and climate change pose a substantial threat to physical, mental, and lack of clean water until changes in disease patterns due to current climate conditions include diseases and injuries due to heat waves, floods, storms, fires, and droughts; burden of diarrheal disease; increased frequency of heart and respiratory disease, due to increased ozone levels; and changes in the spatial distribution of several infectious disease vectors. Increased heat and rainfall caused by climate change can help the process of spreading disease. Increased temperatures will potentially support the lives of insects (mosquitoes and flies), which in turn can accelerate the spread of vector-borne diseases, such as malaria, dengue, or diarrhea/cholera. The occurrence of climate change is also thought to be able to trigger the presence of various new infectious diseases such as SARS, bird flu, and virus attacks (in the form of disasters and typical diseases in the tropics). The Climate Change Adaptation Strategy in the health aspect refers to the provisions based on Minister of Health Regulation Number 1018 of 2011, including (1) Prepare legislation; (2) Increase affordability of health services; (3) Socialization and advocacy; (4) Mapping vulnerable populations and regions; (5) Improve climate change response systems; (6) Improve partnerships; (7) Improve community empowerment; (8) Improve disease control and prevention; (9) Improve surveillance and information systems; (10) Increase the capacity of human resources; and (11) Increase affordability of health services.

(5) Settlement aspects. The condition of settlements in the sea will certainly be susceptible to strong winds which will make it easier for their homes to be swept away by strong winds. In general, Bajo people occupy wooden houses built on the sea; the shape of a house is in the form of a square or rectangle with a pyramid or saddle roof, generally still using thatched roofs or zinc roofs. For walls and floors, boards are used; some houses also use walls from silar leaves (Corypha utan Lamk.) and sago fronds (Metroxylon sp.) or palm (Arenga sp.). Bajo tribal houses in the form of stilt houses mostly use wood foundations. The

pile wood used in the Bajo tribe houses generally uses local wood with the use of round wood that still has skin.

(6) Aspects of Utilizing Coastal Resources (SDP) Mangrove Forests. Adaptation of Coastal Resources (SDP) is a form of economic strategy through the use of SDPs to produce various economically valuable commodities without requiring fishing. One potential resource is mangroves. Cultivation under mangrove stands with silvofishery pattern, namely the integrated use of mangroves between economic and ecological interests in terms of developing aquaculture is very potential to be developed so that the use of mangroves for possible economic interests will be widely developed. The development of mangrove areas is intended to maintain the economy of fishermen who experience shocks due to the effects of climate change in coastal areas. Because, in certain conditions in the area of mangrove cultivation, this can still provide fishery products with high economic value as well as substitution commodities to cover daily necessities of life, which benefit the Bajo fishermen such as mangrove crabs or others, especially when the impact of climate change is coming. The tools used to search for crabs are certain crab and bait nets. Searching for crabs in the mangrove area tends to be safer, not too weather dependent and does not require large production costs. The search for mangrove crabs is also an alternative pattern of adaptation carried out by Bajo fishermen when weather conditions are not possible or uncertain to go to sea.

(7) Increased Aspects of Human Resources. The pattern of adaptation of the allocation of Human Resources especially household members is a form of economic strategy that is very important in saving the economy of fishermen affected by climate change. This strategy includes the optimization of household workforce and the development of a double income pattern. The development of this dual livelihood strategy is intended to ensure that the fishermen/Bajo people do not depend on the results of the capture. This needs to be done because the limited facilities available cause fishermen not to be able to go to sea all year long.

Based on the summary results of the description of the conditions of the Bajo people and the occurrence of climate change, as a form of concrete solutions in providing support capacity for them and their environment to adapt and defend from the effects of extreme climate change that occurs. Because, understood together that climate change caused by extreme weather factors has had a big impact on the productivity of catches due to the inactivity of the Bajo fishermen. Reduced fish stocks and the difficulty in determining fishing areas cause economic losses for coastal/Bajo fishermen when forcing them to do activities at sea, but the results obtained are not worth the production costs incurred. The risk of high seas due to the threat of increasing storms and extreme waves causes a reduction in the frequency of Bajo tribe fishermen looking for catches, which then also impacts on economic conditions. Whether they want it or not, the Bajo people have to look for other alternatives to sustain and defend their lives from the threat of climate change that is happening at this time. Thus, an economic strategy is needed to enable fishermen/Bajo families to obtain additional income through the exploitation of household labor and a dual livelihood strategy so that the household economy does not depend solely on the results of catching at sea.

The local wisdom of the Bajo people regarding the use of resources in the sea has values and norms that govern their lives. This local wisdom makes the Bajo able to survive until now. Wisdom and local knowledge are the result of a very long process from generation to generation. Discussion of the Bajo tribe adapts to changes in the environment in meeting their daily needs. In utilizing natural resources, the Bajo people still apply traditional methods that are environmentally friendly and proven to be able to meet their subsistence needs. They still appreciate the various directions from their ancestors. However, the Bajo are not against new things. They adapt to other resource utilization activities outside of their traditional ways.

Based on some examples of the above cases, it can be seen that the Bajo tribe adapted to changes in the environment in meeting their daily needs. In utilizing natural resources, the Bajo people still apply traditional methods that are environmentally friendly and are proven to be able to meet the needs of their subsystems. However, the Bajo tribe also turned out not to be anti-new. They adapt to other resource utilization activities outside of their traditional ways. Therefore, the implementation of appropriate adaptation strategies for the Bajo people in dealing with climate change needs to be prepared so that people are not vulnerable to the impact of the rate of change that occurs. All forms of appropriate adaptation strategies in response to climate change need to be anticipated by all actors, including the government, the private sector, and the community themselves, especially the Bajo.

The Human Rights of the Bajo Tribe and Its Protection Against the Impact of Climate Change

Human Rights (HAM) are considered as a right that has been inherited from birth and is collective in nature. Also a right that is owned because someone is human, but those rights are very important rights, which are formed to protect human interests that are morally and fundamentally valid. The concept of human rights is actually a mechanism for connecting between individuals "participation in rights is participation in relations with others" so that what is expected of this relationship will bring influence to the pattern of relations and fundamental values as respect for fellow human beings and peace.

Global climate change has directly affected the microclimate, which has been a reference for indigenous peoples and coastal communities/Bajo tribes to plan activities in order to fulfill their needs. The real impact of climate change is felt when they are facing the planting season, harvest season, and the season to go fishing. The uncertainty of the seasonal calendar due to a time shift has caused crop failures and fishermen from the coastal/Bajo tribes have to cancel due to extreme weather conditions which could endanger their lives if forced. Another impact that will be caused by changes in global climate is the increasing frequency of droughts, floods, and the severity of extreme weather occurring at sea. As a community group that depends directly on the quality and quantity of natural resources, they are the groups most vulnerable to climate change and global warming. That is, Indonesia is experiencing a very severe climate crisis. The form is in the form of anomalies in sea surface temperature which warms up at least 26.5 °C in a number of Indonesian waters. The Meteorology, Climatology, and Geophysics Agency (BMKG) calls this phenomenon a tropical cyclone. So, it is only natural that the state guarantees human rights for their protection against the extreme effects of climate change.

Fenemona, for Indonesia as a country that recognizes the universal value of human rights, has obligations and must protect (to protect), respect (to respect), and fulfill (to fulfill) basic rights of its citizens, namely education, welfare, health, employment, security, clothing, and a good and healthy environment. Therefore, human rights cannot be separated from the talk of justice. Justice according to Cicero is to keep someone from harming others, except others who have made mistakes. Whereas, nature has given every type of living thing an instinct to defend its life, avoid loss, and nature unites humans and other human beings in a common bond through words (language) and life. This opinion is also in line with the legal teachings of Thomas Aquinas (1225–1275) that human beings are endowed with three basic obligations which are the main core of natural law, namely maintaining personal life (personal and descendants), knowing the truth about God, and maintaining life together in society. This principle was reiterated by Subhi Mahmassani that human rights essentially refer to the principles of social responsibility and social justice which are composed of the principle of justice and are complemented by the principle of balance and virtue, which ultimately meets the highest values or ideas taught by religion and morals. Based on the conception of human rights described above, it is very relevant if the rights to the environment are part of natural human rights which is a gift from God to mankind. So, it is very relevant if the theory or doctrine of human rights is associated with various occurrences of disasters that occur due to climate change and global warming, as actions rather than humans themselves who over-treat nature and overtake the content contained in the earth just to fulfill wealth satisfaction and with reasons to increase the economy and the original income of the country/region.

Regarding the issue of climate change, it is also necessary to pay attention to the results of the Sustainable Development Summit held in Johannesburg, South Africa in 2002. The principles of sustainable development listed in the UNCED consist of (1) intergenerational equity; (2) justice in one generation (intragenerational equity); (3) the precautionary principle; (4) biodiversity protection (conversation of biological diversity); and (5) internalization of environmental costs (internalization of the environmental cost and incentive mechanism). Then, one of the results agreed to support sustainable development is an integrated approach, taking into account various aspects of danger (multihazard) and inclusion to deal with vulnerability, risk assessment, and disaster management, including prevention, mitigation, preparedness, response, and recovery which are an important element for a safer world in the twenty-first century.

Therefore, related to human rights to the environment referred above, giving and placing the state as a bastion of human rights in the management of natural resources and the environment, the people must be placed as the main interests, including the existence of the Bajo tribe which resides and inhabits coastal areas in the archipelago. This is also in line with the principle that indigenous peoples or certain communities (Bajo tribe) which still up to now carry out the tradition of local wisdom have rights that must be respected, protected, and fulfilled by every country where the community is located. In documents published by NZ Human Rights. 37 (thirty-seven), rights of indigenous peoples were stated, namely, 2 (two) rights were mentioned, namely (1) Recognition and protection of their lands and resources; (2) Fair processes for dealing with their rights to lands and resources. As written in the document NZ Human Rights entitled, "The Rights of Indigenous Peoples: What you Need to Know" yang menyatakan "Around the world, indigenous peoples may be known by names such as: tangata whenua, aboriginal, first nations, 'native' or 'tribal' peoples."

Thus, the position of the state fully acts as an instrument of management and organizers of policies aimed at protecting and advancing human rights to the environment. The interests of the people or people's rights, especially in terms of access to natural resources and the environment, must be used as the main means and ultimate goal of the state's right to control, as referred to in Article 33 paragraph (3) 1945 Constitution of the Republic of Indonesia. towards human rights to the environment becomes a very important and fundamental aspect because the environment has all its limitations, so that control over human behavior over the environment becomes absolute. One of the controls is through instruments, mechanisms, and policies, both at the local, national, and international levels, to achieve a balance called sustainable development. Because, its existence has a connection with the 1945 Constitution of the Republic of Indonesia Article 28 paragraph 3 which states that "cultural identity and the rights of traditional communities are respected in accordance with the development of the times and civilization."

The conception of human rights to the environment began to be discussed at a meeting of the United Nations Conference on Environment and Humans in Stockholm, Sweden, on 5–6 June 1972, which sparked the Stockholm Declaration. This conference is the starting point of awareness of the international community on the importance of environmental sustainability as a fundamental part of fulfilling human rights. In Principle 21 and Principle 11, Declaration on the Human Environment from the Stockholm Conference stated that the state has the sovereign right to use its natural wealth in accordance with the policies of safeguarding and maintaining its environment. In this utilization, the state is responsible for any activity that harms the environment or territory of another country that is outside its national jurisdiction.

Violations of Human Rights (HAM) are not only carried out by dictators, but also by climate change and the country that triggered this phenomenon. Therefore, interpreting environmental issues is not only identical to the problems of waste disposal and pollution, forest fires, or the continued increase in the list of endangered species that are destroyed. In the environment, there is material life about basic rights (basic rights) of humans and the principles of environmental justice (environmental justice) and equal access to the sources of life. All three are a unity in a development policy (city/state) that puts forward the principles of sustainability in development. The failure of a country to translate the environment will directly affect the environmental elements themselves. That is, the state has a moral responsibility to provide protection from transparency in dealing with climate change impacting disaster and loss. A long series of environmental tragedies that occur and threaten the life of the earth and humans always come in turns, as if it becomes an annual routine schedule that can always be predicted or not for its presence and its consequences and seems to not provide important lessons to be anticipated. Earth that will naturally be turbulent is exacerbated by the low carrying capacity/ability of humans to anticipate it through direct destructive activities as well as through political conspiratorial efforts in exploitative environmental policies.

Along with the things mentioned above, the disaster management described in Law No. 24 of 2007 aims to provide protection to the public from the threat of disasters and respect local culture. The government is responsible for its implementation by means of disaster risk reduction and the integration of disaster risk reduction with development programs, as well as community protection from the impact of disasters. The integration of disaster risk reduction with development programs in disaster management will be effective if studies have been carried out in the aspect of disaster as a basis for consideration of regional planning and development. The knowledge base is then used as a reference in planning strategies and disaster management programs in each spatial planning and regional development document.

In the fact that has happened so far, the detailed spatial plan down to the building plan and environment does not adequately examine the aspects of disaster management from detailed community knowledge, including disasters arising from global climate change. Planning without knowledge that is studied in depth then results in often unused planning or producing planning whose application cannot be sustainable. Planning also often does not pay attention to coastal village reviews or the existence of the Bajo people, but a planning program is prepared that emphasizes and focuses on urban development and is related to the development of the city, while the growth of settlements in coastal areas is the existence of the Bajo community which is always marginalized and unplanned in adult development planning patterns.

Conclusion

Global climate change has implications for various forms of adaptation in the livelihoods of the Bajo people as an effort to sustain their lives. This form of adaptation can come from local knowledge and the creative power of the community itself. Local knowledge and creativity of the Bajo people is important because it is often associated with environmental protection and at the same time can be a sustainable supporting factor.

The Bajo have local wisdom to preserve the coastal and marine environment in utilizing natural resources. Because the Bajo are people who live in the sea, they must maintain and care for the sea as a source of livelihood, and generally as an environment. The local wisdom of the Bajo is the concept of pamali, namely the prohibition of disposing of certain wastes into the sea and the prohibition of catching fish that still do not have adult size. The concept of pamali is not only a mechanism of local wisdom to protect the environment, but, for the Bajo, the concept is the way of life, cosmology, and spiritual beliefs of the Bajo people. Without realizing it, by obeying the rules or pamali, the preservation of marine and coastal ecosystems will be maintained. Thus, indirectly the Bajo tribe has made efforts to preserve the marine and coastal ecosystems.

Climate change faced by Bajo fishermen is shifting and changing seasons, rising tide of seawater or rob, rising temperatures, and increasing storm intensity. Climate change has also had an impact on the livelihoods of Bajo fishermen in determining the time and location of the sea and the reduced number of catches. Climate change that occurs has an impact on the economic aspects that lead to the reduced level of welfare of fishermen. Bajo fishermen try to deal with these changes by adopting adaptation strategies that aim to keep them alive. The adaptation strategy that is carried out is by reconstructing the knowledge back about nature and maximizing the use of lempara fishing gear which is considered in accordance with existing environmental conditions.

References

The full reference list for this chapter can be found via https://doi.org/10.1007/978-3-030-55536-8_16.

Part II
Climate Change Mitigation and Other Cross Cutting Issues (Finance, Communication, Law)

Chapter 10
Carbon Stocks from Peat Swamp Forest and Oil Palm Plantation in Central Kalimantan, Indonesia

Nisa Novita, J. Boone Kauffman, Kristell Hergoualc'h, Daniel Murdiyarso, Dede Hendry Tryanto, and Joni Jupesta

Abstract Conserving high carbon density tropical peat forests is one of the most cost-efficient strategies for climate change mitigation at national and global levels. Over past decades, large areas of tropical peat forests have been converted to oil palm plantation in Indonesia resulting in significant carbon emissions into the atmosphere. Here, we quantified the ecosystem carbon stocks in a total of six sites: two primary peat swamp forest sites, one secondary peat swamp forest site, and three young oil palm plantation sites in Tanjung Puting, Central Kalimantan, Indonesia. We further determined potential carbon emissions from vegetation change due to peat swamp forest conversion to oil palm plantation. The mean total ecosystem carbon stock of primary and secondary peat swamp forests were 1770 ± 123 Mg C/ha and 533 ± 49 Mg C/ha, respectively. In contrast, the mean carbon stock of oil palm plantations was 759 ± 87 Mg C/ha or 42% of peat swamp forests. The ratio of the aboveground to belowground C stock was highest in secondary forest with an estimated value of 0.48, followed by primary forest at 0.19 and oil palm plantation at 0.04. Using a stock difference approach, we estimated potential carbon emissions from vegetation change resulting from the conversion of primary peat swamp forest to oil palm plantation

The original version of this chapter was revised: This chapter has been changed to open access under a CC BY 4.0 license. The correction to this chapter is available at https://doi.org/10.1007/978-3-030-55536-8_16

N. Novita (✉) · K. Hergoualc'h · D. Murdiyarso · D. H. Tryanto
Center for International Forestry Research, Jalan CIFOR, Situgede Sindang Barang, Bogor Barat 16680, Indonesia
e-mail: nisa.novita.278@gmail.com

N. Novita · J. B. Kauffman
Oregon State University, Nash Hall, 2820 SW Campus Way, Corvallis, OR 97331, USA

D. H. Tryanto
Badan Restorasi Gambut, Gedung Graha Mandiri Lt. 21, Jl. Imam Bonjol No. 61, Menteng, Jakarta Pusat 10310, Indonesia

J. Jupesta
PT. SMARt Tbk., SMART Research Institute, Jln Sukarno Hatta No 19, Pekanbaru, Riau 28294, Indonesia

R. Djalante et al. (eds.), *Climate Change Research, Policy and Actions in Indonesia*, Springer Climate, https://doi.org/10.1007/978-3-030-55536-8_10

of 640 ± 114 Mg CO_2/ha. Finally, while restoring peatlands is important, avoiding peat conversion is imperative for Indonesia's climate change mitigation effort.

Keywords Tropical peat forest · Land-use change · Oil palm · Indonesia · Climate change mitigation

Introduction

Tropical peat swamp forests provide a broad array of important ecological functions and environmental services (Page et al. 2006). This ecosystem is known as a biodiversity hotspot supporting the existence of endemic and endangered flora and fauna (Posa et al. 2011). Tropical peat swamp forests are imperative for the socioeconomics of local communities, as resource for food, shelter, plant-based medicine, and cultural activities at local and global scale (Hergoualc'h et al. 2016). They also exert hydrological controls on water flow during wet and dry seasons (Wösten et al. 2006). In addition, the prominent role of tropical peatlands as carbon (C) reservoir and sink has been described in many scientific studies (Page et al. 2011; Murdiyarso et al. 2009; Yu et al. 2011; Warren et al. 2017). Approximately 84% of peat carbon in Southeast Asia is found in Indonesia (Page et al. 2011) which equals to 18% of volume peat globally (Gumbricht et al. 2017), where extensive deforestation and degradation of peat swamp forests has occurred since the 1990s.

Recently, a half of tropical peat swamp ecosystem have been converted to managed plantation (particularly oil palm (*Elaeis guineensis* Jacq.) at smallholders and industrial scales, then followed by logged-over areas (Miettinen et al. 2015). In addition, fires emissions during land preparation and peat drainage result in high greenhouse gas emissions to the atmosphere (Langner and Siegerts 2009; Saharjo 2007). Currently, Indonesia is extensively avoiding peat conversion by extending moratorium on the conversion of peatland, rewetting peatlands, restoring degraded peatlands, and improving socioeconomic or community livelihood. Thus, reducing emissions from peat swamp forest deforestation is important relative to the commitment to reduce national emissions by 497 MtonCO_2e from forestry and land-use change sector. Including peat swamp forest conservation and restoration as main activity to achieve national NDC target is a promising step to cut land-based emissions by an estimated 29% compared to the business as usual scenario by 2030.

Given that tropical peat swamp forests have been recognized as having exceptionally high carbon stocks coupled with rapid rates of deforestation in the past, it is important to quantify the changes in carbon stocks and potential carbon emissions due to forest conversion. Limited data exists for ecosystem carbon stocks in tropical peatlands particularly for soil carbon, and even fewer for carbon emissions that would arise from vegetation cover change of peat swamp forest. The lack of data becomes a major problem for stakeholders at national level to develop policy related to reduction of national greenhouse gas emissions from land-use change sector on peatlands. This data can be used to enrich global and national forest inventory data to revise second forest reference emissions level document.

A preliminary carbon stock inventory in peat swamp forest over shallow peats (<3 m) in Tanjung Puting reported stocks to range from 863 to 1048 Mg C/ha (Murdiyarso et al. 2009). We expanded our measurements to slightly deeper peat forests and oil palm plantations located in close proximity to the peat swamp forest sites. The objectives of this study were to quantify the carbon stocks of intact peat swamp forests, secondary forests, and oil palm plantations and estimate potential emissions from vegetation carbon change due to forest to oil palm conversion. To accomplish these objectives, we (1) quantified the structure and composition of primary forest, secondary forest, and oil palm plantation and (2) measured the total ecosystem C stocks in each ecosystem.

Materials and Methods

Study Site

Study sites were located in the Tanjung Puting region, which is ~25 km from the city of Pangkalan Bun in Central Kalimantan Indonesia (Fig. 10.1). The climate of Central Kalimantan is classified as humid-tropic, with annual rainfall of 2000–4000 mm/yr (Ministry of Environment and Forestry 2015). The annual pattern of rainfall in the study area is influenced by a southeast dry monsoon and a northeast wet monsoon. Most precipitation occurs during the wet season between the months of September

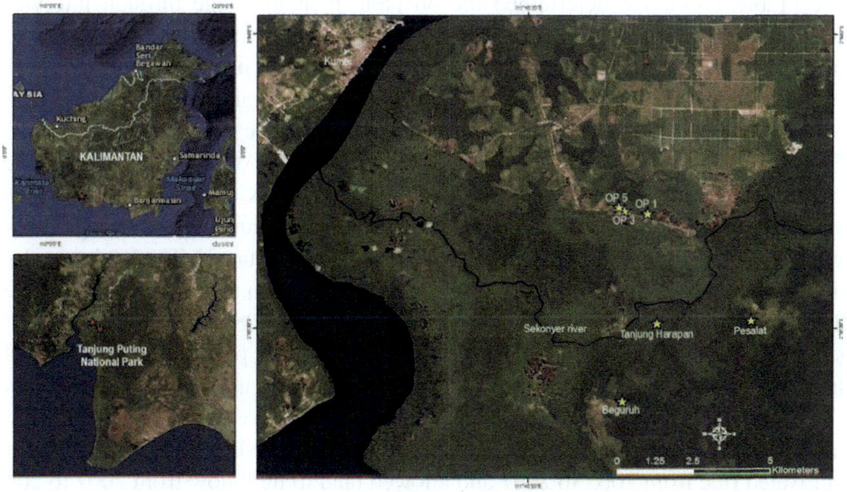

Fig. 10.1 Location of study areas in Tanjung Puting, Indonesia

Table 10.1 Location and peat depth of sampling locations in Tanjung Puting, Indonesia

Ecosystem	Site	Latitude	Longitude	Mean Peat thickness (cm)
Primary forest	Beguruh	S 2°50′51″	E 111°48′11″	290 ± 23
Primary forest	Pesalat	S 2°49′21″	E 111°50′24″	155 ± 6
Secondary forest	Tanjung Harapan	S 2°49′24″	E 111°48′46″	27 ± 2
Oil palm	OP1	S 2°47′33″	E 111°48′36″	20 ± 2
Oil palm	OP3	S 2°47′18″	E 111°48′10″	47 ± 3
Oil palm	OP5	S 2°47′28″	E 111°48′7″	47 ± 4

to May and less during a relatively dry season from June to August (Ministry of Environment and Forestry 2015). Based upon the weather stations installed at the research sites, the annual rainfall recorded from the weather stations during our sampling period was below-average annual precipitation (1870 mm/yr). Mean air temperature was 31.4 °C. The peat in the area of Tanjung Puting is classified as ombrogenous (Ministry of Environment and Forestry 2015) where precipitation is the major supply of water and nutrients to the ecosystem (Page et al. 2006). The peat in this region was estimated to form beginning about 22,120 ^{14}C yr BP (Weiss et al. 2002), with peat accumulation rate is 0.54 mm/year (Dommain et al. 2011) .

Two of the forest sites were relatively intact (Beguruh and Pesalat) and one site was a secondary forest (Tanjung Harapan). The secondary forest was located about 1 km from an abandoned village and was disturbed at least 30 years ago and most likely much earlier than what we expected. The forest sites were located inside the National Park of Tanjung Puting, while the oil palm plantations were located on the opposite side of the Sekonyer River nearby the village of Bedaun. Global Positioning System (GPS) readings were utilized to determine geographic coordinates for each site (Table 10.1). We determined the oil palm plantation age through interviews with farmers/landowners and Bedaun and Sekonyer residents. Oil palm management practices were similar across the sites and were managed by smallholders.

We sampled three oil palm plantations and all had drainage canals constructed around their perimeter. The oil palm plantation sites were (1) One-year old (OP1): This was a small-scale plantation with total area of 1 ha. At OP1, the primary forest clearing occurred in 1989 (about 27 years previously) with slash and burn of the standing forest. The valuable trees were cut and removed, and the remaining biomass was dried for about 1–2 months during dry months then burned. The land was abandoned in one year and was utilized for dry rice cultivation for two years until 1991. The farmer allowed land to lie fallow from 1993 to 2008. In 2008, the land was slashed and burned again for rice cultivation, which was practiced until 2010. The land was fallowed for one year, and in late 2011 the farmer again cleared the fallowed land through slash and burn for oil palm plantation. (2) Three-year-old oil palm plantation or OP3: The forest clearing occurred in 2005 by via slash and burn. The farmer cultivated rice from 2005 to 2007 and allowed land to lie fallow from 2007 to 2009. The oil palm plantation was established in 2009. (3) Five-year-old oil palm plantation or OP5. The primary forest was slashed and burned in 1989 for cultivation and then fallowed. The secondary forest formed on this site was cleared in 2006 via slash and burn. Before the oil palm plantation was established in 2008, the land was utilized for rice and vegetable crops for 2 years.

Field Sampling

We determined the composition, structure, and ecosystem carbon stocks following methods outlined by Kauffman et al. In the forest sites, six 10 m fixed-radius (0.0314 ha) plots were established 50 m apart along a 250 m transect as depicted in Fig. 10.2. Due to area limitation in the plantations, three 10 m fixed-radius plots were established 20 m apart along a 40 m transect as shown in Fig. 10.3. In addition, palm height and density were measured in a square plot (50 m × 50 m) established in each plantation site.

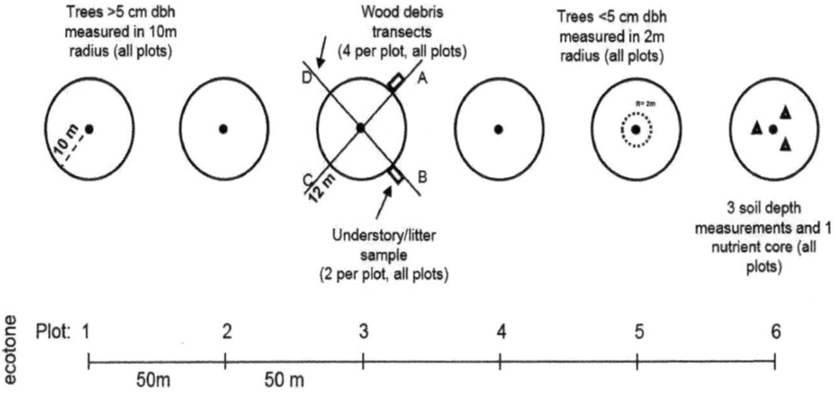

Fig. 10.2 The plot layout for C stock assessment of peat swamp forests in Tanjung Puting, Indonesia (modified from Kauffman et al. 2011)

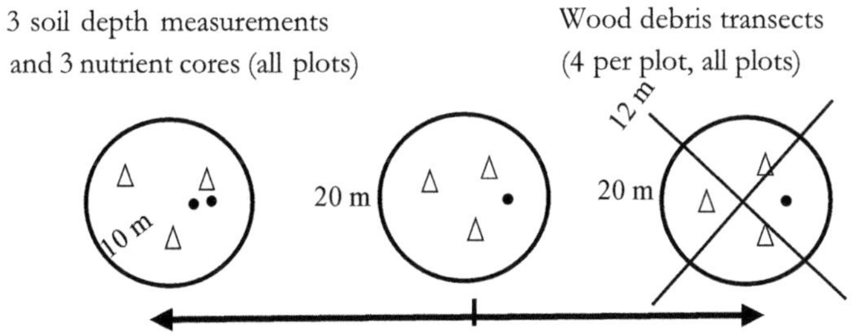

Fig. 10.3 The plot layout for soil C stock assessment of oil palm plantations in Tanjung Puting, Indonesia

Forest Structure and Composition

Structure and species composition of trees were determined through tree identification and measurement of the diameter at breast height (dbh; 1.30 m height or 10 cm above the area of buttresses or other flared tree bases) of all trees in the circular plots. The tree size class distribution in forest sites was separated into 10 cm diameter classes. We also recorded tree status (dead or alive). Live trees with dbh >2 cm were identified in all circular plots. Samples from plants not identified in the field were determined at the Indonesian Institute of Science or the *Lembaga Ilmu Pengetahuan Indonesia* (LIPI) in Bogor, Indonesia. Based on the field data, we calculated the mean basal area, tree density, and Importance Values Index (IVI) to describe relative abundance of each tree species. Species diversity indices were also calculated, including Shannon'n (H'), Simpson's reciprocal, and Shannon's evenness (H'E) (Sheehan 1984).

Overstory C Stocks: Trees and Oil Palm Vegetation

Allometric equations were used to calculate tree and oil palm biomass. The allometric equation by Manuri et al. (2014) was selected to estimate tree biomass in forest sites (Table 10.2). Dead tree biomass was estimated by applying to the allometric equation used for living tree biomass a correction factor based on dead tree status (BSN 2011). The allometric equation based on palm height presented by Khasanah et al. (2015) was chosen to determine oil palm biomass. The palm height was defined as the trunk height from ground level to the lowest frond in palm canopy (Khasanah et al. 2015). Root biomass in all types of ecosystems was estimated based upon a general belowground allometric equation provided by Cairns et al. (1997). This was based

Table 10.2 Equations to estimate aboveground biomass in forest and oil palm plantation sites in Tanjung Puting, Indonesia

C pool	Equation	Factor C conversion
Live tree biomass (kg/tree)	$AGB_{tree} = 0.136 * dbh^{2.513}$	0.48
Oil palm biomass (Mg/palm)	$AGB_{OP} = 0.0939 * H + 0.0951$	0.48
Deadwood biomass (kg)		0.48
Status 2	$AGB_{dw} = 0.8 * (0.136 * dbh^{2.513})$	
Status 3	$AGB_{dw} = 0.7 * (0.136 * dbh^{2.513})$	
Root biomass	$AGB_{root} = \exp(-1.085 + 0.9256 * \ln(AGB_{OP/tree})$	0.39
Woody debris biomass (Mg)		0.50
>7.5 cm diameter	$AGB_{wd} = sg * [(\mu^2\ \Sigma\ D^2)/(8 * L)]$	
2.5–5.5 cm	$AGB_{wd} = sg * [\mu^2 * (NQMD^2)/(8 * L)]$	

on an assumption that root: shoot ratio of oil palm plantation is equal to another tree vegetation (Jourdan and Rey 1997). Vegetation C mass was calculated from biomass by multiplying by a factor of 0.48 and 0.39 for vegetation and root biomass, respectively (Kauffman and Donato 2012).

Symbols used in the equations area: AGB = Aboveground biomass, dbh = diameter at breast height (cm), H = height of the palm (m), sg = specific gravities (g/cm^3) to convert volume to downed wood biomass are from Chao et al. (2008). Sg for solid and rotten woody debris are 0.5 and 0.23, respectively. N = number of pieces of woody debris pieces, QMD = quadratic mean diameter of size class (cm), and L = the transect length (m).

Litter and Understory

Litter and understory biomass were determined through destructive sampling. In each circular plot, we harvested litter and understory vegetation from two 56 cm × 56 cm microplots. Litter consisted of leaves, flowers, fruits, bark fragments, seeds, and small twigs/woody stems. All plant samples were placed in a bag and transported to the laboratory where they were oven-dried to a constant mass at 60 °C and then weighed. The organic carbon content of the litter and understory were determined by the induction furnace method with a Carbon Nitrogen Analyzer (LECO Corporation, St. Joseph, Michigan, USA) in the Soil Biotechnology laboratory, Bogor Agricultural University, Indonesia. The litter C content in the oil palm plantation was estimated based on Khasanah et al. (2015) study.

Downed Wood

The mass of downed wood was measured using the planar intersect technique (Kauffman et al. 2016; Kauffman and Donato 2012). At the center of each circular plot, four 12 m transects were established diagonally at 45° off the main transect (Figs. 2 and 3). Along each transect, coarse downed wood (diameter >7.5 cm) was measured from 2 to 12 m and fine downed wood (diameter 2.5–7.5 cm) was measured from 2 to 7 m. Coarse downed wood was separated into sound and rotten classes. Fine downed wood was only counted. A quadratic mean diameter (QMD) was used to estimate volume of the fine downed wood (Kauffman et al. 2016). We used data of the specific gravity of downed wood reported by Chao et al. (2008). Volume and specific gravity of downed wood were multiplied to estimate biomass. The biomass was multiplied by a factor of 0.50 to estimate downed wood C mass (Kauffman and Donato 2012).

Soil Carbon

Peat depth and sample cores were extracted from three sampling points near the center of each circular plot. We measured peat depth by forcing a steel open-faced peat auger downward through peat horizon to the organic-mineral transitional horizon. An extension handle was used if peat depth was greater than 1 m. Measurements of soil bulk density and C concentration at various peat depths were taken to determine the soil C stocks (Donato et al. 2011; Kauffman et al. 2011). The core was systematically divided into the depth intervals of 0–15, 15–30, 30–50, 50–100, and >100 cm (if mineral layers were not detected before 100 cm depth). Samples of a known volume were carefully placed in *whirlpak* bags and transported to the Bogor Agricultural University for laboratory analysis. Soil samples were dried to a constant mass at 60 °C and weighed to obtain soil bulk density. The organic carbon content of soil samples was determined via the induction furnace method using a carbon–nitrogen analyzer (LECO Corporation, St. Joseph, Michigan, USA). The induction furnace with an elemental carbon analyzer is considered the most reliable method compared to other C determination methods, e.g., wet combustion and loss of ignition (Kauffman and Donato 2012) in tropical peat soils (Farmer et al. 2014). The soil carbon stocks were calculated as the products of bulk density and carbon concentration combined with plot-specific peat depth.

Statistical Analyses

Microsoft Excel and IBM SPSS were used for data analysis with a probability threshold (*p* value) of 5%. The residual values of soil properties (bulk density, C, N, and C/N ratio), ecosystem C stocks, and their log transformations were not normally distributed based on the Shapiro–Wilk test. Differences in soil properties, biomass, and C among ecosystems (primary forest, secondary forest, and oil palm plantation) were tested using the non-parametric Kruskal–Wallis test. If the result was significant, a pairwise comparison test was applied to determine which means were significantly different.

Results

Ecosystem Structure and Composition

The distribution of tree diameters described a typical reverse J-shaped curve in all forest sites where small trees were most abundant (Fig. 10.4). Within forest sites, there was considerable site-to-site variation in the distribution of tree diameter classes and therefore forest structure. In general, the range in diameters was larger in Beguruh and

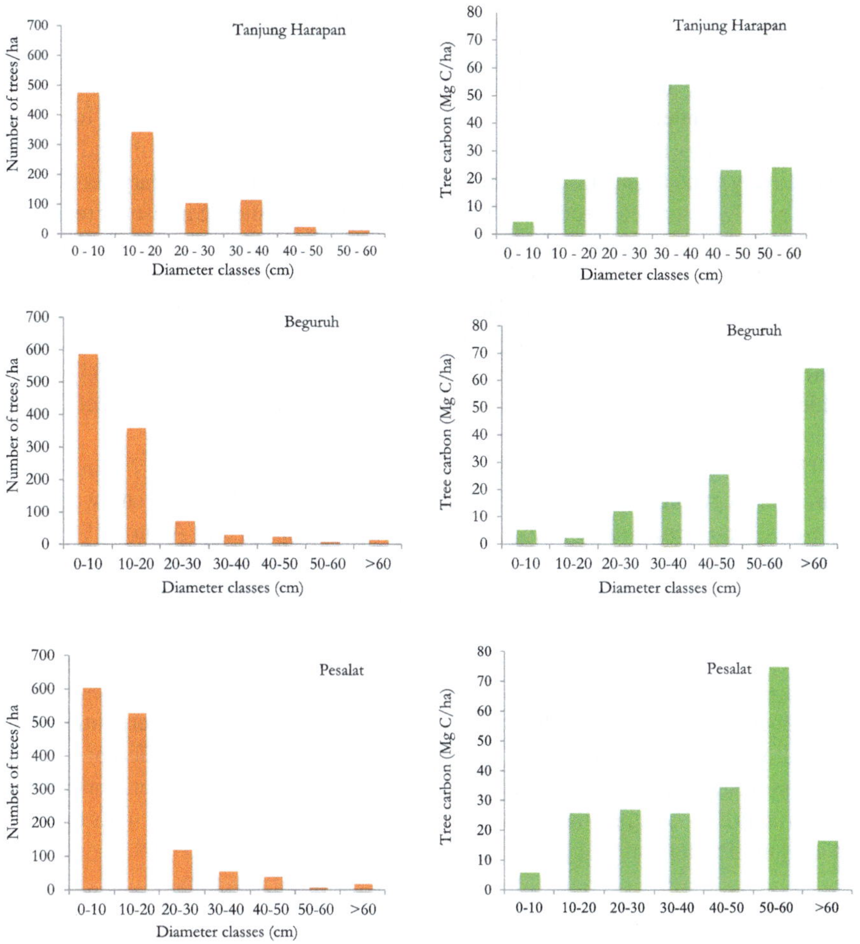

Fig. 10.4 Distribution of tree density and aboveground C stock by 10 cm diameter classes in Tanjung Harapan, Beguruh, and Pesalat forest sites in Tanjung Puting, Indonesia

Pesalat sites (primary forest) compared to Tanjung Harapan site (secondary forest). Living trees with dbh >60 cm were present in primary forests (Pesalat and Beguruh) while in the secondary forest, Tanjung Harapan, they were not present. The small trees (dbh <20 cm) were also lower in abundance at Tanjung Harapan than Beguruh and Pesalat sites (Fig. 10.4). The small trees comprised 76, 86, and 82% of the total distribution of trees in Tanjung Harapan, Beguruh, and Pesalat, respectively. Overall, Tanjung Harapan site differed from Pesalat and Beguruh sites, particularly in having a lower density of small (dbh <20 cm) and large trees (dbh >60 cm).

We also found difference between secondary and primary forests in tree carbon stock distribution among diameter classes (Fig. 10.4). Tree C stock in secondary forest was mostly dominated by tree with dbh <40 cm, which comprised 68% of the total vegetation C stocks. In contrast, tree with dbh >40 cm dominated C vegetation in primary forest sites, comprising 74 and 60% of the total vegetation C stocks in Beguruh and Pesalat, respectively.

Shannon's index (H') of forest structure was lower at the secondary forest site (1.13) compared to primary forest sites (1.35 and 1.36 for Beguruh and Pesalat, respectively), thus reflecting a more even-aged stand in the secondary forest. This index poorly described structural diversity among forest sites. Simpson's reciprocal index of the secondary forest Tanjung Harapan was lower (8.44) than Beguruh (14.98) and Pesalat (14.62) sites. Simpson's reciprocal index better illustrated the diversity of the forest structure in Tanjung Harapan, and was different from the Beguruh and Pesalat sites. Furthermore, Pesalat and Beguruh sites had similar Shannon's evenness indices (0.37), which were slightly higher than Tanjung Harapan site (0.34). Lower values of species diversity and evenness indices in Tanjung Harapan reflect signs of early seral forest succession.

The mean basal area of the forest sites was 31.3, 28.4, and 39.6 m^2/ha in Tanjung Harapan, Beguruh, and Pesalat, respectively. The Tanjung Harapan site was dominated by *Macaranga motleyana*, a pioneer species (Davies and Ashton 1999) with the highest basal area (15 m^2/ha) and importance value index (82%; Fig. 10.5). The high dominance by a single species in Tanjung Harapan site was consistent with the low values described by Simpson's reciprocal index. Unlike Tanjung Harapan site, no single species dominated in the primary forests (Fig. 10.5). It is interesting to see the presence of *Artocarpus heterophyllus* (jackfruit), a crop cultivated species, had the 5th highest IVI value (12.4), showing that Tanjung Harapan was formerly utilized for agriculture. Inevitably, the Tanjung Harapan site clearly showed a different structure and composition compared to Beguruh and Pesalat sites. The different composition and structure at Tanjung Harapan reflects past intensive land use prior to its inclusion in Tanjung Puting national park territory.

Aboveground C Stocks

Ecosystem C stocks are presented in Table 10.3. In primary forests, the aboveground C stock was higher in Pesalat (206 ± 32 Mg C/ha) than in Beguruh (130 ± 30 Mg C/ha; $p < 0.001$). The mean of vegetation C stock in primary forest (168 ± 31 Mg C/ha) was not significantly different from the stock in secondary forest (140 ± 16 Mg C/ha; $p = 0.291$). The tree carbon pools closely followed basal area across forest plots. In the oil palm plantations, vegetation C stocks ranged from 7 to 19 Mg C/ha. There were significant differences among sites in the oil palm plantation ($p < 0.001$), where the highest palm C stock was found at the oldest plantation. Mean aboveground C stocks in the primary forests were 14 times higher than in the oil palm plantations sites ($p < 0.001$).

The downed wood was a significant C pool, contributing 12 and 16% to the total aboveground C stocks in the primary forests and secondary forest, respectively. This was especially true for oil palm plantations, where the downed wood C pool

Fig. 10.5 Basal area (m^2/ha) and important value index in Tanjung Harapan, Beguruh, and Pesalat forest sites in Tanjung Puting, Indonesia

consisting of residual wood from the forests contributed more than half (55%) to the total aboveground carbon. The dead trees were an insignificant C pool, accounting for only 3 and 1% of the total aboveground biomass in the primary forests and secondary forests, respectively. There were no standing dead trees in the oil palm plantation sites. The dead frond piles C stocks were found in the oil palm plantations but this C pool is assumed to be insignificant or may be underestimated.

In forests, mean carbon concentrations of litter and understory were 48.43 ± 0.46% and 45.02 ± 0.68%, respectively. Litter and understory C stocks were 4 Mg C/ha in forest sites and contributed 2% to the total aboveground C stocks. There was less understory and litter in the oil palm plantations because herbicides were regularly applied.

Table 10.3 Ecosystem C stocks (Mg C/ha) in the forest and oil palm plantation sites in Tanjung Puting, Indonesia. AGC and BGC indicate aboveground carbon and belowground carbon, respectively

Site	Vegetation	Forest floor	Dead tree	Woody debris	Root	Soil	AGC	BGC	Total ecosystem C stocks
Beguruh	130 ± 30	5 ± 0.4	6 ± 0	26 ± 0.1	23 ± 6	2311 ± 103	168 ± 30	2334 ± 109	2502 ± 139
Pesalat	206 ± 32	4 ± 0.4	8 ± 0	22 ± 0.1	29 ± 2	769 ± 72	240 ± 33	798 ± 74	1038 ± 107
Primary forest	168 ± 31	4 ± 0.4	7 ± 0	24 ± 0.1	26 ± 4	1540 ± 88	204 ± 32	1566 ± 92	1770 ± 123
Secondary forest	140 ± 16	4 ± 0.5	1 ± 0	28 ± 0.1	25 ± 4	336 ± 29	172 ± 17	361 ± 32	**533 ± 49**
OP1	7 ± 0.1	1 ± 0.1	–	12 ± 0.2	2 ± 0.0	545 ± 33	20 ± 0.3	547 ± 33	567 ± 34
OP3	10 ± 0.1	1 ± 0.1	–	14 ± 0.2	2 ± 0.0	788 ± 50	26 ± 0.3	791 ± 50	816 ± 51
OP5	19 ± 0.3	1 ± 0.1	–	22 ± 0.1	4 ± 0.1	847 ± 23	42 ± 0.4	851 ± 23	893 ± 23
Oil palm	12 ± 0.2	1 ± 0.1	–	16 ± 0.2	3 ± 0.0	727 ± 35	29 ± 0.3	729 ± 35	**759 ± 36**

Data are mean ± standard error

Aboveground C stock was the highest in the primary forests, followed by secondary forest and oil palm plantations. The mean aboveground C stock was seven times higher in the primary forests than in oil palm plantation sites ($p < 0.001$) (Table 10.3).

Soil

The peat was deeper in the primary forests compared to the oil palm plantations or secondary forest (Table 10.4). The mean peat depth of primary forest sites was 223 ± 10 cm. In contrast, the mean peat depth of three oil palm plantation sites was significantly lower compared to primary forests, with a mean of 38 ± 3 cm ($p < 0.001$). The lowest peat depth across sites was observed in secondary forest Tanjung Harapan (27 cm).

There was a significant difference in soil bulk density between primary forest and oil palm plantation sites ($p < 0.001$, Table 10.4), which was apparent throughout all depths. The mean bulk density across depths was twice higher in oil palm plantations sites (0.37 ± 0.05 g cm^{-3}) than in primary forest sites (0.17 ± 0.02 g cm^{-3}; $p < 0.001$). Bulk density of secondary forests (0.30 ± 0.02 g cm^{-3}) was also twice the value in primary forests. Interestingly, we did not observe a significant difference in bulk density between secondary forest and oil palm plantation sites ($p = 0.222$), suggesting a clear effect of past forest clearing and agricultural cultivation in the secondary forest site. Moreover, the higher bulk density and the shallow peat depth in the secondary forest suggest a slow recovery on the peat layers following abandonment.

In general, bulk density tended to decrease with depth in primary forest and oil palm plantation ecosystems, but not in the secondary forest. In oil palm plantations sites, the bulk density at 0–15 cm peat depth was 0.41 g cm^{-3} or 1.5 times higher than the 0.27 g cm^{-3} average for peat layer >30 cm. Furthermore, in the primary forest sites, bulk density at 0–15 cm was 0.25 g cm^{-3} or two times higher compared to 0.13 g cm^{-3} at peat layer 50–100 cm.

The mean C concentration across depths was similar in the primary forest (49.42 ± 0.83%) and oil palm plantation sites (48.73 ± 5.05%) ($p = 1.0$). In both ecosystems, soil C concentration exceeded 43% C at all depths. Peat in the secondary forest had a significantly lower C concentration but still relatively high at 32% ($p < 0.001$). In general, soil C concentration was not affected by depth in all sites (Table 10.4).

Primary forest conversion to oil palm plantations affected soil N concentration at all soil depths (Table 10.4). There was no significant difference in *N* concentration between primary and secondary forests ($p = 0.299$). The mean N concentration in the primary forest, was higher (1.23 ± 0.07%) than at oil palm plantation sites (0.95 ± 0.13%) ($p < 0.001$). The difference in N concentration between forest and oil palm plantation sites was apparent at all depths. Generally, we observed that soil N concentration decreased with depth in all sites. In the primary forest sites, N concentration at the 0–5 cm peat depth (1.61%) was about 1.5 times higher than in the >100 cm peat layer (1.10%). A similar trend was found in the oil palm plantations

Table 10.4 Soil properties (carbon, nitrogen, C/N ratio, bulk density, and carbon mass) by sampling depth in forest and in oil palm plantation sites in Tanjung Puting, Indonesia

Ecosystem	Site	Sample depth (cm)	Carbon (%)	Nitrogen (%)	C/N ratio	Bulk density (g/cm^3)	Carbon mass (Mg/ha)
Primary forest	Beguruh	0–15	43.92 ± 1.04	1.60 ± 0.07	28.48 ± 1.59	0.30 ± 0.03	200 ± 19
		15–30	48.36 ± 1.30	1.22 ± 0.05	40.77 ± 1.95	0.26 ± 0.06	186 ± 18
		30–50	51.36 ± 0.56	1.13 ± 0.04	46.28 ± 1.79	0.16 ± 0.04	167 ± 24
		50–100	51.01 ± 0.61	1.10 ± 0.02	46.66 ± 1.27	0.18 ± 0.04	459 ± 79
		100–290	49.66 ± 1.10	1.05 ± 0.04	48.87 ± 1.33	0.13 ± 0.03	1298 ± 345
							2310 ± 485
	Pesalat	0–15	48.50 ± 1.18	1.62 ± 0.08	31.52 ± 2.27	0.20 ± 0.05	139 ± 22
		15–30	52.37 ± 1.19	1.27 ± 0.04	39.81 ± 3.09	0.15 ± 0.04	110 ± 22
		30–50	52.25 ± 0.03	1.09 ± 0.05	49.04 ± 0.05	0.13 ± 0.03	114 ± 30
		50–100	50.03 ± 1.19	1.04 ± 0.04	48.92 ± 1.64	0.09 ± 0.02	217 ± 42
		100–155	46.75 ± 1.59	1.15 ± 0.09	43.51 ± 2.50	0.08 ± 0.02	162 ± 31
							742 ± 147
Mean primary forest		222 ± 12	49.42 ± 0.83[a]	1.23 ± 0.07[a]	42.29 ± 2.29[a]	0.17 ± 0.02[a]	**1526 ± 316**
Secondary forest	Tanjung Harapan	0–15	33.85 ± 1.05	1.31 ± 0.10	28.47 ± 2.38	0.27 ± 0.03	144 ± 19
		15–30	32.63 ± 1.28	0.97 ± 0.11	37.84 ± 3.09	0.29 ± 0.03	119 ± 17
		30–39	30.00 ± 2.12	0.85 ± 0.16	38.07 ± 7.93	0.34 ± 0.04	73 ± 21
							336 ± 57

(continued)

Table 10.4 (continued)

Ecosystem	Site	Sample depth (cm)	Carbon (%)	Nitrogen (%)	C/N ratio	Bulk density (g/cm^3)	Carbon mass (Mg/ha)
Mean secondary forest		27 ± 2	32.16 ± 0.90^{b}	1.04 ± 0.07^{ab}	34.79 ± 2.95^{b}	0.30 ± 0.02^{b}	**336 ± 57**
Oil Palm Plantation	OP1	0–15	45.79 ± 1.86	1.10 ± 0.10	43.77 ± 3.56	0.46 ± 0.05	326 ± 38
		15–30	47.35 ± 5.10	0.80 ± 0.09	59.83 ± 3.14	0.47 ± 0.07	219 ± 56
							545 ± 94
	OP3	0–15	50.80 ± 2.30	1.33 ± 0.11	40.53 ± 3.90	0.36 ± 0.05	273 ± 33
		15–30	58.08 ± 2.06	1.11 ± 0.07	54.33 ± 4.42	0.35 ± 0.04	302 ± 36
		30–47	47.03 ± 13.76	0.84 ± 0.35	59.70 ± 12.08	0.27 ± 0.09	214 ± 158
							789 ± 227
	OP5	0–15	47.38 ± 5.27	0.92 ± 0.14	57.05 ± 9.63	0.40 ± 0.05	273 ± 46
		15–30	51.81 ± 3.81	0.86 ± 0.07	61.59 ± 4.17	0.35 ± 0.03	264 ± 25
		30–47	41.64 ± 6.26	0.60 ± 0.10	70.51 ± 4.35	0.27 ± 0.02	196 ± 48
							733 ± 119
	Mean OP		48.73 ± 5.05^{a}	0.95 ± 0.13^{b}	55.91 ± 5.66^{c}	0.37 ± 0.05^{b}	**689 ± 147**

Data are mean ± standard error. Different letters in superscript following values describe statistical significance across primary forest, secondary forest, and oil palm plantation

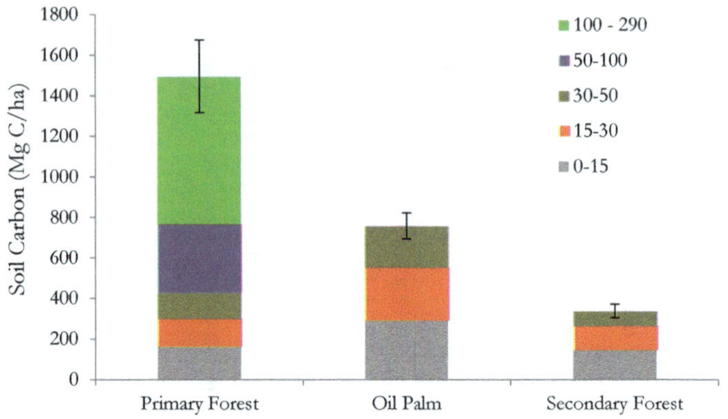

Fig. 10.6 Soil C stocks (Mg C/ha) in different peat layers (cm) in primary forest, secondary forest, and oil palm plantation sites in Tanjung Puting, Indonesia. Vertical bars are standard errors

where soil N content in the 0–15 cm peat depth (1.12%) was about 1.6 times higher than in the peat layer >30 cm (0.72%).

The relatively deep peat and high C concentration throughout the profile in the primary forest sites resulted in high soil carbon stocks. The shallow peat depth coupled with lower C concentration resulted in lower C soil carbon stock in the secondary forest that in primary forests. The mean soil C stock was higher in the primary forest sites than in the oil palm plantation sites ($p < 0.001$; Fig. 10.6). The mean soil C stock of oil palm plantations was about 47% of the mean of primary forests. However, carbon density (carbon mass per unit of soil) remained high due to high bulk density in the oil palm plantation sites.

Ecosystem C Stocks

The mean total ecosystem C stock of the primary forests was 1770 ± 123 Mg C/ha. Total ecosystem C stock was 1038 Mg C/ha at Pesalat site and 2502 Mg C/ha in Beguruh site (Fig. 10.7). In contrast, the mean total ecosystem C stock in the oil palm plantations was 759 ± 36 Mg C/ha, less than half of the mean of the primary forests ($p < 0.001$). Total ecosystem C stock for oil palm plantations showed relatively low variation from a minimum of 567 Mg C/ha at OP1 to a maximum of 893 Mg C/ha at OP5 (Fig. 10.7). The lowest ecosystem C stock was found at the secondary forest.

Soil was a primary carbon pool in all land uses. Soil C pools contribute 3% of the total ecosystem C stock in the primary forests, ranging from 74% at Pesalat site to 92% at Beguruh site. In the oil palm plantations, soil comprised >95% of the total ecosystem C stock in all sites. Lower contribution of soil C pool to the total ecosystem C stock (63%) was observed at Tanjung Harapan. Consequently, the ratio

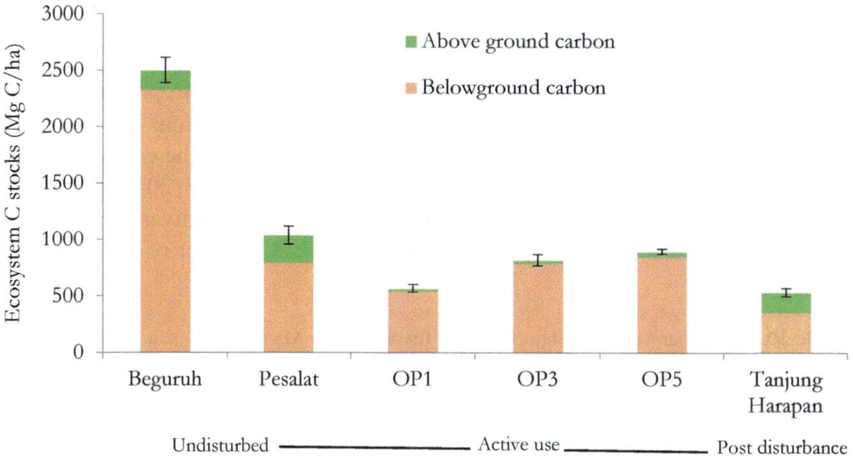

Fig. 10.7 Total ecosystem C stocks (Mg C/ha) of peat swamp forests (primary and secondary) and oil palm plantation sites in Tanjung Puting, Indonesia. Vertical bars are standard errors

of the aboveground to belowground C stock was highest in secondary forest site with an estimated value of 0.48, followed by primary forest (0.19) and oil palm plantation (0.04). Peat swamp forest conversion resulted in substantial C losses from vegetation cover change, of 640 Mg CO_2/ha.

Discussion

Composition and Structure of Forest

Tanjung Harapan as secondary forest clearly shows differences on diameter distribution, basal area, and important value index compared to two other primary forest sites; Beguruh and Pesalat. It clearly indicates that Tanjung Harapan is at early to medium stage growth of secondary forest and recovering from human disturbances in particular encroachment in the past. The secondary forest carbon pool at Tanjung Harapan site suggests a rapid rate of carbon sequestration when sites are abandoned and native vegetation is allowed to grow (in the absence of fire and other human disturbances).

Ecosystem C Stocks

Limited studies on aboveground C stocks have been conducted in tropical peat swamp forests (Table 10.5) and even fewer exist for oil palm plantations and secondary forests (but see Murdiyarso et al. 2009; Warren et al. 2012). The estimate of total aboveground C stocks of primary forest sites in this present study (204 Mg C/ha) was slightly higher than the forest sampled in the same area by Murdiyarso et al. (2009) who reported mean aboveground C stocks of 143 Mg C/ha. Our results fall within

Table 10.5 Summary of ecosystem C stocks (expressed in Mg C/ha) of peat swamp forests in Indonesia (except Micronesia and Mexico)

Location	Peat depth (cm)	AGC (Mg C/ha)	BGC (Mg C/ha)	Ecosystem C (Mg C/ha)	Sources
Mexico	150	95 ± 16	–	722 ± 64	Adame et al. (2015)
Kosrae, Micronesia	–		165–388		Chimner and Ewel (2005)
Various sites	450	–	2774 ± 595	–	Jaenicke et al. (2008)
South Sumatra		144			Novita (2010)
Jambi	–	141	–		Krisnawati et al. (2012)
Jambi	–	153	–	–	Krisnawati et al. (2012)
Riau	–	131			Istomo (2002)
Riau[a]	–	56			Istomo and Wibisono (2009)
Central Kalimantan	–	157 ± 21			Krisnawati et al. (2014)
Central Kalimantan[a]	–	140 ± 2			Krisnawati et al. (2014)
Central Kalimantan	–	118			Simbolon (2015)
Central Kalimantan	–	214			Simbolon (2015)
West Kalimantan	650–1050	122	3899	4021	Basuki (2017)
Tanjung Puting	46			894 ± 112	Murdiyarso et al. (2009)
Tanjung Puting[a]	27	172 ± 17	361 ± 32	533 ± 49	This study
Tanjung Puting	223	204 ± 32	1566 ± 92	1770 ± 123	This study

[a]secondary forest/degraded forest

range of aboveground C stocks in other Indonesian primary peat swamp forests of 118–214 Mg C/ha (Table 10.5). Contrastingly, a peat swamp forest in Mexico had lower aboveground C stocks (Adame et al. 2015). Although the peat swamp forest in Mexico had two times higher tree density (2469 trees/ha) compared to this study, the area was dominated by only a single species of *Pachira aquatic* (Adame et al. 2015).

Our analysis showed that aboveground carbon contributed about 10% to the total ecosystem C stocks in the primary forests (Table 10.3) which is comparable to results from previous studies in the tropics which reported that 10–13% of the total ecosystem C stocks were allocated above ground (Draper et al. 2014; Murdiyarso et al. 2009; Adame et al. 2015; Bhomia et al.2019). In areas where the peat is deep, a smaller contribution of aboveground carbon to total ecosystem C stocks can be expected. For example, Basuki (2017) reported that aboveground carbon only contributed less than 4% to total ecosystem C stocks in West Kalimantan, Indonesia where the peat thickness was more than 9 m and contained 3,800 Mg C/ha.

Globally, carbon stocks of recently established oil palm plantations 1–5-year old fall between 1.3 and 16.2 Mg C/ha. The oil palm C vegetation stocks in this study were comparable to that reported by Corley et al. (1971) for OP1, and Ng et al. (1986) for OP3 and OP5. Our study reported aboveground carbon stocks from young plantation, hence the estimates are lower than the average carbon stocks in oil palm plantation over a full rotation cycle. The variation of oil palm C stocks among studies may be related to differences in soil condition, management practices (e.g., palm density and fertilizer application) (Syahrinuddin 2005; Henson 2003), or sampling approaches (destructive or allometric). A review analysis from data of aboveground carbon of oil palm plantations in Malaysia and Indonesia showed that oil palm C stocks were strongly correlated with the age of the stands, which was best explained by a power model (Oil palm C stocks $= 3.4713 * \text{age}^{0.8382}$; $r^2 = 0.82$). The ecosystem C stocks in the final stage of oil palm plantations (798 Mg C/ha) were 45% of total ecosystem in primary forests, suggesting that increasing of aboveground C stocks over time is insignificant in oil palm plantations and will never compensate the expense of AGB loss before forest conversion. the time-averaged aboveground C managed by smallholder oil palm plantation is 37.76 Mg C/ha for one cycle in Indonesia where almost half of above ground carbon stocks from oil palm plantation in Peruvian Amazonia reported by Malaga et al (2020) (78.2 ± 2.0 Mg C/ha).

Conversion of peat swamp forest to oil palm plantations resulted in substantial carbon losses to the atmosphere from changes in vegetation C stock. As hypothesized, we found that the conversion of forest to oil palm plantations significantly reduced the vegetation carbon by approximately 93% (Table 10.3 and Fig. 10.8). This was similar to the conclusion of Basuki (2017) who also found more than 99% of plant carbon reduction after conversion of primary forest on deep peats to oil palm plantation in West Kalimantan, Indonesia. Our estimate was higher than the one by Kho and Jepsen (2015), who reported about 81–88% of the plant C stocks were lost due to forest conversion to oil palm plantations in Malaysia.

In a chronosequence scenario, the secondary forest in Tanjung Harapan had the rapidly accumulate aboveground C stocks but not peat. It is important to point out

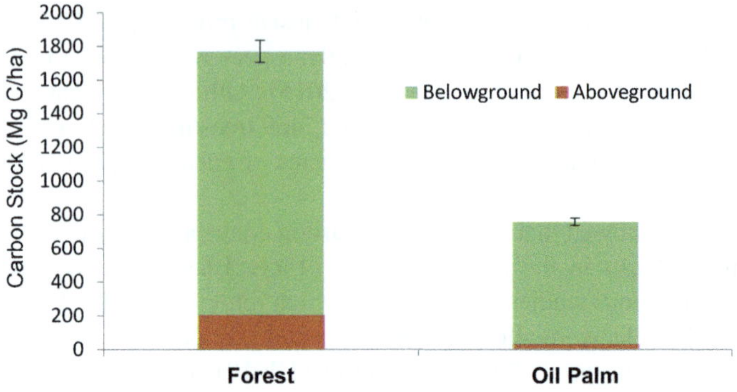

Fig. 10.8 Ecosystem C stocks (Mg C/ha) in the primary peat swamp forest and oil palm plantation sites in Tanjung Puting, Indonesia

that the slow carbon accumulation on peat in secondary forests reveals that peat is a non-renewable source in that it takes centuries to accumulate and is lost quickly during land-use conversion (slow in, fast out). Tanjung Harapan, as secondary forest site, may reach an aboveground carbon stock equal to that in primary forest sites, but not in belowground carbon. Finally, the probabilities of permanent C loss following forest disturbance are not necessarily equal to carbon gain during regrowth period. This suggests that restoration of peat swamp forest is important but conservation is the most effective way to conserve more carbon on peat in the tropics.

Aboveground C stocks of the Tanjung Harapan site were 172 Mg C/ha, which was 84% to the total aboveground C stock of primary forests in the Pesalat and Beguruh sites after at least 30 years disturbance. This sequestration rate of about 5.7 Mg C ha/yr is comparable to results from other secondary forests such as results by Hughes et al. (1999, 2000) who found rates to be about 2.5 Mg C ha/yr in secondary in the heavily deforested sites in Mexico. These results imply that reforested peat swamp forests may provide a new opportunity for climate change mitigation in tropical peatlands. However, there are some critical gaps in our knowledge, and further scientific studies are needed to investigate the role of secondary peat swamp forests in carbon re-accumulation aboveground after long-term disturbance.

Root C stocks in primary and secondary forest sites were estimated to range from 23 to 29 Mg C/ha. These findings are comparable to root carbon stocks from primary peat swamp forests in Riau, Indonesia (33 Mg C/ha) (Istomo 2002). Our results were consistent with root C stocks reported by Krisnawati et al. (2014); 24 and 21 Mg C/ha for primary and secondary peat swamp forests in Central Kalimantan, Indonesia, respectively. Further, root: shoot ratios from peat swamp forests in the present study were 18%, which in agreement with Persch et al. (in prep) who reported 12% of the root: shoot ratios for peat swamp in Jambi. In the oil palm plantation, we found root: shoot ratio was 28%, which was very close to that of 30% for non-forest vegetation evaluated by Fearnside (2000).

We compared our results with Murdiyarso's et al. (2009) study to estimate ecosystem C stocks of peat swamp forests at a landscape-scale from Tanjung Puting area. They reported smaller total ecosystem C stocks (894 ± 112 Mg C/ha) than our estimates (1770 ± 123 Mg C/ha) due to shallower organic soils in their study sites (46 ± 7 cm). We calculated the total ecosystem C stocks of the primary peat swamp forests in Tanjung Puting were 1332 ± 117 Mg C/ha (including 5 sites of primary forest from Murdiyarso's et al. (2009) and 2 sites of primary forest sites from our study). Using the difference of 174 Mg C/ha from the weighted average of 3 sampled sites of primary peat swamp forests and the mean of sampled oil palm plantations, we calculated an estimate of 640 Mg CO_2e/ha are potentially lost due to primary peat swamp forest conversion to oil palm plantations due to vegetation loss. By combining all of the available data on ecosystem C stocks from peat swamp forests in Tanjung Puting, the results consistently show that the losses in C stocks were exceptionally high when peat swamp forest was converted to oil palm plantations.

This result can provide scientific evidence to improve regulations related to peat swamp forest conservation. For example, the Indonesia Presidential Decree No. 32/1990 and the Ministry of Agriculture Decree No. 14/2009 allow the use of Indonesian peatlands if peat thickness is less than 3 m. These policies do not consider carbon emissions from the shallow peat swamp forest that may exceed the emissions from the deep peat swamp forest ecosystem. An improvement of Indonesia's peat map where currently estimated storing about 21.6 Gt C carbon stocks is also needed to reduce uncertainty and national peat carbon data (Warren et al. 2017).

Conclusion

This study provides robust scientific information for considerations of the role of peat swamp forest conservation in climate change mitigation strategies. The large carbon emissions resulting from their conversion clearly show the role of tropical peat swamp forests as the largest pools of soil organic carbon. Thus, avoiding deforestation from peat swamp forests should be regulated as a permanent policy instead of moratorium for a relatively short period of time. This is particularly important to achieve targeted or more ambitious national emission reduction as well as to measure the implementation progress agreed in Paris Agreement. Specifically stated in Article 6 of Paris Agreement that allows Indonesia to use carbon market as an international cooperation for climate change mitigation strategy.

Further, the European Union just enacts the Delegated Act in 2019 which is aimed to phase out palm oil as biofuel feedstock by 2030 since palm oil caused deforestation and conflict with food security (EU 2019). This is could be perceived as a good momentum for Indonesia to increase the research to intensify the yield of oil palm without land expansion and to utilize the degraded land instead of forest. By doing so, dual benefit might be attained, the economic incentives such as REDD+ incentive and eligible to export to EU for biofuel feedstock due to its effort on reducing deforestation and land degradation. In the end, avoiding peat forest conversion into palm oil might hamper the economic growth. Hence the solution is to intensify the yield of oil palm without land expansion or to develop palm oil on degraded dryland. These actions could bring double benefits: the economic incentives such as REDD+ incentive and eligible to export to EU for biofuel feedstock due to its effort on reducing deforestation and land degradation.

In this study, we have quantified ecosystem C stocks of forests and oil palm plantations but have not examined the land-use transitions, including critical initial phases of land conversion entailing the losses associated with slash and burn. Future studies are needed to assess long-term chronosequences of ecosystem carbon dynamics associated with land use land cover change in the tropical peatlands.

Funding This study was possible through funding provided by the United States Agency for International Development (USAID) to the Center for International Forestry Research. This work was carried out as part of the Sustainable Wetland Adaptation and Mitigation Program (SWAMP).

Literature Cited

Adame MF, Santini NS, Tovilla C, Vázquez-Lule A, Castro L, Guevara M (2015) Carbon stocks and soil sequestration rates of tropical riverine wetlands. Biogeosciences 12:3805–3818

BSN [Badan Standardisasi Nasional] (2011) Pengukuran dan penghitungan cadangan karbon-Pengukuran lapangan untuk penaksiran cadangan karbon hutan (ground based forest carbon accounting). Indonesia

Basuki I (2017) Carbon stocks and emission factors of tropical peat swamp forests in respond to land cover changes in West Kalimantan, Indonesia. Dissertation. Oregon State University, Corvallis, Oregon, USA

Bhomia RK, van Lent J, Grandez-Rios JM, Hergoualc'h K, Honorio Coronado EN, Murdiyarso D (2019) Impacts of Mauritia flexuosa degradation on the carbon stocks of freshwater peatlands in the Pastaza-Maranon river basin of the Peruvian Amazon. Mitig Adapt Strateg Glob Change 24(4):645–668

Cairns MA, Brown S, Helmer EH, Baumgardner GA (1997) Root biomass allocation in the world's upland forests. Oecologia 111:1–11

Chao KJ, Phillips OL, Baker TR (2008) Wood density and stocks of coarse woody debris in a northwestern Amazonian landscape. Can J For Res 38:795–805

Chimner RA, Ewel KC (2005) A tropical freshwater wetland: II. Production, decomposition, and peat formation. Wetlands Ecol Manage 13:671–684

Corley RHV, Gray BS, Kee NS (1971) Productivity of the oil palm (Elaeis guineensis Jacq.) in Malaysia. Exp Agric 7:129–136

Davies SJ, Ashton PS (1999) Phenology and fecundity in 11 sympatric pioneer species of Macaranga (Euphorbiaceae) in Borneo. Am J Bot 86:1786–1795

Dommain R, Couwenberg J, Joosten H (2011) Development and carbon sequestration of tropical peat domes in south-east Asia: links to post-glacial sea-level changes and Holocene climate variability. Quatern Sci Rev 30(7–8):999–1010. https://doi.org/10.1016/j.quascirev.2011.01.018

Donato DC, Kauffman JB, Murdiyarso D, Kurnianto S, Stidham M, Kanninen M (2011) Mangroves among the most carbon-rich forests in the tropics. Nat Geosci 4:293–297

Draper FC, Roucoux KH, Lawson IT, Mitchard ET, Coronado ENH, Outi L, Montenegro LT, Sandoval EV, Baker TR (2014) The distribution and amount of carbon in the largest peat land complex in Amazonia. Environ Res Lett 9:124017

EU (2019) Commision Delegated Act No 2055. Supplementing Directive (EU) 2018/2001 as regards the determination of high indirect land-use change- risk feedstock for which a significant expansion of the production area into land with high carbon stock is observed and the certification of low indirect land use change risk biofuels, bioliquids and biomass fuels

Farmer J, Matthews R, Smith P, Langan C, Hergoualc'h K, Verchot L, Smith JU (2014) Comparison of methods for quantifying soil carbon in tropical peats. Geoderma 214:177–183

Fearnside PM (2000) Global warming and tropical land-use change: greenhouse gas emissions from biomass burning, decomposition and soils in forest conversion, shifting cultivation and secondary vegetation. Clim Change 46:115–158

Gumbricht T, Roman-Cuesta RM, Verchot L, Herold M, Wittman F, Householder E, Herold N, Murdiyarso D (2017) An expert system model for mapping tropical wetlands and peatlands reveals South America as the largest contributor. Glob Change Biol 23:3581–3599

Henson IE (2003) The Malaysian national average oil palm: concept and evaluation. Oil Palm Bull 46:15–27

Hergoualc'h K, Carmenta R, Atmadja S, Martius C, Murdiyarso D, Purnomo H (2016) Managing peatlands in Indonesia: challenges and opportunities for local and global communities. CIFOR Infobrief no. 205

Hughes RF, Kauffman JB, Jaramillo VJ (1999) Biomass, carbon, and nutrient dynamics of secondary forests in a humid tropical region of Mexico. Ecology 80:1892–1907

Hughes RF, Kauffman JB, Jaramillo VJ (2000) Ecosystem-scale impacts of deforestation and land use in a humid tropical region of Mexico. Ecol Appl 10:515–527

Istomo (2002) Phosporus and calcium contents in the soil and biomass of peat swamp forest (A Case study at the concession area of PT. Diamond Raya Timber, Bagan Siapi-api, Riau Province). Dissertation. Bogor Agricultural University, Bogor, Indonesia.

Istomo C, Wibisono ITC (2009) Plant diversity and biomass content in relation to wise use of tropical peatland. In: Proceedings of the Bogor symposium and workshop on tropical Peatland management, Indonesia, 14–15 July

Jaenicke J, Rieley JO, Mott C, Kimman P, Siegert F (2008) Determination of the amount of carbon stored in Indonesian peatlands. Geoderma 147:151–158

Jourdan C, Rey H (1997) Modelling and simulation of the architecture and development of the oil-palm (*Elaeis guineensis* Jacq.) root system. Plant Soil 190:235–246

Kauffman JB, Heider C, Cole TG, Dwire KA, Donato D (2011) Ecosystem carbon stocks of Micronesian mangrove forests. Wetlands 31:343–352

Kauffman JB, Donato D (2012) Protocols for the measurement, monitoring and reporting of structure, biomass and carbon stocks in mangrove forests. Working Paper. Center for International Forestry Research (CIFOR), Bogor, Indonesia

Kauffman JB, Arifanti VB, Basuki I, Kurnianto S, Novita N, Murdiyarso M, Donato D, Warren MW (2016) Protocols for the measurement, monitoring, and reporting of structure, biomass and carbon stocks in tropical peat swamp forest. Working Paper. Center for International Forestry Research (CIFOR), Bogor, Indonesia

Kauffman JB, Trejo HH, Garcia MDCJ, Heider C, Contreras WM (2016) Carbon stocks of mangroves and losses arising from their conversion to cattle pastures in the Pantanos de Centla, Mexico. Wetlands Ecol Manage 24:203–216

Khasanah NM, van Noordwijk M, Ningsih H (2015) Aboveground carbon stocks in oil palm plantations and the threshold for carbon-neutral vegetation conversion on mineral soils. Cogent Environ Sci 1:1119964

Kho LK, Jepsen MR (2015) Carbon stock of oil palm plantations and tropical forests in Malaysia: a review. Singap J Trop Geogr 36:249–266

Krisnawati H, Adinugroho WC, Imanuddin R, Hutabarat S (2014) Estimation of forest biomass for quantifying CO_2 emissions in Central Kalimantan: a comprehensive approach in determining forest carbon emission factors. Research and development center for conservation and rehabilitation. Forestry Research and Development Agency, Bogor, Indonesia

Krisnawati HWC, Adinugroho R, Imanuddin (2012) The dynamic of aboveground carbon stock in peat swamp forest. Proceedings of international symposium on wild-fire and carbon management in peat forest in Indonesia, 13–14 Sept. Bogor, Indonesia

Langner A, Siegert F (2009) Spatiotemporal fire occurrence in Borneo over a period of 10 years. Glob Change Biol 15:48–62

Manuri SC, Brack NP, Nugroho K, Hergoualc'h N, Novita H, Dotzauer L, Verchot C, Agung S, Putra E, Widyasari (2014) Tree biomass equations for tropical peat swamp forest ecosystems in Indonesia. Forest Ecol Manag 334:241–253

Miettinen J, Shi C, Liew SC (2015) Land cover distribution in the peatldns of Peninsular Malaysia, Sumatra and Borneo in 2015 with changes since 1990. Glob Ecol Conserv 6:67:68

Ministry of Environment and Forestry (2015) Tanjung Puting national park. https://tanjungputing.dephut.go.id/index.php/tentang-kawasan/letak-dan-luas. Retrieved 6 Jan 2015

Murdiyarso D, Donato D, Kauffman JB, Kurnianto S, Stidham M, Kanninen M (2009) Carbon storage in mangrove and peatland ecosystems: a preliminary account from plots in Indonesia. Working Paper 48. Center for International Forestry Research (CIFOR), Bogor, Indonesia

Málaga N, Hergoualc'h K, Kapp G, Martius C (2020) Variation in vegetation and ecosystem carbon stock due to the conversion of disturbed forest to oil palm plantation in Peruvian Amazonia. Ecosystems 1–19

Ng SK, Thamboo S, de Souza P (1968) Nutrient contents of oil palms in Malaysia. II. Nutrients in vegetative tissues. Malaysia Agric J 46:332–391

Novita N (2010) Above ground biomass of logged-over forests in Merang, South Sumatra. Thesis. Graduate School. Bogor Agricultural University. Bogor, Indonesia

Noway Government (2019) News: https://www.norway.no/en/indonesia/norway-indonesia/news-events/news2/indonesia-reports-reduced-deforestation-triggering-first-carbon-payment-from-norway/

Page SE, Rieley JO, Banks CJ (2011) Global and regional importance of the tropical peatland carbon pool. Glob Change Biol 17:798–818

Persch S, Hergoualc'h K, Hölscher D, Verchot L (In prep) Large above- and below-ground biomass carbon losses from a tropical peat swamp forest transition into an oil palm plantation: a case study in Sumatra, Indonesia

Page SE, Rieley JO, Wüst R (2006) Lowland tropical peatlands of Southeast Asia. Peatlands: evolution and records of environmental and climate changes. Elsevier

Posa MRC, Wijedasa LS, Corlett RT (2011) Biodiversity and conservation of tropical peat swamp forests. Bioscience 61:49–57

Saharjo BH (2007) Shifting cultivation in peatlands. Mitig Adapt Strat Glob Change 12:135–146

Sheehan PJ (1984) Effects on community and ecosystem structure and dynamics. Effects of pollutants at the ecosystem level. Wiley

Simbolon H (2015) Above ground biomass changes in the peat swamp forests of Central Kalimantan https://www.forestday.org/fileadmin/tropical-workshop/Plenary-3/17A_SimbolonH_Above%20ground%20biomass.pdf. Accessed 13 July 2015

Syahrinudin (2005) The potential of oil palm and forest plantations for carbon sequestration on degraded land in Indonesia. Dissertation. University of Goettingen, Germany

Warren MW, Kauffman JB, Murdiyarso M, Anshari G, Hergoualc'h K, Kurnianto S, Purbopuspito J, Gusmayanti E, Afifudin M, Rahajoe J, Alhamd L, Limin S, Iswandi A (2012) A cost-efficient method to assess carbon stocks in tropical peat soil. Biogeosciences 9:4477–4485

Warren M, Hergoualc'h K, Kauffman B, Murdiyarso D, Kolka R (2017) An appraisal of Indonesia's immense peat carbon stock using national peatland maps: uncertainties and potential losses from conversion. Carbon Balance Manage 12:12

Weiss D, Shotyk W, Rieley J, Page S, Gloor M, Reese S, Martinez-Cortizas A (2002) The geochemistry of major and selected trace elements in a forested peat bog, Kalimantan, SE Asia, and its implications for past atmospheric dust deposition. Geochim Cosmochim Acta 66:2307–2323

Wösten JHM, Van Den Berg J, Van Eijk P, Gevers GJM, Giesen WBJT, Hooijer A, Idris A, Leenman PH, Rais DS, Siderius C, Silvius MJ (2006) Interrelationships between hydrology and ecology in fire degraded tropical peat swamp forests. Water Resour Develop 22:157–174
Yu Z, Beilman DW, Frolking S, MacDonald GM, Roulet NT, Camill P, Charman DJ (2011) Peatlands and their role in the global carbon cycle. EOS 92: 97-108.

Chapter 11
Transforming Exploitative Land-Based Economy to Reduce Terrestrial Carbon Stock Loss: The Case of Kalimantan, Indonesia

Chun Sheng Goh and Ser Huay Janice Teresa Lee

Abstract Large-scale land exploitation has been regarded as a quick way to jump-start backward agricultural economies, but it also brought about profound impacts on climate change through terrestrial carbon stock change. Kalimantan is a major site of terrestrial carbon stock loss. Major carbon stock loss happened in the 1970–1980s due to predatory logging activities under Suharto's regime. Since 2000, the conversion of forest to oil palm plantation has become a prominent driver. In 2006–2010, the total carbon stock loss in Kalimantan amounted to 53 Tg CO_2/yr. Boosting upstream productivity of cash crops, mobilising under-utilised low carbon and degraded land resources, creating value for carbon stock (e.g. REDD+) and enhancing resilience to natural and human-made disasters were regarded as the four major strategies proposed for transforming exploitative land-based activities. However, the implementation of each strategy has faced various challenges. The first two strategies with wealth creation as the centre of policymaking may prevent further degradation but are inadequate to repair the previous environmental damage. Similarly, the last two strategies that emphasise restoration have limited contribution to economic growth. To better understand the dynamics of such transformation, careful attention must be paid to the territorial-specific characteristics and on-the-ground realities. This must also include the historical background of land-based development and its continuity.

Keywords Kalimantan · Land-use · Carbon stock · Forest · Bio-economy

C. S. Goh (✉)
Center for Government and International Studies, Harvard University, Asia Center, 1730 Cambridge St, Cambridge, MA 02138, US
e-mail: chunshengg@sunway.edu.my

Jeffrey Sachs Center on Sustainable Development, Sunway University, Malaysia, Malaysia

S. H. J. T. Lee
Asian School of the Environment, College of Science, Nanyang Technological University (NTU), Subang Jaya, Singapore
e-mail: chunshengg@sunway.edu.my

R. Djalante et al. (eds.), *Climate Change Research, Policy and Actions in Indonesia*, Springer Climate, https://doi.org/10.1007/978-3-030-55536-8_11

Introduction

In Indonesia, large-scale land exploitation has been regarded as a quick way to jump-start backward agricultural economies, especially in regions with vast land resources. It has, however, brought about profound impacts on climate change through terrestrial carbon stock change, including deforestation and peat loss. Kalimantan is a typical example of such disasters. For decades, Kalimantan has been suffering from extensive environmental degradation due to rampant timber extraction, uncontrolled fire and rapid oil palm expansion. In terms of terrestrial carbon stock loss, the island contributed roughly 400–700 Tg CO_2/year or 10–17% of global land-use emission in 2000–2010 (Abood et al. 2015; Agus et al. 2013). These have been accompanied by not only transboundary disaster like haze, but also mushrooming social conflicts stemmed from land resource exploitation.

From an economic point of view, over-reliance on export-oriented resource exploitation for fiscal revenues will unavoidably lead to economic bottlenecks. In fact, the regional economy has always been exposed to periodic economic crisis due to fluctuations of commodity prices, such as the recent sharp decline of Crude Palm Oil (CPO), largely preventing them from building their own secondary and tertiary industries in a steady manner. While such exploitative activities have generated quick revenues for the economy, the livelihood of people has indeed been threatened in all aspects from immediate local health risk to long-term global climate change. Furthermore, it has been plagued by poor governance, corruption, ineffective law enforcement and limited growth in skilled labourers. Low population density in Kalimantan also encourages large-scale primary land-based activities and inhibits industrial advancement. These large-scale activities are often accompanied by social conflicts, largely attributed to rapid (sometimes forceful) changes of local lifestyle and inequitable distribution of wealth created from resource exploitation (e.g. Potter 2016; Scheidel et al. 2018). Exploring alternative development strategies for economic growth to prevent the exacerbation of environmental degradation is urgently needed.

In this context, advocates for productivity or so-called 'productivitists' have been pushing for the use of cutting-edge biological knowledge and technological innovation to increase overall productivities instead of furthering unsustainable large-scale land exploitation. These concepts have caught global imagination in producing more agricultural and forestry products while dealing with exacerbated environmental and developmental issues of conventional land-based economies (Bugge et al. 2016). Meanwhile, alternative economic strategies with priorities over conservation have also been proposed by conservationists. These strategies stress the multifunctionality of land-based activities, advocating the needs to observe the bio-capacity of the Earth system when optimising the human use of nature (Marsden and Farioli 2015). These broad ranges of strategies can lead to different impacts, synergies and trade-offs.

Taking Kalimantan as the case study, this chapter first presents a brief history of carbon stock change linking to economic development in the region in

Sect. "Land-Use History in Kalimantan". Then, four strategies proposed for transforming exploitative land-based activities are elaborated in Sect. "Reconciling Development and Conservation: Proposals, Progress and Problems". The chapter ends with a discussion in Sect. "Discussion".

Land-Use History in Kalimantan

Large-scale deforestation in Kalimantan began in the 1980s. Figure 11.1 illustrates the changes in land-use from 1970 to 2010s. The deforestation rate was the steepest in mid-1980s to early 1990s. It has slightly slowed down since then. At the same time, large expansion of oil palm plantations has been undertaken throughout the Kalimantan region. Meanwhile, Fig. 11.2 displays the types of forest conversion. While large areas of forests were converted to oil palm and pulpwood, much larger areas of forests were cleared or destroyed without any follow-up productive activities. These areas were largely deforested in the 1980s. The carbon stock loss resulted from these changes was displayed in Fig. 11.3.

Under the Suharto regime in 1968–1998, land resources were under a highly centralised governance by the central government in Jakarta. It drove large-scale logging throughout the Kalimantan provinces under the claim of national development. At the same time, illegal logging activities by local elites (including local

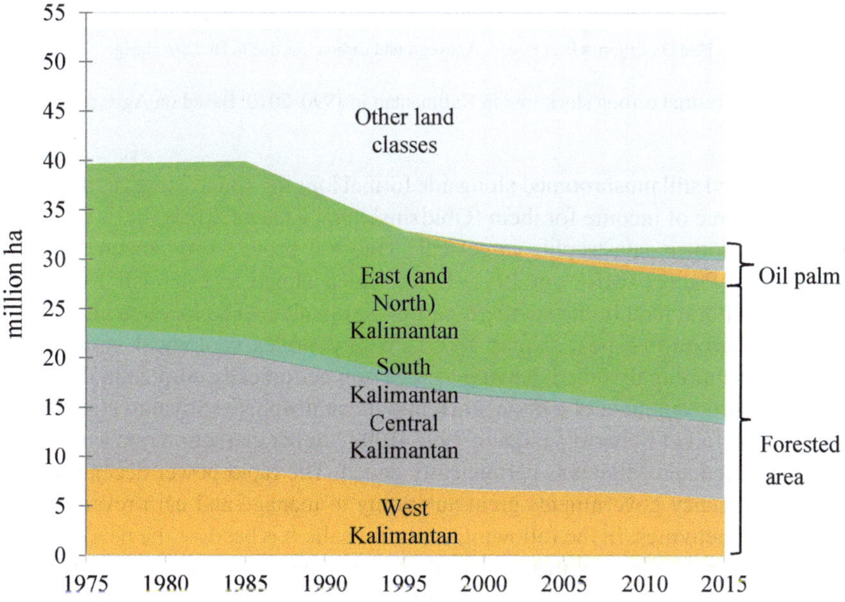

Fig. 11.1 Land cover change in Kalimantan

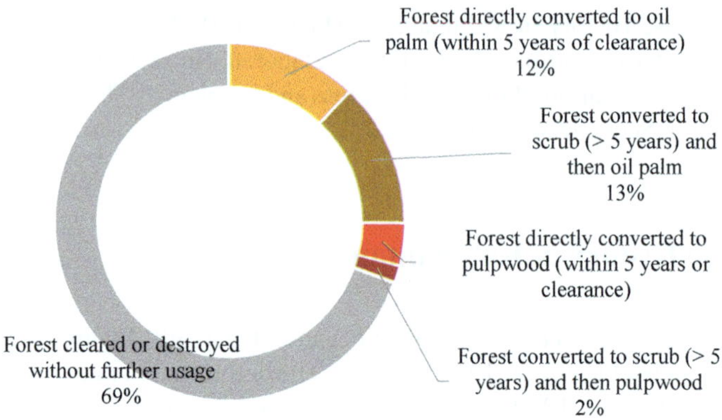

Fig. 11.2 Conversion of forest in Kalimantan in 1973–2016. Data taken from CIFOR Atlas (https://www.cifor.org/map/atlas/). Accessed on May 30, 2019

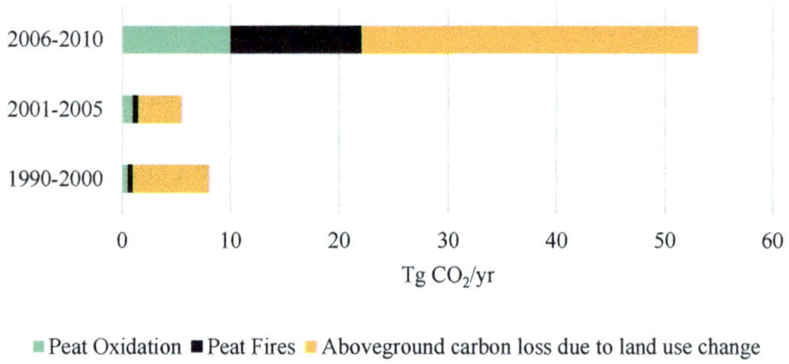

Fig. 11.3 Terrestrial carbon stock loss in Kalimantan in 1990–2010. Based on Agus et al. (2013)

governments) still mushroomed alongside formal logging concessions, as these were a major source of income for them (Obidzinski and Kusters 2015).

Then, another catastrophic large-scale land-use project implemented was the Mega Rice Project (MRP) in 1995–1998, which aimed to establish 1.4 Mha of new agricultural land in the southern part of Kalimantan. The location chosen was, however, carbon-rich peat swamp forests with complex ecological systems. The project has thoroughly failed, leaving a mega ruin periodically erupted in flames and releasing large amounts of carbon stock into the atmosphere (Suyanto et al. 2009).

With the fall of Suharto's reign in 1997–1998, timber extraction rate was dramatically reduced and MRP was permanently ended. The rapid power decentralisation gave the regency governments great autonomy to manage and earn revenues from land-based activities. In the following years, oil palm has become the new economic paradigm. Until today, Kalimantan is still experiencing deforestation albeit in smaller scale and periodic peat fire, although the province has seen some recoveries in certain areas from the past environmental damages with extra-local intervention. Continuous pressure from NGOs has forced both the governments and the big oil palm companies to seek ways for enhancing economic productivity while not causing socio-environmental issues.

Reconciling Development and Conservation: Proposals, Progress and Problems

Transformation of exploitative land-based economies has gained great attention due to its urgency in face of rapid environmental destruction, global warming effects, unavoidable economic bottlenecks and mushrooming social conflicts. Conventional wealth generation from timber extraction followed by large-scale cash crop expansion cannot be continued as a sustainable way of development. Multiple strategies have been proposed, however, with priorities given to environment, economy and society to a different extent and received various responses from stakeholders. The following sections describe the four proposals for transforming the economic activities into more sustainable ones, their progress and problems faced.

Boosting Upstream Productivity of Cash Crops

Pushing for higher production per unit of land has been deemed the most direct measure to reduce further unsustainable expansion (Garnett et al. 2013). The most widely cultivated crop as well as the major cash generator in Kalimantan, oil palm, has always been characterised as the most productive oil crop in terms of land area used. Large industrial players together with public entities have been investing in technology breakthroughs for boosting upstream productivity of oil palm, motivated by the limited land availability in the country for future expansion (Corley and Tinker 2015; Ali Nordin et al. 2017). With substantial financial inputs in upgrading crop breeding and genomics, as well as better agricultural management, the breakthrough in yield was reported from time to time in certain experimental plots—the most recent peak was reported at 12 t/ha/yr (Woittiez et al. 2017).

Yet, the historical data in Borneo shows that there has no substantial improvement in recent years as the CPO yields from matured plantations fluctuate in the range of 2.7–4.4 t/ha/yr in 2011–2017 (Fig. 11.4a). It seems that the yield improvement is stuck within this range. Furthermore, the crop's performance can be quite uneven across the island. The agro-ecological characteristics like soil type and latitude, which vary from place to place, are the major constraints for yield breakthrough (Mulyani and Sarwani 2013). For example, although the economic outcome of oil palm cultivation on marginal soils can be greatly improved from very poor to satisfying with intensive agro-inputs and proper mitigating practices, the yield can hardly

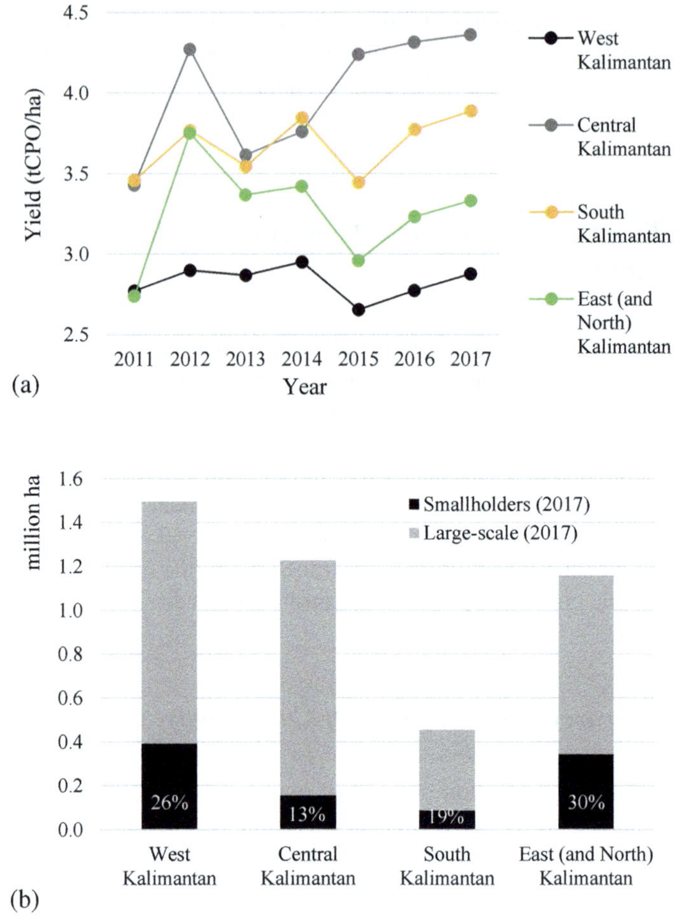

Fig. 11.4 **a** Yield of matured oil palm plantations in Borneo (DG Estate Crops Indonesia 2012–2016); **b** Percentage of oil palm plantation managed by smallholders

grow further as compared to those planted on better soil (Goh et al. 2018). The other reducing factors like droughts, fires and diseases can also be critical to the overall productivity (Woittiez et al. 2017; Mohd Hassan et al. 2018). The lower points in Fig. 11.4a in 2015–2016 are indeed the results of the severe water deficit brought by El Niño in that period (Oettli et al. 2018). There is still no evidence on average yield improvement in face of these agro-ecological and climatic constraints.

Another factor that has often been raised is the underperformance of the smallholders. Currently, this group contributes to 22% of total oil palm area in Kalimantan, ranging from 13–30% in different territories (Fig. 11.4b). Unlike the industrial players, the smallholders usually lack support in terms of agricultural inputs (fertilisers and pest control), machineries and knowledge about best management practices (Ali Nordin et al. 2017). While in their Malaysian neighbours in Borneo,

Sabah and Sarawak have designed several schemes to technically support the small farmers, the situation in Kalimantan is less progressive as the governments generally lack capacity in implementing measures like these (Goh et al. 2018). While lacking direct financial assistance like seedling assistance schemes, measures like establishing small farmers' cooperatives were created among the farmers themselves as observed in West Kalimantan (own fieldwork). Such engagement has shown to be crucial in overcoming the constraints for productivity. The previous 'plasma scheme' designed to assist small farmers by attaching them to large companies in early 2000s was proven to be quite unsuccessful with numerous cases of dispute between both parties reported (Potter 2016; Goh et al. 2018).

Labour shortage is yet another factor that drags down the overall performance (Murphy 2014). Lack of labourers implies sub-optimal management of plantation with longer harvesting round (Sheil et al. 2009; Sayer et al. 2012). Most plantations in Kalimantan relying heavily on extra-local workers from other Indonesian islands for daily operation. With the availability of other opportunities at home due to the booming economic development across Indonesia, the plantation jobs away from home have become less attractive (Selvadurai et al. 2018). Increasing wages and partly substituting with machines are some immediate measures, but this largely depends on the overall profitability, i.e. market price of CPO in the long run.

Considering these tough challenges from both natural and human aspects, monocultural intensification on the upstream to maximise economic productivity does not show high potential as expected from experimental breakthrough in yield. Rather, it seems to be more about combating the multiple emerging problems to prevent yield declining in the future (Rasmussen et al. 2018).

Mobilising Under-Utilised Low Carbon and Degraded Land Resources

Shifting future production away from high carbon and biodiversity land is another direct way to keep pace with demand growth yet not adding pressure on the environment. Land resources with the following criteria, or so-called Under-utilised Low Carbon (ULC) land, may potentially be used: (i) the current economic productivity of the land is insignificant or low compared to its optimal potential and (ii) the level of carbon stock is low so that land utilisation is unlikely to incur additional carbon stock loss and negative ecological impacts (e.g. forest and wetland must be excluded) (Goh et al. 2017). In this direction, possible scenarios of further oil palm expansion on ULC land has been extensively investigated by various studies (Austin et al. 2015; Mosnier et al. 2017; Sumarga and Hein 2016; WRI 2012). Figure 11.5 illustrates the extent of low carbon land areas that do not belong to the high carbon or functional land classes. By 2015, the Kalimantan provinces have significant areas of low carbon land, amounted to roughly 18 Mha.

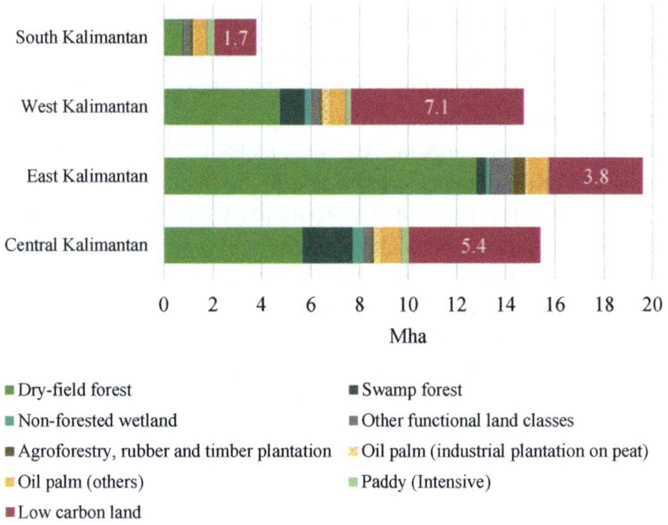

1. Oil palm data were adapted from DG Estate Crops Indonesia (2016)
2. Wetland and peatland data were adapted from Miettinen et al. (2016)
3. Dry-field forest data were adapted from NRE (2015) and MoE (2016)
4. Other functional land classes were adapted from Gunarso et al. 2013 and Goh et al. 2017, assuming insignificant changes in area from 2010-2015

Fig. 11.5 Extend of low carbon land in Borneo by territories in 2015

At the first sight, it seems that the region still has massive areas of ULC land that might be potentially used for production if only land cover and carbon stock are put into consideration. However, as revealed by Goh et al. (2017), such physical area estimations need to be further evaluated from various perspectives, such as land suitability and land-use intensity. This can considerably reduce the actual area of land that can be practically turned into productive land. Currently, accurate spatial data of agro-ecological characteristics is still largely missing (Goh et al. 2017). Furthermore, mobilisation of ULC land resources is not that straightforward. Underlying socio-economic dynamics, such as labour availability, can impose great limitation to the actual use (Goh et al. 2018). These are further intertwined with subtle institutional and cultural elements, such as fragmentation and uncertainties of land ownership that can inhibit productive use of land (Sklenicka 2016). A proposal about swapping such ULC land with high carbon land in oil palm concession was made but it has not been successful due to immense legal complexity in Indonesia (Rosenbarger et al. 2013).

It is questionable for the further expansion of monoculture to max out productivity considering the aforementioned constraints. Generally, oil palm cultivation has been regarded the best economic opportunities with no comparable competitor, and thus is still strongly promoted by many local governments (Goh et al. 2018). This is, however, shadowed by the past experience of improper planning throughout Kalimantan—large tracts of 'idle' land (which may still be covered by forests) to private companies which could involve seizure of land from native people under the name of development (Cramb 2016). Forest encroachment may also happen in the absence of effective forest governance, as some of these ULC areas are located just next to the remaining forests. Also, the fragmented ULC areas may actually contribute significantly to connectivity of forest patches and thus intensive use may cause impacts on important ecosystem services (Evans et al. 2017).

Despite the multiple risks and tough challenges, activating ULC land for production is still a better option than converting forest or other high carbon lands for production especially in face of growing demand for food and materials. In addition, proper management of these land may help to avoid further land degradation and replenish lost carbon stock. Reviving abandoned timber plantation for both production and restoration is a good example. However, there is currently no specific incentives for activating ULC land resources in sustainable manners.

Creating Value for Carbon Stock: The Case of REDD+

The concept of capitalising Ecosystem Services (ES) is not new but exists for decades already. It advocates incorporating ecosystem services with contemporary market economy through creating values for 'nature' that are compatible with economic accounting practices (Missemer 2018). The basis is to link 'natural capital' to human benefits in terms of provisioning, regulating, supporting and cultural services (Millennium Ecosystem Assessment 2005). Quantification of ES is deemed crucial for this strategy as it provides manageable attributes of natural capital stocks for interventions to take place (Maseyk et al. 2017). However, it has been very challenging especially for the last two categories (Gunton et al. 2017).

The concept was experimented since early 2000s in Kalimantan with carbon stock which is among the most conceivable and measurable components and can be banked on carbon trading. Reducing Emissions from Deforestation and Forest Degradation (REDD+) is the major programme widely promoted in Kalimantan (see Fig. 11.6 for the locations of the sites) (Sills et al. 2014). It aims to raise billions of US dollars from international donors to compensate the stakeholders who avoid degradation and deforestation for the opportunity costs of converting these lands for agricultural production. The Kalimantan Forests and Climate Partnership (KFCP) that covers 120,000 ha of intervention area in Central Kalimantan was regarded the most established REDD+ programme among the others in the country (Atmadja et al. 2014). The programme was designed to reduce ongoing GHG emissions from peatland degradation using various measures (Aldhous 2004). The programme also includes the four other projects throughout Kalimantan: Katingan Peatland Restoration and Conservation Project, Rimba Raya Biodiversity Reserve Project, Ketapang Community Carbon Pools (KCCP) and the Berau Forest Carbon Program (Anandi et al. 2014; Indriatmoko et al. 2014a, b; Intarini et al. 2014).

These projects suffer from many challenges. First, there have been stiff competitions between productive-use and conservation of land which were demonstrated

1. Ketapang Community Carbon Pools (KCCP)
2. Rimba Raya Biodiversity Reserve Project
3. Katingan Peatland Restoration and Conservation Project
4. Kalimantan Forests and Climate Partnership (KFCP)
5. Berau Forest Carbon Program

*Google Map was used as the base map

Fig. 11.6 REDD+ sites in Kalimantan

by the polarised views between villagers and villages and the inconsistencies in government policies as seen in the case of Ketapang and Berau (Anandi et al. 2014; Intarini et al. 2014). The immediately following question would be who has the right to benefit from the 'sales' of 'ecosystem services' (in this case carbon credits)? The Rimba Raya case that was dominated by a foreign company based in Hong Kong raises concerns about the real implications of such schemes for local people (Indriatmoko et al. 2014b). Even for the other programmes, the actual implementations were troubled by uneven distribution of benefits among the local people (Joshi et al. 2010; Howson and Kindon 2015). Third, tenure uncertainties are another long-standing issue to be solved. The Indonesian government itself has a lot of concerns in giving out license of managing a large area of land for such projects due to legal uncertainties and extreme difficulties in managing potential land conflicts, not to mentioned the underlying political complexity (Indriatmoko et al. 2014a). Last but not least, monitoring of carbon stock would be a difficult technical challenge. In addition to various arguments about verification methods (e.g. carbon leakage issue), difficulties in understanding the system also cause hesitance of officials to get involved (Thompson 2018).

With these multiple challenges, creating values for carbon and ES may not truly benefit the livelihoods of local people at least in short term (Suyanto et al. 2009). Either top-down prescriptions or bottom-up approaches have received objections from local stakeholders. Furthermore, it is also unclear how to sustain large-scale payment in long term once the local communities begin to rely heavily on this for their livelihoods—can developed countries continue to pay 'sufficiently' for the ES in Borneo for several decades to come? Naturally, long-term economic development cannot rely solely on voluntary payment from the other countries.

Enhancing Resilience to Natural and Human-Made Disasters

Disasters like fire and haze will inflict enormous damage to environment, economy and society. Just taking the example of transboundary haze caused by uncontrolled land fire in Kalimantan and Sumatera in 2015—the amount of CO_2 released into the atmosphere in just one month is equivalent to the annual emission of Germany and the direct economic loss is estimated to be USD 16 billion for Indonesia alone (not yet included are losses for Malaysia and Singapore as well as health damages due to transboundary haze). These unwanted consequences, with fire and drought as the main themes for Borneo, are likely to become more pervasive threats due to drastic alteration of the earth system. In recent years, resilience to these disturbances has been proposed to be an important economic indicator in addition to productivity due to the numerous evidences of impacts from harmful disasters on existing production system (Walker et al. 2010).

A landscape that can supply a broad range of products and ecosystem services (with rich biodiversity) is deemed less vulnerable to external shocks. It was proven that the decline of biodiversity as a result of excessive land exploitation in the past decades has substantially crippled the agro-ecological resilience (Newbold et al. 2016). A more holistic landscape management, i.e. forging synergies between different land-use and services throughout the landscape, may avoid unwanted environmental impacts and also create opportunities for restoration (Sayer et al. 2013). Practically for Kalimantan, this can include for example re-activation of abandoned monoculture timber plantations by introducing new species in combination with natural forest regeneration depending on the suitability (Brockerhoff et al. 2008; Crouzeilles et al. 2017). Having said that, it is not entirely clear how such a landscape approach can be applied to meet different needs optimally. Past experience shows that its effectiveness is difficult to be monitored and measured (Sayer et al. 2017).

Some had advocated that reforming of the current system with decentralisation and empowering of indigenous and rural communities may significantly help anchoring resilience (Marshall 2009). Important to recognise is that different stakeholders hold different views on land-use that can vary widely even within a smaller administrative territory like districts and villages (Goh et al. 2018). While value judgement is unavoidable, a 'balanced' configuration that satisfies the majority may just not be truly resilient. In certain occasions it can be more effective with a stronger move from the higher-level governments, such as the strict enforcement of 'no burning' policies to prevent the use of fire for land clearing in Kalimantan. This is of vital importance to avoid the exacerbation of extreme drought events. Although the degree of compliance is unclear, decline in fire activities was observed in the recent years with efforts from multiple parties (Lambin et al. 2018; Noojipady et al. 2017).

The more crucial consideration would be the capability of local actors and their understanding in the diverse mechanisms in establishing resilient agro-ecological systems. On many occasions, the stakeholders simply do not have enough financial means to equip themselves with proper tools in countering unforeseen changes. For example, key infrastructure like irrigation and firefighting systems are still largely missing throughout Kalimantan (Goh et al. 2018). In addition, leveraging available resources and local strengths is also necessary. Local communities may also possess traditional ecological knowledge which they learnt from decades of human-environmental interactions (van Oudenhoven et al. 2011). Incorporating these can further enrich the 'knowledge bank' or 'tool box' to improve the agro-ecological resilience.

To make this strategy works, innovative ways of managing landscape are needed. While investments in both people capacity and key infrastructure are necessary, effective governance would be the key to steer the transformation as it involves cross-sectorial and cross-scale coordination. Despite the emergence of multitudinous theoretical and conceptual frameworks, on-ground implementation remains an arduous challenge for Borneo, especially in Kalimantan. Evidence from many parts of the world shows that implementing such strategies heavily depends on external funding if not other forms of strong interventions (Ortiz et al. 2018).

Discussion

The conventional exploitative land-based economies are facing a predicament: how to maintain economic growth not only without causing further environmental impacts but also repairing the damage done in the past. This overview identified and discussed four key transformative strategies with elaboration of the experience in Kalimantan as well as the associated opportunities and challenges. The first two strategies with wealth creation as the centre of policymaking may prevent further degradation but are inadequate to repair the previous environmental damage. Similarly, the last two strategies that emphasise restoration have limited contribution to economic growth.

The interconnected nature of economic productivity and conservation means that no single strategy is a perfect solution, although some can be more practical and effective than the others in different places and periods of time, or more or less acceptable by different stakeholders. These inadequacies demand optimally combining the different strategies to reach both ends. On the positive side, these strategies may create synergies between each other. For example, shifting agricultural expansion onto non-forested land with alternative incentives can be viewed as a big synergy to the REDD+ programme. However, due to the fundamental differences in nature, it is difficult to define 'optimal' combinations as not all outcomes can be easily measured and linked to the policy actions for analytical evaluation. For example, the attempt to create values for nature has found to be very challenging even for carbon stock which is relatively easy to measure compared to other functions like biodiversity. In this sense, the choice of strategies may rely more on value judgement.

The existence of multiple stakeholders with different interests, values and education level means that the 'optimal' set of strategies would more likely be a result of political negotiations that can be practically implemented. One vivid example has been the proposal to divert agricultural expansion onto ULC land. Exploiting ULC land for productivity may seem acceptable purely from an environmental point of view, especially when the governments are under strong economic pressure. However, it may significantly alter local socio-cultural dynamics through, e.g. lifestyle changes and migration, and potentially victimised certain groups of people. For example, a large part of local communities may be left out from the discussion about large-scale investments in converting ULC land to productive plantations as shown in some cases in the past. Some can be very eager to integrate into market economies, but some may opt to go for traditional self-sufficient farming. Crucially, people change their minds from time to time. For example, the author's recent visit to West Kalimantan revealed that some villagers who were strongly against oil palm a decade ago are now planting oil palm themselves, while some former oil palm smallholders who failed in planting have either switched to other crops or went for off-farm jobs.

It is thus vital to recognise that there are no straightforward solutions considering these profound differences in both spatial and temporal dimensions. To better understand the dynamics of such transformation, careful attention must be paid to the territorial-specific characteristics and on-the-ground realities. This must also include the historical background of land-based development and its continuity. It is necessary to recognise the differences in values that deeply rooted in the mindsets of the people, as well as the dynamic socio-economic changes. These 'human factors' require in-depth understanding of a territory beyond just physical and monetary analyses when one intends to understand how the transformation of land-based economies was driven. While generalised knowledge can help to layout the framework, in-depth area studies would play a much larger role when it comes to actual implementation and execution of transformative strategies. That would require a more holistic development thinking beyond just 'land' but the entire territory.

(Note that this chapter is a summary of the other two full research papers by the author submitted elsewhere.)

References

Abood S, Lee JSH, Burivalova Z, Garcia-Ulloa J, Koh LP (2015) Relative contributions of the logging, fiber, oil palm, and mining industries to forest loss in Indonesia. Conserv Lett 8(1):58–67

Agus F, Gunarso P, Sahardjo BH, Harris N, van Noordwijk M, Killeen TJ (2013) Historical CO_2 emissions from land use and land use change from the oil palm industry in Indonesia, Malaysia and Papua New Guinea. In: Killeen TJ, Goon J (eds) Reports from the technical panels of the second RSPO GHG working group, roundtable for sustainable Palm Oil, Kuala Lumpur, pp 65–88

Aldhous P (2004) Land remediation: Borneo is burning. Nature 432:144–146. https://doi.org/10.1038/432144a

Ali Nordin AZ, Ahmad SM, Sahidan AS; Abdullah N, Ain H (2017) An economic study on technical efficiency among independent oil palm smallholders in Sabah and Sarawak. Oil Palm Indus Econ J 17(1):16–31. Available at: https://palmoilis.mpob.gov.my/publications/OPIEJ/opiejv17n1-ali.pdf. Accessed 5 Aug 2018

Anandi CAM, Resosudarmo IDP, Komalasari M, Ekaputri AD, Intarini DY (2014) TNC's initiative within the Berau Forest Carbon Program, East Kalimantan, Indonesia. In: Sills EO, Atmadja SS, de Sassi C, Duchelle AE, Kweka DL, Resosudarmo IAP, Sunderlin WD (eds) REDD+ on the ground: a case book of subnational initiatives across the globe. CIFOR. Bogor, Indonesia. Available at: https://www.cifor.org/redd-case-book/. Accessed 29 Mar 2018

Atmadja SS, Indriatmoko Y, Utomo NA, Komalasari M and Ekaputri AD (2014) Chapter 17. Kalimantan Forests and Climate Partnership, Central Kalimantan, Indonesia. In: Sills EO, Atmadja SS, de Sassi C, Duchelle AE, Kweka DL, Resosudarmo IAP, Sunderlin WD (eds) REDD+ on the ground: a case book of subnational initiatives across the globe. CIFOR. Bogor, Indonesia. Available at: https://www.cifor.org/redd-case-book/. Accessed 29 Mar 2018

Austin KG, Kasibhatla PS, Urban DL, Stolle F, Vincent J (2015) Reconciling oil palm expansion and climate change mitigation in Kalimantan Indonesia. PLoS ONE 10(5):e0127963

Brockerhoff EG, Jactel H, Parrotta JA, Quine CP, Sayer J (2008) Plantation forests and biodiversity: oxymoron or opportunity? Biodivers Conserv 17(5):925–951

Bugge MM, Hansen T, Klitkou A (2016) What is the bioeconomy? A review of the literature. Sustainability 8(7):691

Corley RHV, Tinker PB (2015) The oil palm, 5th Edn. Wiley. ISBN: 978–1–405–18939–2

Cramb RA (2016) Political economy of large-scale oil palm development in Sarawak. In: Cramb R, McCarthy JF (eds) The oil palm complex: smallholders, agribusiness and the state in Indonesia and Malaysia. NUS Press, Singapore, pp 189–246

Crouzeilles R, Ferreira MS, Chazdon RL, Lindenmayer DB, Sansevero JB, Monteiro L et al (2017) Ecological restoration success is higher for natural regeneration than for active restoration in tropical forests. Sci Adv 3(11):e1701345

DG Estate Crops Indonesia (2012) Komoditas Kelapa Sawit di Indonesia 2011–2013. Available at: https://ditjenbun.pertanian.go.id/tinymcpuk/gambar/file/statistik/2013/SAWIT.pdf. Accessed 9 Apr 2018

DG Estate Crops Indonesia (2013) Komoditas Kelapa Sawit di Indonesia 2012–2014. Available at: https://ditjenbun.pertanian.go.id/tinymcpuk/gambar/file/statistik/2014/SAWIT%20Ok.pdf. Accessed 9 Apr 2018

DG Estate Crops Indonesia (2014) Komoditas Kelapa Sawit di Indonesia 2013–2015. Available at: https://ditjenbun.pertanian.go.id/tinymcpuk/gambar/file/statistik/2015/SAWIT%202013%20-2015.pdf. Accessed 9 Apr 2018

DG Estate Crops Indonesia (2015) Komoditas Kelapa Sawit di Indonesia 2014–2016. Available at: https://ditjenbun.pertanian.go.id/tinymcpuk/gambar/file/statistik/2016/SAWIT%202014-2016.pdf. Accessed 9 Apr 2018

DG Estate Crops Indonesia (2016) Komoditas Kelapa Sawit di Indonesia 2015–2017. Available at: https://ditjenbun.pertanian.go.id/tinymcpuk/gambar/file/statistik/2017/Kelapa-Sawit-2015-2017.pdf. Accessed 9 Apr 2018

Evans LJ, Goossens B, Asner GP (2017) Underproductive agriculture aids connectivity in tropical forests. For Ecol Manage 401:159–165

Garnett T, Appleby MC, Balmford A, Bateman IJ, Benton TG, Bloomer P et al (2013) Sustainable intensification in agriculture: premises and policies. Science 341(6141):33–34

Goh CS, Wicke B, Potter L, Faaij APC, Zoomers A, Junginger M (2017) Exploring under-utilised low carbon land resources from multiple perspectives: case studies on regencies in Kalimantan. Land Use Policy 60:150–168

Goh CS, Junginger M, Potter L, Faaij A, Wicke B (2018) Identifying opportunities and barriers to mobilising under-utilised low carbon land resources: a case study on Kalimantan. Land Use Policy 70:198–211

Gunarso P, Hartoyo ME, Agus F and Killeen TJ (2013) Oil palm and land use change in Indonesia, Malaysia and Papua New Guinea. In: Killeen TJ, Goon J (eds) Reports from the technical panels of the second RSPO GHG working group, roundtable for sustainable Palm Oil, Kuala Lumpur, pp 29–64

Gunton RM, van Asperen EN, Basden A, Bookless D, Araya Y, Hanson DR et al (2017) Beyond ecosystem services: valuing the invaluable. Trends Ecol Evol 32(4):249–257

Howson P, Kindon S (2015) Analysing access to the local REDD+ benefits of Sungai Lamandau, Central Kalimantan Indonesia. Asia Pacific Viewpoint 56(1):96–110

Indriatmoko Y, Atmadja SS, Utomo NA, Ekaputri AD, Komalasari M (2014a) Katingan Peatland restoration and conservation project, Central Kalimantan, Indonesia. In: Sills EO, Atmadja SS, de Sassi C, Duchelle AE, Kweka DL, Resosudarmo IAP, Sunderlin WD (eds) REDD+ on the ground: a case book of subnational initiatives across the globe. CIFOR. Bogor, Indonesia. Available at: https://www.cifor.org/redd-case-book/. Accessed 29 Mar 2018

Indriatmoko Y, Atmadja SS, Ekaputri AD, Komalasari M (2014b) Rimba Raya biodiversity reserve project, Central Kalimantan, Indonesia. In: Sills EO, Atmadja SS, de Sassi C, Duchelle AE, Kweka DL, Resosudarmo IAP, Sunderlin WD (eds) REDD+ on the ground: a case book of subnational initiatives across the globe. CIFOR. Bogor, Indonesia. Available at: https://www.cifor.org/redd-case-book/. Accessed 29 Mar 2018

Intarini DY, Resosudarmo IAP, Komalasari M, Ekaputri AD, Agustavia M (2014) Ketapang community Carbon pools, West Kalimantan, Indonesia. In: Sills EO, Atmadja SS, de Sassi C, Duchelle AE, Kweka DL, Resosudarmo IAP, Sunderlin WD (eds) REDD+ on the ground: a case book of subnational initiatives across the globe. CIFOR. Bogor, Indonesia. Available at: https://www.cifor.org/redd-case-book/. Accessed 29 Mar 2018

Joshi L, Janudianto VN, Pradhan U (2010) Investment in carbon stocks in the eastern buffer zone of Lamandau River Wildlife Reserve, Central Kalimantan province, Indonesia: a REDD+ feasibility study. Project Report. World Agroforestry Centre (ICRAF) Southeast Asia Regional Office, Bogor. Available at: https://theredddesk.org/sites/default/files/lamandau_report_final_version.pdf. Accessed 2 Apr 2018

Lambin EF, Gibbs HK, Heilmayr R, Carlson KM, Fleck LC, Garrett RD, de Waroux YL, McDermott CL, McLaughlin D, Newton P, Nolte C (2018) The role of supply-chain initiatives in reducing deforestation. Nature Climate Change 29:1

Marsden T, Farioli F (2015) Natural powers: from the bio-economy to the eco-economy and sustainable place-making. Sustain. Sci. 10:331–344

Marshall GR (2009) Polycentricity, reciprocity, and farmer adoption of conservation practices under community-based governance. Ecol Econ 68(5):1507–1520

Maseyk FJ, Mackay AD, Possingham HP, Dominati EJ, Buckley YM (2017) Managing natural capital stocks for the provision of ecosystem services. Conservation Letters 10(2):211–220

Missemer A (2018) Natural Capital as an Economic Concept, History and Contemporary Issues. Ecol Econ 143:90–96

Mohd Hassan NA, Ahamad SM, Balu N (2018) Relationship between severe El Niño Phenomena and Malaysia's Palm Oil production a VECM approach. Oil Palm Indus Econ J 18(1):1–8. Available at: https://palmoilis.mpob.gov.my/publications/OPIEJ/opiejv18n1-ain.pdf. Accessed 5 Aug 2018

Mosnier A, Boere E, Reumann A, Yowargana P, Pirker J, Havlík P, Pacheco P (2017) Palm oil and likely futures: assessing the potential impacts of zero deforestation commitments and a moratorium on large-scale oil palm plantations in Indonesia. CIFOR Infobrief 177. CIFOR. Bogor. Indonesia. Available at: https://www.cifor.org/library/6468/palm-oil-and-likely-futures-assessing-the-potential-impacts-of-zero-deforestation-commitments-and-a-moratorium-on-large-scale-oil-palm-plantations-in-indonesia/. Accessed 29 Mar 2018

Mulyani A, Sarwani M (2013) The characteristic and potential of sub optimal land for agricultural development in Indonesia. Prosiding Seminar Nasional Lahan Suboptimal Intensifikasi Pengelolaan Lahan Suboptimal Dalam Rangka Mendukung Kemandirian Pangan Nasional. Palembang, Indonesia, pp 20–21

Murphy DJ (2014) The future of oil palm as a major global crop: opportunities and challenges. J Oil Palm Res 26(1):1–24

Newbold T, Hudson LN, Arnell AP, Contu S, De Palma A, Ferrier S et al (2016) Has land use pushed terrestrial biodiversity beyond the planetary boundary? A global assessment. Science 353(6296):288–291

Noojipady P, Morton DC, Schroeder W, Carlson KM, Huang C, Gibbs HK et al (2017) Managing fire risk during drought: the influence of certification and El Niño on fire-driven forest conversion for oil palm in Southeast Asia. Earth Syst Dyn 8(3):749

Obidzinski K, Kusters K (2015) Formalizing the logging sector in Indonesia: Historical dynamics and lessons for current policy initiatives. Soc Nat Res 28(5):530–542

Oettli P, Behera SK, Yamagata T (2018) Climate based predictability of Oil Palm tree yield in Malaysia. Sci Rep 8(1):2271

Ortiz W, Vilsmaier U, Osorio ÁA (2018) The diffusion of sustainable family farming practices in Colombia: an emerging sociotechnical niche? Sustain Sci 13(3):829–847

Potter L (2016) How can the people's sovereignty be achieved in the oil palm sector? Is the plantation model shifting in favour of smallholders. In: McCarthy J, Robinson K (eds) Land and development in Indonesia: searching for the People's Sovereignty. ISEAS–Yusof Ishak Institute, Singapore, pp 293–320

Millennium Ecosystem Assessment (2005) Ecosystems and Human Well-being: Synthesis. Island Press, Washington, DC. Available at: https://www.millenniumassessment.org/documents/document.356.aspx.pdf. Accessed 7 Aug 2018

Rasmussen LV, Coolsaet B, Martin A, Mertz O, Pascual U, Corbera E et al (2018) Social-ecological outcomes of agricultural intensification. Nat Sustain 1(6):275

Rosenbarger A, Gingold B, Prasodjo R, Alisjahbana A, Putraditama A, Tresya D (2013) How to change legal land use classifications to support more sustainable Palm Oil production in Indonesia. World Resource Institute. Available at: https://www.wri.org/publication/how-to-change-legal-land-use-classifications-to-support-sustainable-palm-oil-in-indonesia. Accessed 22 Aug 2018

Sayer J, Ghazoul J, Nelson P, Boedhihartono AK (2012) Oil palm expansion transforms tropical landscapes and livelihoods. Global Food Sec 1(2):114–119

Sayer J, Sunderland T, Ghazoul J, Pfund JL, Sheil D, Meijaard E (2013) Ten principles for a landscape approach to reconciling agriculture, conservation, and other competing land uses. Proc Natl Acad Sci 110(21):8349–8356

Sayer J, Margules C, Boedhihartono AK, Sunderland T, Langston JD, Reed J et al (2017) Measuring the effectiveness of landscape approaches to conservation and development. Sustain Sci 12(3):465–476

Scheidel A, Temper L, Demaria F, Martínez-Alier J (2018) Ecological distribution conflicts as forces for sustainability: an overview and conceptual framework. Sustain Sci 13(3):585–598

Selvadurai S, Lyndon N, Ilelmi M, Rahim A (2018) Economic problems faced by Oil Palm smallholders and its solution: the case of Malaysia. Int Inf Inst (Tokyo). Inf 21(6):1789–1800

Sheil D, Casson A, Meijaard E, van Noordwijk M, Gaskell J, Sunderland-Groves J et al. (2009) The impacts and opportunities of oil palm in Southeast Asia: what do we know and what do we need to know? Working paper no. 51. Center for International Forestry Research. Bogor, Indonesia. Available at: https://www.cifor.org/library/2792/the-impacts-and-opportunities-of-oil-palm-in-southeast-asia-what-do-we-know-and-what-do-we-need-to-know/. Accessed 15 Aug 2018

Sills EO, Atmadja SS, de Sassi C, Duchelle AE, Kweka DL, Resosudarmo IAP, Sunderlin WD (2014) REDD+ on the ground: a case book of subnational initiatives across the globe. CIFOR. Bogor, Indonesia. Available at: https://www.cifor.org/redd-case-book/. Accessed 29 Mar 2018

Sklenicka P (2016) Classification of farmland ownership fragmentation as a cause of land degradation: a review on typology, consequences, and remedies. Land Use Policy 57:694–701

Sumarga E, Hein L (2016) Benefits and costs of oil palm expansion in Central Kalimantan, Indonesia, under different policy scenarios. Reg Environ Change 16:1011–1021

Suyanto S, Sardi I, Buana Y, van Noordwijk M (2009) Analysis of local livelihoods from past to present in the Central Kalimantan ex-mega rice project area. Working Paper 94. World Agroforestry Centre. Bogor, Indonesia. Available at: https://www.worldagroforestry.org/publication/analysis-local-livelihoods-past-present-central-kalimantan-ex-mega-rice-project-area. Accessed 29 Mar 2018

Thompson BS (2018) Institutional challenges for corporate participation in payments for ecosystem services (PES): insights from Southeast Asia. Sustain Sci 13:919–935

van Oudenhoven FJ, Mijatović D, Eyzaguirre PB (2011) Social-ecological indicators of resilience in agrarian and natural landscapes. Manage Environ Quality Int J 22(2):154–173

Walker B, Sayer J, Andrew NL, Campbell B (2010) Should enhanced resilience be an objective of natural resource management research for developing countries? Crop Sci 50(Supplement_1):S-10

Woittiez LS, van Wijk MT, Slingerland M, van Noordwijk M, Giller KE (2017) Yield gaps in oil palm: a quantitative review of contributing factors. Eur J Agron 83:57–77

WRI (2012) WRI Suitability Mapper. World Resource Institute. Available at: https://www.wri.org/resources/maps/suitability-mapper. Accessed on 10 Mar 2015

Chapter 12
Innovative Financing for Peatland Restoration in Indonesia

Agus P. Sari, Alue Dohong, and Budi Wardhana

Abstract Indonesia harbors the largest area of tropical peatland globally, with a total area of approximately 15 million hectares (Mha) and store about 45–65 G_tC which is strategically important for controlling global climate change. However, Indonesia's peatland experiences an alarming rate of degradation in recent decades due to anthropogenic activities such as logging, conversion to industrial timber and agriculture, fires, and drainage associated with these land use changes. Peatland degradation imposes socio-economic, environmental consequences, and health costs to the country. In order to reverse the peatlands degradation pace, President Joko Widodo enacted a Presidential Regulation Number 1 of 2016 concerning the establishment of Peatland Restoration Agency or Badan Restorasi Gambut (BRG). The newly ad hoc governmental agency is tasked to coordinate and facilitate stakeholders in implementing peatlands restoration of 2.0 Mha target in seven provinces up to 2020. The 2.0 Mha target is then revised and updated based on certain criteria to a new target of 2.49 Mha. Of the 2.49 Mha target, about 1.40 Mha is located within private concessionaire licensed areas; meanwhile, the remaining 1.09 Mha is allocated in non-private licensed areas. A substantial budget is needed to achieve restoration target aforementioned. The currently available funding from both state budget and bilateral donor countries remains inadequate to fill in the funding needed, which is an estimate of total US$ 1.70 billion for the whole 2.49 Mha target. Hence, strategic and innovative funding sources need to be created and tapped in order to address the funding gap. This chapter aims to present business cases for peatland restoration and to propose a potential innovative financial instrument to finance the restoration activities in Indonesia. The creation of an investment instrument is necessary by means of structuring a peat bond where a combination of cash and carbon asset returns can be gained by investors.

A. P. Sari
Landscape Indonesia, Jakarta, Republic of Indonesia

A. Dohong (✉)
Indonesian Ministry of Environment and Forestry, Jakarta, Republic of Indonesia
e-mail: aluedohong@gmail.com

B. Wardhana
Peatland Restoration Agency, Jakarta, Republic of Indonesia

R. Djalante et al. (eds.), *Climate Change Research, Policy and Actions in Indonesia*, Springer Climate, https://doi.org/10.1007/978-3-030-55536-8_12

Fig. 12.1 Map of wetlands and peatlands in Indonesia. Adopted from Sustainable Wetlands Adaptation and Mitigation Program (SWAMP), Global Wetlands, public-domain map of wetlands and peatlands, https://www.cifor.org/global-wetlands

Keywords Peatlands · Peatlands restoration · Restoration cost · Peat bonds

Introduction

Indonesia hosts the largest area of tropical peatlands in the world, occupying a vast total area of nearly 15 million hectares (Mha) (Wahyunto et al. 2016; Dohong 2017),[1] with a carbon content of approximately between 45 and 65 G_tC (Jeanicke et al. 2008), of which around 20–30 G_tC are stored at below ground (Agus et al. 2011). Destruction and degradation of peatlands have become a major concern in Indonesia and elsewhere in the world. Reduction of forest covers and uncontrolled construction of drainage canals in peatlands has led to peat over-drain and in turn will pose significant risks such as peat oxidation, subsidence, forest, and peat fires that produce a massive release of carbon dioxide (CO_2) and other climate change inducing greenhouse gases to the atmosphere (Agus 2015) (Fig. 12.1).

The extraction and felling of commercial timbers from peat swamp forests for both export and domestic consumptions in the 1970s–1980s, and then later on followed by the conversion of peat swamp forest for farmlands and industrial plantations, have been perceived as the main drivers of peatlands deforestation and degradation in Indonesia (Dohong et al. 2017a, b). Another factor that drives forest degradation is

[1]There is a contentious discussion whether Indonesia hosts the largest or the second largest area of tropical peatlands in the world. Dohong (2017) suggests that Indonesia hosts the largest tropical peatlands area in the world. However, a recent study carried out by Gumbricht et al. (2017) used new maps and expert system approach concludes that Brazil actually is the largest area (but this study used a minimum 30 cm depth in defining peatland area; meanwhile, in Indonesia, a minimum threshold 50 cm depth has been used in defining peatland) (e.g. Dohong 2017; MOEF 2018b).

mainly because there is no other alternative value for the peatlands other than being utilized as farmlands (MOEF 2018a, b).

Emissions of climate change-inducing greenhouse gases are especially of big concern. Under the 2015 Paris Agreement of the United Nations Framework Convention on Climate Change (UNFCCC), Indonesia has committed itself to an unconditional emission reduction of 26 and 29% from its business as usual trajectories by 2020 and 2030, respectively, and 41% reduction is fledged when overseas financial assistance is made available in those years. Of the unconditional target, almost 90 and 60% of the emission reductions in 2020 and 2030, respectively, will come from the reduction of deforestation and degradation of forests and peatlands. Peatlands contain rich carbon concentration (e.g. Jaenicke et al. 2010). They contain about one-third of about 550 G_tC of the world's land-based carbon stocks (Murdiyarso et al. 2017a, b; Parish et al. 2007). Indonesia's peatlands may contain between 28.1 G_tC (Warren et al. 2017) and 33.6 G_tC (Suryadiputra et al. 2018).

The 2015 catastrophic forest and peatland fires put Indonesia as the fourth largest greenhouse gas emitting country (1.62 G_tC) (Harris et al. 2015a). In a number of days of the fire period, Indonesia's daily emissions were even larger than those in the United States (Harris et al. 2015b). The local impacts of peat degradation are also significant. The World Bank estimates that the 2015 forest fires burned around 2.6 Mha of Indonesia's forests, costing the country more than US$16 billion (World Bank 2016). The 1998–1999 fires were estimated to cost the country approximately US$8.5 billion (Tacconi 2003).

In the meantime, avoiding emissions from restoring degraded peatlands has become a strategic action to reduce emissions in Indonesia. In Indonesia's NDC, the forestry sector contributes 17 out of 29% of the unconditional emission reduction commitment below its BAU trajectory by 2030, or 23% out of the conditional commitments. Emissions from peat may increase by almost 13% due to decomposition and 30% due to fires by 2030, compared with 253.6 and 307 M_tCO_2 in 2010 (e.g. GoI 2016a).

To address peatland degradation and its resulting emissions, hence, newly elected President of the Republic of Indonesia, Joko Widodo, issued Presidential Regulation No. 1 of 2016 concerning the establishment of Badan Restorasi Gambut (Peatland Restoration Agency). This ad hoc governmental agency is tasked to coordinate and facilitate the implementation of peatland restoration activities in seven provinces, namely Riau, South Sumatra, Jambi, West Kalimantan, Central Kalimantan, South Kalimantan, and Papua, with total area targeted of 2.0 Mha during the period of 2016–2020 (Peatland Restoration Agency 2016). However, having re-examined and synchronized the maps on fire burnt area, artificial drainage networks, concessions permit areas, and conservation as well as remaining pristine peat forest cover areas (see Table 12.1), BRG came up with even bigger peatland restoration target (almost 2.5 Mha) in the seven provinces. Additionally, four districts (kabupatens) consist of Pulang Pisau in Central Kalimantan, Musi Banyuasin and Ogan Komering Ilir in South Sumatera, and Meranti in Riau have explicitly mentioned in the Presidential Regulation as priority districts of peatlands restoration (e.g. GoI 2016b).

Table 12.1 Selected criteria for mapping out peatlands restoration priority areas

Priority 1	Areas burnt in the 2015 fires
Priority 2	Protected areas not burnt in the 2015 fires with existing drainage canals
Priority 3	Protected areas not burnt in 2015 fires without drainage canals
Priority 4	Other areas with drainage canals

Table 12.2 Classification of area permits under the BRG contingency plan

Area permit classification	Size area (ha)
Cultivation areas with permit (concessions)	1,410,943
Cultivation areas without a permit	396,945
Protected areas	684,638
Total	2,492,527

By the end of this year (2018), BRG is expected to accomplish 70% of its restoration target, whereas the remaining 20 and 10% will be implemented in 2019 and 2020, respectively.

The BRG Contingency Plan shows a plan to restore a vast area of almost 2.5 Mha up to the year 2020 (e.g. Chief of BRG (Badan Restorasi Gambut, the Peat Restoration Agency).

Out of this 2.5 Mha target, around 1.4 Mha is located under cultivation area with a permit (concessionaire areas), about 0.40 Mha of cultivation area without a permit, and approximately 0.70 Mha of protected areas, respectively. Restoration target based on area permit classification is shown in the following Table 12.2 (e.g. Harris et al. 2015a, b).

Since its establishment in early 2016, BRG has carried out many activities ranging from coordinating and facilitating stakeholders on planning, implementing restoration measures through the 3Rs approach, socialization, education, partnership, research, and development. Additionally, BRG has also commenced active collaboration with many donor agencies, universities, and so forth both domestic and overseas.

Moreover, to expedite its restoration target, BRG introduced the 3Rs approach. The 3Rs stand for Rewetting of drained peat (R1), Revegetation of bare and secondary fragmented peatland (R2), and Revitalization of local livelihoods (R3) (Dohong et al. 2017a, b).

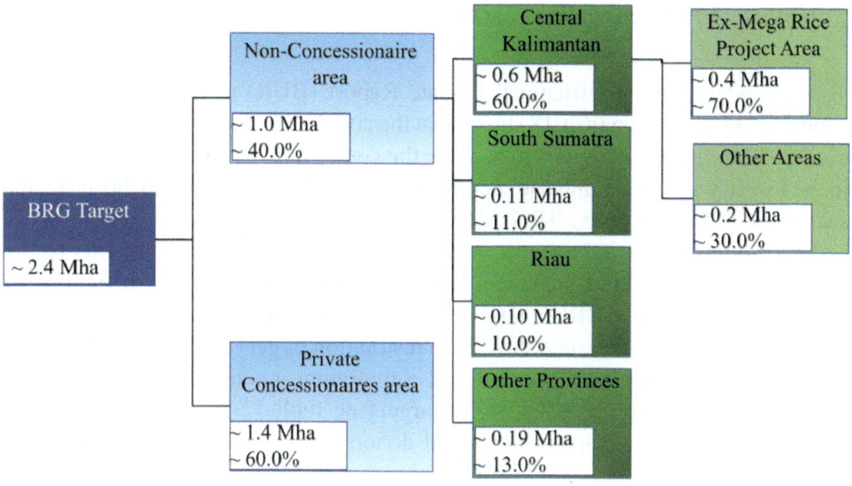

Fig. 12.2 Grouping of areas within the 2.5 Mha of BRG mandate areas

Financing Requirements for Peat Restoration

Concession Versus Non-concession Dichotomy

Out of the 2.5 Mha of the BRG mandate, about 1.4 Mha is located within private concessionaires' jurisdiction. Since the Ministry of Environment and Forestry (MOEF) is the issuer of the concession permits, the ministry holds the responsibility for facilitating the planning and oversighting of peatlands restoration activities implemented by those concessionaire permit holders. Meanwhile, BRG has focused its effort and prioritized to coordinate and facilitate restoration activities in the remaining 1.1 Mha target (see Table 12.2).

About 0.60 Mha (66%) of the 1.0 Mha non-concessionaire target area is located in Central Kalimantan, which is around 0.4 Mha (70%) is occurred in the ex-mega rice project area. The remaining 0.40 Mha (36%) of the non-concessionaire restoration target area is located in other provinces such as 0.1 Mha (11%) in South Sumatera; 0.1 Mha (10%) in Riau; and another 0.1 Mha (13%) in other provinces (Fig. 12.2).[2]

[2]Ex-mega rice is a failed government project that attempted to develop one million ha of peatlands as rice-farming fields in Central Kalimantan. Commenced in 1996, the project already opened up a large portion of peat forests and built canals, only to be abandoned later on. It remains a problem until today.

Estimates of Restoration Costs

The second Indonesia's Biannual Update Report (BUR) to the Paris Agreement submitted by Indonesia in 2018 shows that the country requires about US$5.5 billion until 2030 to finance activities to achieve the conditional emission reduction targets in the forestry and other land use sectors and around US$2.2 billion in the agriculture sector under the NDC (e.g. BRG 2016).

Few studies have tried to estimate funding needs for BRG to deliver its restoration mandate. These studies present facts that a substantial amount of funding gap still remains to achieve BRG's target. The World Bank estimates that an amount of US$2 billion is needed to fund the 2 Mha restoration target (Beat 2016). Meanwhile, Hansson and Dargusch (2018) find that a minimum amount of US$4.6 billion is required to restore the 2.49 Mha BRG's target (see Table 12.3). The existing funding pledges from overseas governments and donors are still insufficient in narrowing down the funding gap.

The figures presented by Hansson and Dargusch (2018) were mainly built upon interviews and discussions with BRG's technical staff and senior leaders. As such, some suggest that the figures are too high—and indeed these are among the higher estimates at the moment. As a result, BRG is currently reviewing these cost figures and commencing a process that is expected to produce a better insight into not only the costs of restoring peatlands but also the strategic means to close down the financing gaps.

By now, the review process is still ongoing, and the figures are entirely preliminary, but a more realistic estimate has come closer. The review is carried out by matching the license types of the land uses within the 2.5 Mha with two principal indicators, namely peat ecosystem regulation and the peat burnt frequency. Based on spatial analysis, it is found that the restoration targets are mainly concentrated in private licensing areas such as industrial timber estate plantations (Hutan Tanaman Industri—HTI) and business utilization rights (Hak Guna Usaha—HGU). Among them, six concessionaire areas made up more than 84% of the total area within the 2.5 Mha.[3] In the meantime, the remaining target is located under non-licensed production and conservation areas.

The cost calculation is estimated based on the 3Rs (rewetting, re-vegetation, and revitalization) intervention approach. Peat rewetting (R1) is implemented through the construction of rewetting infrastructures such as canal blocking, canal backfilling, and deep well. Meanwhile, re-vegetation (R2) is carried out by providing seedlings production and seed banks, seeds transplanting, enrichment planting as well as promoting natural succession. Finally, revitalization of local livelihoods (R3) is developed on three principal bases, namely those based on land, water, and environmental services. Apart from the 3Rs, other costs for socialization and educating local people are also being considered.

[3]Preliminary results to determine the costs required for BRG to fulfill its mandate in the 2.5 million ha restoration area carried out by SystemIQ and Landscape Indonesia.

Table 12.3 Cost estimates for restoration of the BRG priority restoration areas (adopted from Hansson and Dargusch 2018)

BRG Area classification	Restoration requirement	Size area (ha)	Lowest estimated cost (US$/ha)	Lowest estimated total cost (US$)
Licensed production areas	Hydrological restoration of primary, secondary, or tertiary canals	1,410,943	2000	2,821,886,000
Non-licensed product areas burnt post-2015	Hydrological restoration of primary, secondary, or tertiary canals with full re-vegetation	396,945	3225	1,280,147,625
Protected areas burnt post-2015	Hydrological restoration of commodity to primary canals with full re-vegetation	226,335	1625	367,794,375
Protected areas with canals	Hydrological restoration of commodity to primary canals with potential enrichment planting	201,457	400	80,582,800
Protected areas with shallow peat and canals	Hydrological restoration of commodity to primary canals with potential enrichment planting	256,846	400	102,738,400
Total		2,492,527	1867	4,653,149,200

According to the nature of intervention, there are two kinds of financing requirements which are those that can yield financial returns and those that are cannot. For those that can produce financial returns, the total amount of initial investments is expected to be about US$675 million, while on the other hands, for those yield no financial returns, but are necessary anyhow to reduce risks so that investments can be made, the total amount of initial investments is expected to be in approximately of US$615 million. The grand total of both combined initial investments is US$1.3 billion. Added with recurring annual costs of US$380 million, the total amount of investments required is about US$1.7 billion.

Table 12.4 Cost estimates in concession and non-concession areas under the 1.5 Mha BRG restoration mandate

Description	Concessions	Non-concessions		Total
		Production	Protection	
Startup costs	143	181	498	822
Initial capital costs for 3Rs	445	102	–	547
Operational costs	231	77	71	379
Total	819	360	569	1748

The following table presents the preliminary cost estimates based on whether or not the areas under private or non-private concessions (note: this preliminary cost estimation is still under review process) (e.g. Harris et al. 2015a, b) (Table 12.4).

The preliminary cost estimation carried out by BRG shows that around US$819 million out of the total US$1,748 million investment required for 1.5 Mha restoration target is under private concessions, meanwhile, the remaining US$929 million investment needed is for 1 Mha target under non-concessionaire areas.

Implications of Regulatory Changes on Restoration Costs

There were changes in the peatland protection and management regulations in Indonesia in recent years. The new approach, using Ministerial Decree (of the Environment and Forestry) No. 130/2017, produces significantly different results compared with the old approach using the Government Regulation (GR) No. 71/2014 and Government Regulation (GR) No. 57/2016. The differences prevail from the different baselines. In addition to having a peat depth of more than 3 m and the prevailing spatial and forestry plans, GR 71/2014 and GR 57/2016 use different criteria for protection, namely whether the area contained strategic biodiversity values such as endemic germplasm and endangered species. Ministerial Decree (MOEF) No. 130/2017, which is the new approach, uses different criteria, namely whether there is any peat dome exceeding critical hectarage, and whether there contains any high conservation value area (HCVA) or high conservation value forest (HCVF). The new approach yields a higher cost as there are bigger protected areas in the new approach. In fact, in the new approach, about 90% of the 2.5 Mha would need to be protected.[4]

[4]This assessment has not included the impacts of the Ministerial Decree No. 17/year that was challenged in court and the successing Ministerial Decree No. 10/2019.

Options for Peat Restoration Financing

There are options to reach the financial amount necessary to fulfill BRG's mandate. They include, for example, the government's own budget, grants, and assistance from bilateral and multilateral public sources, philanthropy, and others. Nevertheless, the potential amount from these public sources may not be sufficient. Business options need to be made available for an even larger amount of financing to be made available so that private investors will come. As such, there need to be business cases in peat restoration.

Public Interest in Financing Peat Restoration

In the latest 2018 Budget, the allocation for environmental protection was IDR11.1 trillion (US$792,000) in the 2017 outlook and budgeted for IDR15.7 trillions (US$1,122,000) for 2018 (e.g. GOI 2018a, b). From 2007 to 2014, GOI spent a total of US$17.48 billions for climate change adaptation, mitigation, and supporting activities. It allocated US$55 billion more for the period of 2015–2019 and would continue to set aside a national budget for climate change in the future (e.g. GOI 2016c).

For peat restoration, the Indonesian government budgeted just US$200 millions by the end of 2017 and as a result, only about 10% of the total area was rewetted. This funding includes US$4.4 million from the Government of Norway, US$134.6 million from the US, the UK, Japan, Germany, and the Netherlands. The Government of Indonesia itself allocated about $60 million from its own budget (e.g. Jong 2018a). Further, the Government of Norway increased its contribution to about $50 million. This amount remains far from the required amount, however. The business case for peat restoration will have to be made available for private investors to participate. This needs to be done when we can find a business case for peat restoration to attract private investors to participate.

The Business Case for Peatlands Restoration

The landscape approach in peatlands restoration allows for an area larger than the areas being restored themselves. BRG embraces this approach through the establishment of about 106 Peatland Hydrological Units (PHU, or Kesatuan Hidrologis Gambut, KHG, in Indonesian). The areas outside and around the peatlands within the PHU can be utilized productively for economic purposes through farming, plantations, and other income-generating activities. In the peat areas, farming can also be done with species that are peat-adaptive or peat-appropriate called paludiculture. There are at least the following business opportunities that can be developed.

- Agriculture development around the restoration areas. A significant portion of peatlands degradation is due to agricultural use, or agricultural development, in and around the peatlands. Development of the agricultural sector around the restoration areas will reduce the risks of encroachments as not only is it increasing agricultural productivity, it also increases the prosperity of the community members, especially the farmers, resulting in less propensity to encroach.
- Other livelihoods are supporting activities to the communities around the restoration areas. In addition to agricultural development, support to other livelihoods in the communities around the restoration areas such as animal husbandry and fisheries will also reduce the propensity to encroach.
- Paludiculture. There are species that can be planted in peatlands, such as sago, jelutong, pineapple, and others. Liberica coffee, although not entirely a species that can be planted on peatlands without hydrological modifications, is considered "pealand-safe" as there are only small modifications required. There are opportunities to develop paludiculture as alternative livelihoods. This, too, will benefit the communities and hence reduce the propensity to encroach.
- Ecosystem Services. Among the most significant impacts of peat conservation and restoration is that it will reduce or avoid emissions from peatlands degradation and peat fires. Through the emergence of carbon markets in the future, including that established under Article 6 of the Paris Agreement, a sufficient amount of funding can be raised. In addition to carbon, there are also valuable ecosystem services such as biodiversity, water supply, and ecosystem-based tourism that can be capitalized.

Carbon Asset: A Business Case for Peatland Restoration

Covering a mere three percent of the world, peatlands store 30–40% of the world's carbon. The extent of the world's peatland areas varies due to some definitional challenges. The extent of peatlands in Indonesia similarly varies, ranging from 206,950 sq km (km^2, or about 20.7 Mha) with about 1138 km^3 volume at 5.5 (m) average depth to 225,420 km^2 (about 22.5 Mha) with about 1388 km^3 at 3.4 m average depth (Page et al. 2011; Gumbrich et al. 2017).

Degradation of peatlands including especially peatland fires leads to massive emissions of greenhouse gases. Conserving and restoring degraded peatlands will lead to a reduction and avoidance of greenhouse gas emissions. According to the Paris Agreement, Article 6, efforts to reduce emissions through conservation and restoration of peatlands may be carried out in cooperation with other countries voluntarily. The resulting emission reductions maybe claimed and used by one of the parties towards meeting their NDCs—usually the investing ones—but not both.[5]

[5]Article 6.1 or the Paris Agreement stipulates that "Parties recognize that some Parties choose to pursue voluntary cooperation in the implementation of their nationally determined contributions to allow for higher ambition in their mitigation and adaptation actions and to promote sustainable

As revealed in a study in the Sebangau area in Central Kalimantan—used for an example—for every million ha of successful rewetting of peatlands, it is estimated about 25 (23.7–27.11) M_tCO_2 per year emissions mitigated, or about 25 ton CO_2 per ha (Jeanicke et al. 2010).[6] Additional activities such as re-vegetation of degraded peatlands will add additional reductions. As such, assuming that the emission reductions can be capitalized at, say, US$5 per ton of carbon dioxide, assuming that rewetting one million ha of peatlands will result in a potential income of US$125 million (IDR1.8 trillion) per year, or US$3.1 billion (IDR44 trillion) in the course of 25 years. This is about US$125 (IDR 1.8 million) per ha per year or about US$3,100 (IDR44 million) per ha over the course of 25 years. This amount is clearly more than sufficient to finance the peat restoration for the one million ha of the non-concessionaire peatlands under the BRG mandate.

The Case for Peat Bonds

In 2018, the first-ever issued green "sukuk" in the world, issued by the Indonesian government, was greatly oversubscribed by almost 2.5 times (Dunkley 2018). It raised US$3 billion (about IDR42 trillion), US$1.25 billion of 5-year tenor with a 3.75% coupon, and US$1.75 billion of 10-year tenor with 4.4% coupon. Sukuk is a form of bond financing that is in compliance with Islamic Shariah's non-interest debt. Proceeds from the green sukuk would be allocated for renewable energy projects, sustainable land use, green tourism, and waste management (e.g. CBI 2018a).

The use of proceeds from the issuance of the green sukuk would be used to finance various projects as follows: IDR501 billion (US$34.5 million) green projects such as flood management, urban drainage management, and coastal conservation under the Ministry of Public Works and Housing; IDR165 billion (US$11.4 million) of supporting facilities for railways under the Ministry of Transportation; and the remaining are for projects under the Ministries of Agriculture, and Energy and Mineral Resources. Interestingly enough, there are no significant environmental projects, including those for peat restoration, mentioned.[7]

Worldwide, green bonds have become an attractive instrument to finance environmental ("green") projects, and the amount continues to grow over time. There were more than US$167 billion of green bonds issued in 2018, up by 3% from about US$162 billion in 2017. The slowdown from the 84% year-on-year growth

development and environmental integrity," whereas Article 6.5 stipulates that "Emission reductions resulting from the mechanism referred to in paragraph 4 of this Article shall not be used to demonstrate achievement of the host Party's nationally determined contribution if used by another Party to demonstrate achievement of its nationally determined contribution." These Articles and others under Article 6 practically establish a global emissions trading mechanism under the Paris Agreement.

[6]The study cites a figure of 1.4–1.6 Mt CO_2 for an area of 590 ha. This figure is then extrapolated for one million ha.

[7]The Minister of Finance, Sri Mulyani Indrawati, was quoted as saying in (e.g. Alika 2018).

in 2017 might be due to more diversified labeling in the forms of sustainability, sustainable development goals (SDGs), and social bonds. Taking all of these labeled bonds together, the market actually grew by 13% from US$199.3 billion in 2017 to US$226.1 billion in 2018 (e.g. CBI (Climate Bond Initiative) 2019). At EUR7 billion (US$7.5 billion) with 22 years tenor, France's green bond was the largest to date (e.g. CBI 2018b).

Green bonds have indeed already become a strategic instrument to raise funds for green purposes. Among others, it could access a broader investor base and, in general, it also has better pricing compared to regular bonds (e.g. CBI 2017). Bonds are attractive for financing peatland restoration because it is inexpensive, and allows for a long tenure.

Bonds are practically an investment in the form of a loan. One invests through buying bonds expecting to get a return on their investments that is higher than the amount they spend buying the bonds. One also expects the security of their investments as bonds are considered a relatively safe investment, although the level of security depends on the rating of the bond issuers. The level of safety and security of the bonds are usually depicted by some sort of rating system, either for government- or corporate-issued bonds. Government-issued bonds are usually safe, or "investment grade," although not all of them.

The Architectural Structure of the Indonesian Peat Bond

A so-called "Peat Bond" works as depicted in Fig. 12.3, below. The bond instrument is guaranteed by the government (sovereign guaranteed) or even issued by the government as a sovereign bond. The sovereign guarantee allows for increased safety and security of the bond, which allows for a better price, longer tenor, and overall marketability of the bonds. Indonesia's credit rating is an investment grade with a stable outlook. This means that Indonesia is a safe place for investment.[8] In which case, Indonesia can sell its sovereign bonds at favorable price. The 2018 green sukuk, for example, could be sold at favorably low price at 3.75% for the 5 years tenor and 4.4% for the long 10 years tenor (e.g. Mediatama 2019). There is no case yet for longer tenor in Indonesian bonds, although it may be possible.

The proceeds from the sales of the bonds will be used to finance the project types above. The first two, paludiculture projects and farms and plantations in mineral lands surrounding the peatland restoration areas as well as other economic activities, maybe more straightforward. These activities will yield incomes that can be used to pay back the financing that eventually will be used to pay the investors (the bond buyers) in coupons.

---

[8]As of today, Indonesia's sovereign (credit) rating hovers around BBB (Standard and Poor's), BBB (Fitch Ratings), Baa2 (Moody's Investor Service), BBB (Japan Credit Rating Agency), and BBB (Rating and Investment Information Inc.), all with stable outlook. This means that Indonesia is a safe place to invest.

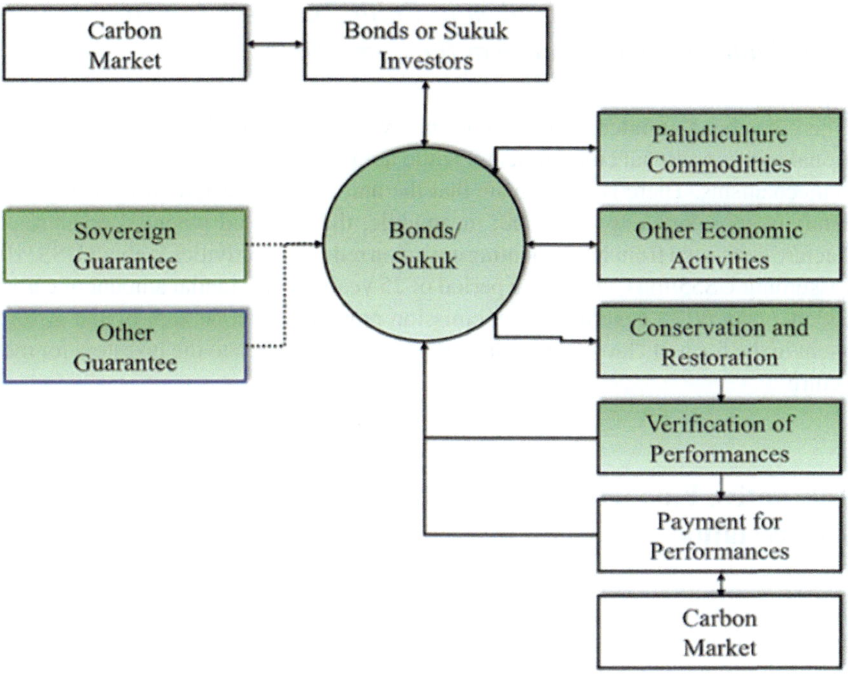

Fig. 12.3 Diagram of the structure of the Indonesian peat bond

For conservation and restoration, however, it is not that straightforward. Conservation and restoration activities will not immediately yield financial income, but they restore environmental services. The restored environmental service, when verified by an authorized party, is considered a "performance." These performances are valuable assets that can be monetized through the so-called "performance-based payment." The most obvious case is to define performance in terms of greenhouse gas emissions reductions. When conservation and restoration activities methodologically and verifiably avoid emissions of greenhouse gases, emission reductions "credits"—also known as carbon credits—will be issued. These credits have monetary value, as every ton of emissions reduced (or avoided) may get paid through the performance-based payment system. In this case, incomes can be defined in terms of the monetary value gained through the performance-based payment, or in terms of the valuable carbon credits, or both. Investors will be able to claim their coupons in cash, partly in cash and partly in carbon credits, or entirely in carbon credits, either to be retired as a fulfillment of their carbon liabilities or traded.

Green Bonds as a Mechanism to Fulfill Indonesia's NDC and Traded Under Article 6 at the Same Time

There has been an intense discussion on how Indonesia fulfills its NDC (unconditional and conditional commitments) while at the same time participating in Article 6 mechanisms. There is an estimate that the unit cost per hectare of restoring peatlands is approximately US\$1600; meanwhile, the expected potential revenue per hectare generated from implementing emission reduction activities is about US\$3100 (assuming US\$5/ton CO_2) over a period of 25 years. A substantial amount of carbon income (say 60%) generated from emission reductions can be kept by the government in the form of "levy," while the remaining can be given to the investors for their return.

Innovative Financing Through Compensation for Past Deforestation

Many commodity sustainability standard certification bodies such as the Roundtable for Sustainable Palm Oil (RSPO) and the Forest Stewardship Council (FSC) require companies with past deforestation histories to compensate their deforestation debts by conserving or restoring other forest areas, using certain formulas. These compensation formulas can be calculated based on hectarage or based on the financial values of the deforestation liabilities.

RSPO has recently strengthened its zero-deforestation standard. The de facto industry standard for sustainable palm oil adopted a standard—principles, and criteria—that is more robust and consistent with the "no deforestation, peat, or exploitation" (NDPE) principle. The previous standard bans companies from cutting primary forests, but still allows cutting of secondary forests and planting in peatlands that are less than 3 m depth. In the new standard, however, there is a total ban for cutting secondary as well as primary forests and no planting is allowed in peatlands of any depth (e.g. Jong 2018b).

In addition to applying tougher zero-deforestation standards, RSPO allows for those certified companies to compensate for their past deforestation. RSPO-certified palm oil companies need to address land clearance and plantation development undertaken since November 2005 without prior High Conservation Value (HSC) assessment by complying with the Remediation and Compensation Procedure (RaCP). The procedure requires palm oil plantation companies to first disclose any new development that was carried out without due HCV assessment, calculate the environmental liabilities through a Land Use Change Analysis (LUCA), and undertake onsite or offsite remediation for the affected sites or with affected parties. Compensation liabilities, accounted in a hectare, are determined utilizing a matrix that combines HCVs (through vegetation coefficients as proxies), the land clearance period, membership status, and areas requiring remediation (e.g. RSPO (undated)). In other words,

RSPO-certified palm oil plantation companies caused deforestation in the process of establishing their plantations to compensate such deforestation liabilities with rehabilitating and conserving forests, hectare to hectare, using a methodological formula.

For the remediation and compensation procedure to benefit peat restoration, a framework can be devised. According to RSPO, there are 82,609 ha of non-compliant land clearance in palm oil plantation establishment in Indonesia that causes deforestation without due HCV assessments since November 2005. Consequently, through a formula, there are 59,563 ha of conservation liability in Indonesia (e.g. Harris et al. 2015a, b). The compensation projects can be directed towards those areas under BRG mandate and therefore, such compensation can contribute to the fulfillment of the restoration mandate.

The so-called compensation projects can be located within BRG's one million ha of non-licensed areas in the seven provinces. By implementing this kind of compensation, the plantation companies can fulfill their compensation liabilities under the sustainability standards while assisting the BRG in fulfilling its mandate.

Blending Financial Sources

Many financial options available can be strategically created to fulfill BRG's financial gap in achieving its restoration mandate. Practically, these options can be implemented at the same time in a concoction that matches financial needs. For example, sources with a high appetite for risks—private equity, hedge funds, etc.—may be able to be matched with financial needs with high risks. Similarly, sources with a low appetite for risks such as private banks may only be matched with financial needs with low risks (Fig. 12.4).

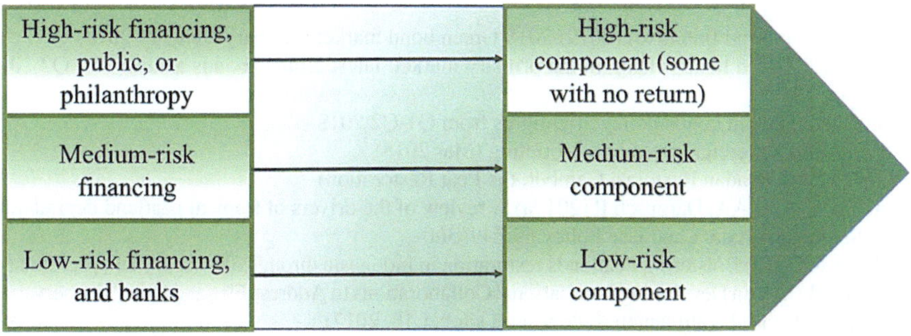

Fig. 12.4 Blending financial sources to match risk appetites of the sources with the risk profile of the financial needs

Conclusion

Peatland restoration is one of the effective land-based emissions reduction activities in Indonesia, and therefore, to achieve better peatland restoration performance notably in avoiding CO_2 emission release, better restoration planning, and proper restoration actions on the ground are required. Financing peatland restoration is often perceived as low hanging fruit investment opportunity due to its cost-effectiveness on a per ton basis.

BRG is mandated to coordinate and facilitate the restoration of 2.5 Mha in seven provinces during the period of 2016–2020. Substantial financial investment is needed to realize the vast area of BRG's restoration target. Hence, efforts have been done to develop innovative financial instruments apart from just rely on a limited state budget. A blended innovative financing that involves bond and ecosystem services is considered as an effective way of addressing financing peatland restoration in Indonesia.

References

Agus F, Hairiah K, Mulyani (2011) Measuring carbon stock in peat soils: practical guidelines, World Agroforestry Centre (ICRAF) Southeast Asia Regional Program, Indonesian Centre for Agricultural Land Resources and Development. 60p

Agus F (2015) Degradation and sustainable management of peat soils in Indonesia. https://www.naro.affrc.go.jp/archive/niaes/marco/marco2015/text/ws3-8_f_agus.pdf

Alika R (2018) Jadi pionir penerbit green sukuk di Asia, Indonesia Raup US$3 Miliar (Pioneering the Issuance of Green Sukuk in Asia, Indonesia raised US$3 billion), Katadata (Feburary 26, 2018)

Beat MH (2016) Indonesia seeks foreign funds to aid peat restoration drive (Mongabay, 26 Sept 2016), https://news.mongabay.com/2016/09/indonesia-seeks-foreign-funds-to-aid-peat-restoration-drive/

BRG (2016) op cit

CBI (Climate Bond Initiative) (2019) 2018 Green bond market summary (January 2019)

CBI (2017) Green bond pricing in the primary market: latest climate bonds analysis for Q2 2017 (9 Nov 2017)

CBI (2018a) Green bonds policy: highlights from Q1-Q2 2018

CBI (2018b) Sovereign green bonds briefing (Mar 2018)

Chief of BRG (Badan Restorasi Gambut, the Peat Restoration)

Dohong A, Azis AA, Dargusch P (2017a) A review of the drivers of tropical peatland degradation in South-East Asia. Land Use Policy 69:349–360

Dohong A (2017) Bolstering Peatlands restoration in Indonesia through 3Rs Approach, presented at the Workshop on Developing International Collaborations to Address Fire and Other Conservation Issues in Central Kalimantan, Indonesia (October 18, 2017)

Dohong A, Cassiophea L, Sutikno S, Triadi BL, Wirada F, Rengganis P, Sigalingging L (2017b) Modul Pelatihan: Pembangunan Infrastruktur Pembasahan Gambut Sekat Kanal Berbasis Masyarakat. Badan Restorasi Gambut Republik Indonesia, Jakarta

Dunkley E (2018) Indonesia issues world's first green sukuk bond. Financial Times (23 Feb 2018) https://www.ft.com/content/e38ea51c-184c-11e8-9376-4a6390addb44

GoI (Government of Indonesia) (2018a) Indonesia second biennial update report under the united nations framework convention on climate change. The Government of Indonesia, Jakarta, Indonesia
GoI (The Government of the Republic of Indonesia) (2016a) First nationally determined contribution, Republic of Indonesia, submitted to the UNFCCC as a contribution to the Paris Agreement. The Government of the Republic of Indonesia, Jakarta, Indonesia.
GoI (The Government of the Republic of Indonesia) (2016b) Presidential regulation No. 1 of 2016 regarding the establishment of Peatland Restoration Agency
GOI (2016c) op cit
GOI (2018b) Informasi APBN 2018: Pemantapan Pengelolaan Fiskal Untuk Mengakselerasi Pertumbuhan Ekonomi Yang Berkeadilan. The Government of the Republic of Indonesia, Jakarta, Indonesia
Gumbrich T, Roman-Cuesta RM, Verchot L, Herold M, Wittmann F, Householder E, Herold N, Murdiyarso D (2017) An expert system model for mapping tropical wetlands and peatlands reveal South America as the largest contributor. Global Change Biol 23:3581–3599
Hansson A, Dargusch P (2018) An Estimate of the financial cost of Peatland Restoration in Indonesia. Case Studies in the Environment January 2018 https://doi.org/10.1525/cse.2017.000695
Harris N, Minnemayer S, Sizer N, Mann SA, Payne O (2015a) With latest fires crisis, Indonesia surpasses russia as world's fourth-largest emitter, world resources Institute. World Resources Institute, Washington, D.C.
Harris N, Minnemeyer S, Stolle F, Payne O (2015b) Indonesia's fire outbreaks producing more daily emissions than entire US economy. World Resources Institute (October 16, 2015). World Resources Institute, Washington, D.C.
Jaenicke J, Woesten H, Budiman A, Siegert F (2010) Planning hydrological restorations of Peatlands in Indonesia to mitigate carbon dioxide emissions. Mitig Adapt Strat Glob Change 15(3):223–239
Jeanicke J, Rieley JO, Mott C, Kimman P, Siegert F (2008) Determination of the amount carbon stored in Indonesian peatlands. Geoderma 147(2008):151–158
Jong HN (2018a) Is Indonesia's ambitious peat restoration initiative severely underfunded?" Mongabay (3 Apr 2018)
Jong HN (2018b) RSPO adopts total ban on deforestation under sweeping new standards. Mongabay (16 Nov 2018)
Mediatama G (2019) Peminat green sukuk melimpah. Retrieved 5 Sept 2019, from https://nasional.kontan.co.id/news/peminat-green-sukuk-melimpah
MOEF (Ministry of Environment and Forestry, Republic of Indonesia) (2018a) Managing Peatlands to cope with climate change: Indonesia's experience, (Mar 2018). Ministry of Environment and Forestry, Jakarta, Indonesia
MOEF (Ministry of Environment and Forestry, Republic of Indonesia) (2018b) The state of Indonesia's forests 2018. Ministry of Environment and Forestry, Jakarta, Indonesia
Murdiyarso D, Hergoualch K, Sasmito S, Hanggara B (2017a) Cadangan Karbon di Lahan Gambut. Center for International Forestry Research, Bogor, Indonesia
Murdiyarso D, Roman-Cuesta RM, Verchot L, Herold M, Gumbricht T, Herold N, Martius C (2017b) New map reveals more peat in the tropics, Info Brief (189 Oct 2017). Center for International Forestry Research, Bogor, Indonesia
Peatland Restoration Agency (2016) Decree No. 5 of 2016 regarding Penetapan Peta Indikatif Restorasi Gambut (peatland restoration indicative map)
Page S, Rieley JO, Bank CJ (2011) Global and regional importance of the tropical carbon pool. Glob Change Biol 17:798–818. https://doi.org/10.1111/j.1365-2486.2010.02279.x
Parish F, Sirin A, Charman D, Joosten H, Minayeva T, Silvius M (2007) Assessment on Peatlands, biodiversity, and climate change: the executive summary. Global Environment Centre, Kuala Lumpur, Malaysia, and Wetlands International, Wageningen, the Netherlands
RSPO (undated) RSPO remediation and compensation procedure. https://www.rspo.org/certification/remediation-and-compensation

Suryadiputra INN, Irwansyah RL, Iwan TCW, Dipa SR (2018) Restorasi Lahan Gambut di HLG Londerang dan Tahura Orang Kayo Hitam, Provinsi Jambi. Wetlands International Indonesia, Bogor, Indonesia

Tacconi L (2003) Fires in Indonesia: causes, costs and policy implications, CIFOR occasional Paper No. 38. Bogor: Center for International Policy Research

Wahyunto, Nugoro K, Agus F (2016) Perkembangan Pemetaan dan Distri-busi Lahan Gam-but di Indonesia. In: Agus F, Anda M, Jamil L, Masganti (Eds) Lahan Gambut Indonesia: Pembentukan, Karakteristik, dan Potensi Men-dukung Ketahanan Pangan. Badan Penelitian dan Pengembangan Pertanian, Kementerian Pertanian, IAARD Press

Warren M, Hergoualc'h K, Kauffman JB, Murdiyarso D, Kolka R (2017) An appraisal of Indonesia's immense peat carbon stock using national peatland maps: uncertainties and potential losses from conversion. Carbon Balance and Management 12:12. 10.1186/s13021-017-0080-2

World Bank (2016) The cost of fire: an economic analysis of Indonesia's 2015 Fire Crisis. The World Bank Jakarta, Indonesia

Chapter 13
Climate Budget Tagging: Amplifying Sub-National Government's Role in Climate Planning and Financing in Indonesia

Zahra Zafira Mutiara, Dede Krishnadianty, Budhi Setiawan, and Joko Tri Haryanto

Abstract The economic growth that Indonesia has experienced relies heavily on the availability of natural resources, and consequently the success to mitigate environmental risks is associated with climate change. Such fact is widely acknowledged in the medium-term national planning process and reflected appropriately in several key national climate policies. The establishment of a far-reaching national plan on climate mitigation has pushed the sub-national governments to set ambitious targets that supposedly guide provincial and district governments to carry out low carbon development plans. Using climate budget tagging as defined by UNDP as a public expenditure assessment tool, insights from East Kalimantan Province and West Kutai District were attained. The results of climate budget tagging in the province and one of its districts consecutively depict that 7% and 47% of budget allocation in 2015 were potentially in support of low carbon development, while in 2016 both allocations in East Kalimantan (24%) and West Kutai (16%) encouraged further discussions on the dynamics of climate mitigation at the sub-national level of governance. Nonetheless, such seemingly high allocations of the regional government budget that were earmarked for achieving climate mitigation targets were not equipped with clear outputs and outcomes that were sensitive to climate mitigation. The findings also show that there was insufficient local government fiscal space to reinforce the newly set Indonesia's NDC target of 29% emission reduction by 2030. In ensuring that climate mitigation is being integrated into the development agenda, as well as in promoting transparency, accountability, comparability and consistency in climate finance, the following policy recommendations are proposed: (1) anticipating and

Z. Z. Mutiara (✉) · D. Krishnadianty
WWF Indonesia, Graha Simatupang, Jakarta Selatan 12540, Jakarta, Indonesia
e-mail: zz.mutiara@gmail.com

B. Setiawan
Faculty of Engineering, Sriwijaya University, South Sumatra, Inderalaya 30662, Indonesia
e-mail: budhi.setiawan@unsri.ac.id

J. T. Haryanto
Centre for Climate Change and Multilateral Policy, Fiscal Policy Agency, Ministry of Finance, Jalan Dr. Wahidin Raya No. 1, Jakarta Pusat 10710, Indonesia
e-mail: djohar78@gmail.com

R. Djalante et al. (eds.), *Climate Change Research, Policy and Actions in Indonesia*, Springer Climate, https://doi.org/10.1007/978-3-030-55536-8_13

preparing for the incoming reforms in functions, responsibilities and resources across levels of sub-national government; (2) clarifying government activities' indicators, outputs and outcomes to better suit the climate mitigation targets; and (3) creating an innovative fiscal transfer financing scheme.

Keywords Climate tagging · Climate finance · Mitigation · Fiscal · NDC

Introduction

Blessed with an abundance of natural resources, Indonesia has been enjoying the economic growth that relies heavily upon them (Alisjahbana and Yusuf 2004; Resosudarmo 2005a, b). Such enjoyment, however, does not last, and even more critically, it may not be able to fuel the growth and welfare that Indonesia always envisions. The exploitation of natural resources brings irreversible and perpetual consequences for the provision of ecosystem services, which are the prerequisites to support communities' day-to-day life and business operations (Moreno-Mateos et al. 2017). Environmental degradation that happens in Indonesia is exacerbated by climate change that significantly impacts the country's development (Measy 2010). Acknowledging that damages to non-sustainable management are permanent, and the impact of climate change to the economy is real, Indonesia may not be able to maintain its growth at 7% while achieving a new status of a high-income country, if it continues doing business-as-usual (MoF 2015). Alternative strategies to anticipate the above-mentioned possibilities are required.

Government of Indonesia (GoI) recognizes potential vulnerability to climate change and other environmental pressures in the national medium-term development plan (RPJMN) 2014–2019 (MoNDP 2014). The plan specifies that "inclusive and sustainable growth; increasing value added of natural resources with a sustainable approach; improving quality of the environment; efforts to tackling climate change" are needed to be done within the 5 years (MoNDP 2014). Post-2020 climate pledge of GoI to reduce greenhouse gas emissions by 29% by 2030 is also reflected through the renewal of Presidential Regulation that declares an avid commitment toward climate mitigation and adaptation (Republic of Indonesia 2016).

Post-Paris Agreement, the challenges faced by countries with plans of contributing toward curbing the impacts of climate change to meet the shared goal remain in mobilizing resources and investments (CPI 2017). Any possible public, private and alternative sources of financing that are utilized to address climate change is referred to as climate finance. Under the current availability and accessibility of climate finance, optimizing the efficacy of climate finance to support the plans is crucial (Nakhooda 2013). This means national and sub-national government should be able to use its authority and organize relatively small amounts of the finance to bring about the greatest possible impact in achieving climate targets and plans (Gordon and Johnson 2017).

In line with the above-mentioned plans, the Ministry of Finance through its Center for Climate Finance and Multilateral Policy launched the Green Planning and Budgeting Strategy for Sustainable Development that further details key priority areas in which policy and public investment can help in mitigating climate change—and to cope with the risks of loss and damage to natural resources. There are 21 top practical focuses clustered in six policy areas highlighted in the document that enable a smooth shift toward a green economy, which in turn maintains the economic growth (MoF 2015). Formal endorsement to a low carbon development plan, which aims at protecting the environment from business-as-usual practices, has slowly become a norm at the national level.

The national government has required each province to set emission reduction targets within the framework of their regional action plans through the issuance of Governor Regulation. Proportionately, provinces thus need their districts to also collectively address the issues. The existing climate and environmental governance in Indonesia mimics multilateral climate governance, where contributions are configured by sub-national governments, accompanied with an effort by the national government to monitor and verify them. In this light, the sub-national government plays an instrumental role, given the nature of the environmental problem that originates from processes that are embedded in specific provincial areas (Betsill and Bulkeley 2006). The success of multilevel governance setting of climate governance hence depends largely on the policy implementation at the sub-national level.

The climate action plan issued by each province in Indonesia will only be meaningful if it is clearly integrated into their development agenda. The accountability of the commitment should then be reflected on budget availability to execute the plans. Ultimately, financing is an elemental enabler of many of climate actions. This paper examines planning and budgeting dynamics at the sub-national level through the lens of climate budget tagging (CBT). CBT offers a framework to understand how planning and budgeting correlate to monitoring and tracking of climate-related expenditures in the budgetary system (Ellis and Moarif 2017; Kissinger et al. 2019; Resch et al. 2017; UNDP 2015). The tool has been widely advocated for in the context of GoI in hope to comprehend relevant climate change spending that allows the government to make informed decisions on climate investments, as well as to strengthen the transparency of public budget use.

This study aims at addressing the main research question: Given the authority and responsibility to reach climate targets, what lessons can be learnt from sub-national governments in mainstreaming climate actions into the development agenda? The question can also be elaborated into several further questions that this study discusses: referring to the case studies, how is the quantity and quality of climate planning and budgeting at sub-national level? what can be done to improve the efficacy of resource use in achieving climate targets at the sub-national level? The rest of the study is organized as follows: Section "Methodology" discusses the framework and methodology of understanding the planning and budgeting at the sub-national level. Section "Results" presents the case studies and actual findings on the ground. Section "Discussion" highlights the analysis of the results laid out in the previous section. Finally, Section "Conclusion" concludes the progress of climate governance at the sub-national level in terms of planning and budgeting.

Methodology

Budget tagging is defined as the process of giving a mark in a budget document to track, identify and compare the output of budget activity against and its budget, which are listed in the government institutions' work plan and budgeting. Climate budget tagging (CBT) is the process of identifying the amount of the budget used to finance specific output aimed for climate change mitigation and adaptation (UNDP 2014). The legal bases of the budget tagging process are as follows:

1. Government Regulation No. 17 Year 2017 about Synchronizing Planning and Budgeting Processes for National Development;
2. Presidential Regulation No. 61 Year 2011 about the National Act of Greenhouse Gas Emission Reduction Plan;
3. Presidential Regulation No. 71 Year 2011 about implementing the National Inventory of Greenhouse Gases;
4. Minister of Home Affairs' Regulation No. 54 Year 2010 about the Execution of the Implementation of Government Regulation No. 8 Year 2008.

In the context of sub-national CBT, there needs to be a diagnostic tool to access opportunities and constraints for integrating climate change concerns within the national and sub-national budget allocation and expenditure process or Climate Public Expenditure Institutional Review (CPEIR). The needs to come up with a framework review of budget allocation in the context of climate expenditure sparked the productions of several guiding references both at the national and sub-national policy field, for instance the one guided this academic exercise by UNDP (UNDP 2014) and WWF Indonesia (WWF 2017).

The CPEIR analytical framework has three key main pillars: policy analysis, institutional analysis and climate public expenditure analysis (UNDP 2014). Following are the steps used to conduct the analysis:

(i) Determine minimum two-point time of interests to check the consistency of policies and expenditures;
(ii) Review regional medium-term development plan, regional strategic environmental assessment (KLHS) to check if province and district officially include responding to climate change as part of their development plan's mission;
(iii) Review relevant regional working agencies' strategic and annual plans to see the extent of planning elaboration into annual programs and activities.
(iv) Thereafter, review the activities' budget and output pursuant to the six key clusters of interventions as shown in Table 13.1 following the suggestion by Ministry of Finance (2015).

Table 13.1 Priority program to address climate change

Cluster of interventions	Programs
Natural resources protection	Forest protection
	Peatland rehabilitation
	Coral and marine protection
Agriculture	Climate change adapted crops
	Plantation crops
	Irrigation
Energy and industry	Energy and resource efficiency
	Renewable energy
	Resource efficiency
	Fuel pricing
	Large-scale power
	Sustainable mining
Transportation and urban planning	Public transportation
	Waste management
	Climate proofing roads/bridges
	Regional infrastructures
Education and health	Green education
	Climate change-sensitive health services
Supporting policies	Disaster reduction and/or management
	Green economy coordination including its monitoring and evaluation

To attest to the climate planning and financing at sub-national level, East Kalimantan Province and one of its districts, Kutai Barat, were selected to be the area of interests. Delving into these two case studies allows comprehension on the climate governance happening at sub-national level. While doing the review, following the above framework and steps, these are guiding questions to assess the two cases presented in this study:

- How much of the total budget allocation can be claimed to be contributing in the achievement of climate targets?
- How is the quality of the planning and budget allocating in achieving climate targets through indirect (budget composition for employee and institutional logistics) and direct spending (budget composition for activities and investment in support of output and outcome of climate policy)?
- Under the existing policies and institutional arrangement, what can be done to improve the role of sub-national level given the authority and resources in achieving climate targets on the ground?

Results

East Kalimantan

East Kalimantan is Indonesia's second largest province after Papua and is the largest revenue sharing fund (RSF) recipient for oil and gas. Its economy depends highly on extractive industries such as oil, gas, coal and mining. Recently, however, East Kalimantan's government has committed to decoupling its economic growth from extractive industries toward more sustainable sectors, as described in their official documents such as the Low Carbon Development in East Kalimantan (2011), Green Growth Compact (2014), proposed Emission Reduction Program Document for Forest Carbon Partnership Facility Carbon Fund (2018). In addition, East Kalimantan has proposed for all of its areas to be entitled to performance-based payment from jurisdictional REDD+program funded by the World Bank (Forest Carbon Partnership Facility).

East Kalimantan's climate budget tagging was conducted by the regional planning agency and related government working units. Documents that became the bases of the budget tagging activities included the Regional Development Plan (RPJMD) 2013–2018 and its regional Environment Strategic Assessment. Three of five mission statements of the East Kalimantan's government are marked in the climate budget tagging process. These are: to promote economic competitiveness based on natural resources and renewable energy (mission 2); to provide high-quality basic infrastructure for the community (mission 3); and to create better and healthy environmental framed by a climate change perspective (mission 5).

Based on the policy reviews, there are three important developments as well as environment issues that are highlighted in both RPJMD and KLHS: (1) pollution and environmental degradation, (2) infrastructure services and sector's competitiveness, and (3) social economic and public health issues. These issues were considered as strategic issues that needed to be addressed by the provincial government by 2018. Looking at correlations among regional planning documents, the abovementioned issues are translated into programs and activities tackled by 11 government working units: Forestry Office, Environment Office, Tourism Office, Transportation Office, Fisheries, Plantation office, Agriculture Office, Food Security Office, Livestock Office, Public Works and Settlement Office, and Energy and Mineral Resources Office.

As Fig. 13.1 illustrates, when those 11 government working units' budgeting sheets of East Kalimantan are reviewed referring to the six clusters of preferred interventions, transportation and regional cities planning cluster received the highest portion of climate budget between 2015 and 2017. Delving into the quality of spending in relation to the impact it brings on the climate mitigation effort, Table 13.2 shows the direct (budget composition for employee and institutions' logistical costs) and indirect spending (budget composition for activities and investment). It also shows the ratio of what considered as direct spending on six cluster key interventions that aim to achieve climate targets of the entire regional budget allocation.

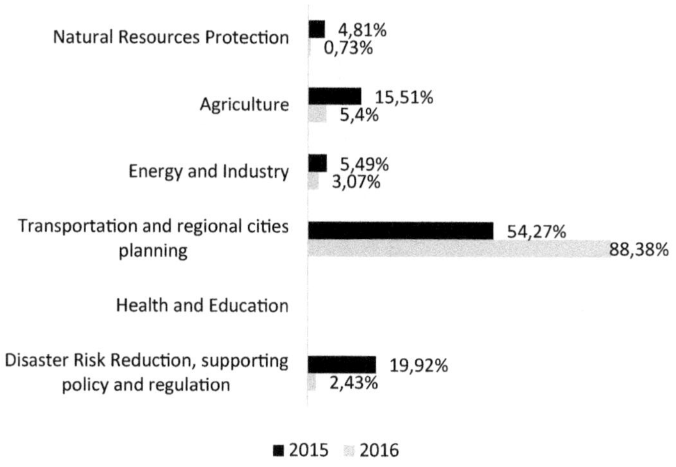

Fig. 13.1 Percentage of East Kalimantan's budget allocation on the path of achieving climate targets to total six clusters budget in 11 OPDs

The results of climate budget tagging in the province in 2015–2016 consecutively reach 7 and 24% of total public budget allocation. The distribution of programs and activities contributed by the 11 government working units is shown in Table 13.3. Some government working units solely contributed to just one cluster of intervention. Of the 11 government working units, only three offices contributed to more than one cluster of interventions. Table 13.3 tells how sectoral the approach is still to climate mitigation at the sub-national level.

West Kutai District

The key programs in West Kutai District that were outlined in the Regional Development Plan (RPJMD) 2011–2016 are: (1) Infrastructure development that supports the improvement of local community welfare; (2) Improving the quality and coverage of education services for all levels of society; (3) Improving the quality and coverage of health services, especially for the poor; (4) Village-based community economic development; (5) Poverty alleviation; (6) Bureaucratic reform and public services; and (7) Environmental preservation. According to West Kutai's RJPMD 2011–2016. Eleven government working units had been identified as offices that directly responded to the achievement of sustainable development and its contribution to climate targets. In reference to the cluster of key interventions, Fig. 13.2 depicts the proportion.

In general, the climate expenditure allocation of 11 government working units in 2015 is greater than the expenditure allocation in 2016. Such reduction of climate spending at the end of the 5-year planning could be the result of other spending on

Table 13.2 Climate spending's composition in East Kalimantan

Year	Six cluster of key interventions	Indirect expenditure of 11 government working units (OPD)	Direct expenditure of 11 government working units (OPD)	Regional budget allocation	Ratio of spending on six cluster key interventions to public budget expenditure
	IDR	IDR	IDR	IDR	%
2015	687.487.210.170	235.260.952.523	2.782.189.962.051	1.02053E + 13	7
2016	1.815.026.589.116,00	246.626.536.578	1.987.631.899.411	7.60124E + 12	24

Table 13.3 Percentage budget allocation in support of six clusters of key interventions in mitigating climate change in East Kalimantan in the year of 2015 and 2016 to total budget of six clusters per year

Cluster of key interventions	Contributors (OPD)	Average of public budget ratio allocated to total climate finance per cluster of key interventions	
		2015	2016
		%	%
Natural resources protection	Forestry offices	1.12	0.30
	Environment office	0.65	0.17
	Fisheries office	3.04	0.27
Agriculture	Agriculture office	–	1.41
	Plantation office	8.15	2.13
	Food security office		0.005
	Livestock office	0.07 7.29	1.85
Energy and industry	Energy and mineral resources office	5.37	2.98
	Environment office	0.13	0.09
Transportation and regional cities planning	Public works and settlement office	23.19	75.35
	Transportation office	29.88	12.55
	Energy and mineral resources office	0.74	0.18
	Environment office	0.35	0.21
	Tourism office	0.11	0.09
Health and education	None		
Disaster risk reduction, supporting policy and regulation	Public works and settlement office	15.88	0.81
	Energy and mineral resources office	1.29	0.37
	Transportation office	1.22	0.15
	Environment office	0.66	0.29
	Plantation office	–	0.23
	Forestry office	0.24	0.20
	Livestock office	–	0.16
	Food security office	0.27	0.10
	Tourism office	0.17	0.04
	Fisheries office	0.15	0.05
	Agriculture office	0.05	0.02

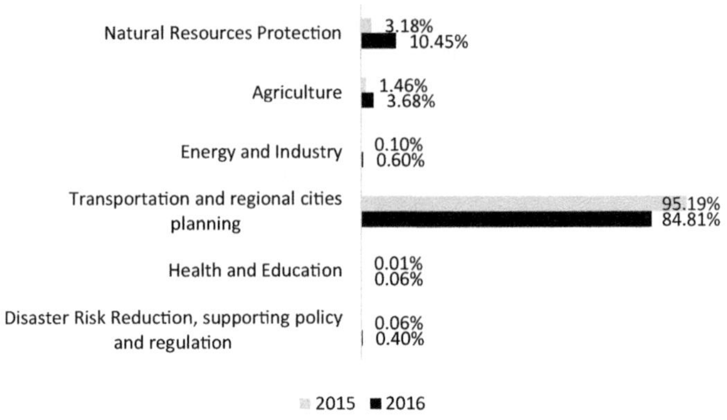

Fig. 13.2 Percentage of West Kutai's budget allocation on the path of achieving climate targets to total six cluster budgets in 11 OPDs

the non-climate targets. It could also be due to that in the last year of being in office, the government had no prospect of being reelected, so that the expenditure pattern was only business-as-usual (Klein and Sakurai, 2015). The problem is that there was a significant reduction in climate expenses from 2016 when compared to 2015. The budget allocation cuts even reached almost 50% in 2016 compared to the previous year.

The results of climate budget tagging in the district in 2015–2016 consecutively reach 47 and 16%, respectively, of total budget allocation in those 11 government working units. The distribution of programs and activities contributed by the 11 government working units is shown in Table 13.4. However, seemingly high number of percentages do not completely reflect the climate target's achievement. This is admitted by the personnel dues to the minimum capacities in defining meaningful climate output and outcome, despite professedly programs' and activities' naming (Table 13.5).

Table 13.4 Climate Spending's Composition in West Kutai

Year	Six clusters of key interventions	Indirect expenditure of 11 government working units (OPD)	Direct expenditure of 11 government working units (OPD)	Total expenditure	Ratio of total six clusters of key interventions to public budget expenditure
	IDR	IDR	IDR	IDR	%
2015	998,456,074,998.88	61.565.448.223	1.307.239.387.366	2.14384E + 12	47
2016	281,162,364,583.68	74.092.203.055	623.001.332.776	1.80827E + 12	16

Table 13.5 Average budget allocation in support of six clusters of key interventions in mitigating climate change in West Kutai between 2015 and 2016

Clusters of key interventions	Contributors (OPD)	Average of public funds ratio allocated to total climate finance per cluster of interventions	
		2015	2016
		%	%
Natural resources protection	Forestry offices[a]	2.94	10.30
	Plantation, agriculture and livestock office	0.02	0.05
	Environment agency[b]	0.01	0.05
	Culture, tourism youth and sport office	0.005	0.01
	Social office	0.01	0.02
	Public works office	0.20	–
Agriculture	Public works office	0.52	–
	Plantation, agriculture and livestock office	0.93	3.65
	Community empowerment and village government office	0.01	0.02
	Food security office	0.01	0.01
Energy and industry	Mining and energy office	0.04	0.59
	Public works office	0.06	–
	Environment office		0.01
Transportation and regional cities planning	Public works office	95.18	82.6
	Industry, trade and cooperative office	0.005	1.88
	Environment office	0.002	0.34
	Community empowerment and village government office	0.005	0.02
Health and Education	Environment office	0.008	0.01
	Industry, trade and cooperative office	–	0.03
	Disaster mitigation office		0.02
Disaster Risk Reduction, supporting Policy and Regulation	Industry, trade and cooperative office	–	0.17
	Social office	0.004	–

(continued)

Table 13.5 (continued)

Clusters of key interventions	Contributors (OPD)	Average of public funds ratio allocated to total climate finance per cluster of interventions	
		2015	2016
		%	%
	Environment office	0.003	0.03
	Disaster mitigation office	0.054	0.20

[a]In 2017 based on Law No. 32 on Local Regional Government's functionality this function is taking over by provincial government
[b]In 2017 has changed in to Environment office

If explained based on the six clusters of interventions, both in 2015–2016, the largest budget for climate mitigation was allocated for indicators of natural resources protection, agriculture and transportation and urban planning. From the composition of expenditures per working units, public works agency receives the largest budget priority. With such allocation, the role of West Kutai District in carrying out development primarily uses linear infrastructures as means of being economically competitive district without much attention on the climate output and outcome during the infrastructure development.

Discussion

The findings tell that priorities for these two sub-national governments are given to mainly infrastructure development, sanitation and other basic needs. This sheds light on what sectors need to incorporate climate-relevant features of activities, which in return bring manifold climate impacts in addition to providing original services to the people. The allocated budget for the key interventions can only be categorized as direct and indirect based on its compositions without further analyses due to the minimum information on activities' indicators, output and outcome on climate policies. Considering the existing authorities and potential resources that can be allocated by sub-national governments, below are several recommendations to amplify the role of sub-national governments in translating political commitment as set by Paris Agreement based on the case studies in East Kalimantan Province and West Kutai District.

Clarifying Activities' Indicators

Mapping the authority to plan and implement climate policies at sub-national level, potential flexibility for climate policy planning and implementation can be amplified better than the existing practices shown through climate budget tagging exercises. It is a matter of integrating one development agenda into other targets governments commit to achieving. Instead of specifically addressing only one problem at a time, climate co-benefits can also be expected from other clusters of development that are closely related to the climate mitigation while maintaining economic growth. As identified by Ministry of Finance, there are five other prioritized sectors that will significantly help the climate; these are: agricultural energy and industry, transportation and urban planning, education and health, and disaster mitigation (MoF 2015). The above-mentioned priorities should be more climate-sensitive, along with conservatively relying on the natural resource protection to gain benefits of climate. Hence, it is suggested that sub-national governments should improve the way planning is conducted by incorporating activities target indicators that are contributing to climate impacts in order to serve more significant outputs and outcomes that help the government in achieving their planned climate targets.

Deliberate on the use of climate budget tagging to see how governments translate the implementation of climate targets, the more comprehensive the activities' indicators, output and outcome that include suitable climate targets are, the easier will be tracking and assessing climate-relevant spending. The information resulting from the tag can be used to notify the state of planning, budgeting as well as implementation to strengthen the transparency and accountability around climate change commitments, both at the sub-national and national governments. Besides, if applied consistently across sub-national governments under similar framework and indicators, the national government can compare outcomes, which will then help the national government to subsequently refocus the expenditures to achieve the targets better.

Anticipating the Incoming Reform at the Sub-National Level

Under the enacted Law No. 23 of 2014 on local government, there is a reform in land-use governance at the sub-national level where authorities on forestry, mining and energy sectors are based at the provincial level, while agriculture, plantation and other land-use allocations are still governed at the district level. Having the situation in hand, sub-national governments need to be more adaptive in anticipating such shift without contravening the Governor's Regulation on their strategic and action plans for climate mitigation. There is a large opportunity for district governments to involve further in forest and land management in accordance with Law No. 23 of 2014. One opportunity that the Law unlocks is the opportunity for district governments to translate and articulate management efforts supported by a robust financing policy and framework, which incentivizes land governance. There is the need of an elaborative attempt to follow-up on targets that can be achieved at the provincial and district levels under the current regime, especially knowing that land-use sector contributes the most to the greenhouse gas emissions in Indonesia (Republic of Indonesia 2016). Having smaller scope of the intervention area and slimmer structure of entities to coordinate with, is beneficial for sub-national governments to adapt faster to changes in circumstances (Oliveira 2009) and to interact with constituencies on the ground.

Introducing Innovative Fiscal Transfer at the Provincial Level

Despite the incoming reform of land governance at the sub-national level, protecting natural resources and mitigating climate change shall be kept unambiguous to achieve climate targets. Bringing benefits of natural resource protection and climate mitigation on the ground requires financing and incentives. To ensure a successful mainstreaming effort for climate targets from provincial to district level, an incentive for district governments becomes a far-reaching instrument. An indicator-based incentive that rewards district governments for their performance within the context of climate mitigation through a fiscal transfer vehicle can become an option.

Putting in place mechanisms in which district governments can benefit financially from climate mitigation and prevention of further land conversion for short-term benefit is needed. The effectiveness of financing in generating climate mitigation outcomes will ultimately depend on the broader political economy of land-use change. One of the major steps leading to that will be to assist local governments in understanding the value of their natural capital. Through this, they will have an understanding or comparison of the potential opportunity cost of when they preserve their natural resources—and of when they merely exploit and extract these resources. This will encourage proper debate and discussions at a policy level that will result in knowledge-based decision-making processes in local governments. This will counter current thinking that income from natural resources relies solely on how much is being extracted instead of how much is being managed and preserved (Nurfatriani 2015). A facility that rewards district government for their efforts to drive low carbon development is the first step to compensate for thinking that climate mitigation is an expenditure, not an investment.

Conclusion

The exercises that were conducted together with the local staff in both East Kalimantan Province and West Kutai District depict the sub-national dynamic of climate governance, and in a broader sense of natural resources management and environmental protection, in Indonesia. Yet, the seemingly high budget allocation in the province (7% in 2015; 24% in 2016) and the district (47% in 2015; 16% in 2016) is not accompanied with climate impacts and benefits. Learning from the case studies, sub-national development agendas have mainly been around overcoming poverty. While doing the poverty alleviation, development agendas should pay attention to the potential climate benefits they can bring on the table. While performing such economic advancement, it is also necessary to comprehend that greening the effort to maintain growth and achieve climate targets should be a non-sectoral attempt. They should not be only the responsibility of one government institution.

As a means to strengthen climate governance and climate commitment compliance, series of analyses and recommendations can become viable options to amplify the role of provincial and district governments. Most importantly, these recommendations can be used as ways to accommodate climate actions through more systematic planning and budgeting from one level of government to the level below it. This includes valuing districts with high ecological value, which perform climate-explicit activities planning and budgeting, with more fiscal support to compensate for costs in conducting climate mitigation. Such settings eliminate the economic limitation that otherwise exists when a region is unable to arbitrage their natural resources but needs to pay for the cost of natural resource protection and climate mitigation. This way, climate financing at the sub-national level can be amplified in terms of transparency, consistency and accountability.

References

Alisjahbana A, Yusuf A (2004) Assessing Indonesia's sustainable development: long-run trend, impact of the crisis, and adjustment during the recovery period. ASEAN Econ Bull 21(3):290–307

Betsill M, Bulkeley H (2006) Cities and the multilevel governance of global climate change. Glob Gov 12:141

CPI (2017) The global landscape of climate finance 2017. From climate policy initiative: https://climatepolicyinitiative.org/wp-content/uploads/2017/10/2017-Global-Landscape-of-Climate-Finance.pdf

de Oliveira JA (2009) The implementation of climate change related policies at the subnational level: an analysis of three countries. Habitat Int 33(3):253–259

Ellis J, Moarif S (2017) Enhancing transparency of climate finance under the Paris agreement: lessons from experience. OECD Publishing, Paris

Gordon DJ, Johnson CA (2017) The orchestration of global urban climate governance: conducting power in the post-Paris climate regime. Environ Polit 26(4):694–714

Kissinger G, Gupta A, Mulder I, Unterstell N (2019) Climate financing needs in the land sector under the Paris agreement: an assessment of developing country perspectives. Land Use Policy 83:256–269

Klein FA, Sakurai SN (2015) Term limits and political budget cycles at the local level: evidence from a young democracy. Eur J Polit Econ 37:21–36

Measey M (2010) Indonesia: a vulnerable country in the face of climate change. Glob Majority E-J 1(1):31–45

MoF (2015) Green planning and budgeting strategy for Indonesia's sustainable development 2015–2019. Centre for Climate Finance and Multilateral Policy, Fiscal Policy Agency, Jakarta

MoNDP (2014) Rencana pembangunan jangka menengah nasional 2015–2019. MonDP, Jakarta

Moreno-Mateos D, Barbier E, Jones P, Jones H, Aronson J, Lopez-Lopez J, Benayas J (2017) Anthropogenic ecosystem disturbance and the recovery debt. Nat Commun 8:14163

Nakhooda S (2013) The effectiveness of international climate finance. Overseas Development Institute

Nurfatriani FD (2015) Redesigning Indonesian forest fiscal policy to support forest conservation. For Policy Econ 61:39–50

Republic of Indonesia (2016) First nationally determined contribution. From UNFCCC: https://www4.unfccc.int/sites/ndcstaging/Pages/Home.aspx

Resch E, Allan S, Alvarez L, Bisht H (2017) Mainstreaming, accessing and institutionalising finance for climate change adaptation. ACT learning paper

Resosudarmo BP (2005a) The politics and economics of Indonesia's natural resources. Insitute of Southeast Asian Studies

Resosudarmo BP (2005b) The politics and economics of Indonesia's natural resources. Institute of Southeast Asian Studies

UNDP (2014) Methodological guidebook: climate public expenditure and institutional review (CPEIR). UNDP, Bangkok

UNDP (2015) Climate budget tagging: country-driven initiative in tracking climate expenditure: studies of Bangladesh, Indonesia, Nepal and the Philippines. UNDP, New York

WWF (2017) Pedoman penandaan anggaran hijau (green budget tagging) di daerah. WWF, Jakarta

Chapter 14
Unveiling the 'Green': Media Coverage on the Aceh Green Vision, Indonesia

Rizanna Rosemary and Darrick Evensen

Abstract Indonesia has had many trials in recent years associated with legal and illegal threats to forests and their biodiversity, including in Aceh's forests—Leuser ecosystem, which are home to one of the key forest reserves. The Aceh government has undertaken a policy and programme of actions to combat the deforestation. Nevertheless, no research has explored how information on the challenges to the forests, or the government's solutions, have been communicated to the public in Aceh. The public will likely need to play a role in realising the protection of Aceh's forests, and more broadly in addressing environmental and climate concerns in Indonesia. In this chapter, we examine the types of information members of the public received on environmental and climate challenges in Aceh from the key local mass media outlet in Aceh. We specifically explore the nature of 'efficacy' information—on what actions can be taken by various societal actors to combat the challenges. Newspaper remains an important platform to address about climate change. This content analysis systematically examines how local newspaper in Indonesia covers important and alarming environmental issues through the Aceh Green Vision (AVG) initiative.

Keywords Forest · Climate change · Biodiversity · News media · Information

Landscape and Climate Challenges Tied to Indonesia's Forests

Indonesia has the third largest amount of forest of any nation. Forests in Indonesia are generally tropical rainforests with high rainfall, which is more than 1200 mm per year, and have a short dry season. According to the Indonesian Forest Watch (IWF) data in 2013, 82 million ha of Indonesia's land area are still covered by forests. However, within the last 12 years, Indonesia's forest lost 6.02 million ha (Mha)

R. Rosemary (✉)
University of Syiah Kuala, Banda Aceh, Indonesia
e-mail: rizanna.rosemary@unsyiah.ac.id

D. Evensen
University of Edinburg, Edinburgh, UK

R. Djalante et al. (eds.), *Climate Change Research, Policy and Actions in Indonesia*, Springer Climate, https://doi.org/10.1007/978-3-030-55536-8_14

overall from 2000 to 2012; this rate of decline steadily rose on average by 47,600 ha per year (Atmadja and Wollenberg 2010).

By 2012, the annual primary forest loss in Indonesia was 0.84 Mha, which was estimated to be higher than in Brazil (0.46 Mha) (Margono et al. 2014; Springate-Baginski and Wollenberg 2010). Fibre plantation, illegal logging, and palm oil industry in Indonesia significantly contributed to the deforestation not just in Indonesia but also within the Southeast Asia region (Abood et al. 2015).

Between the mid-1990s and 2015, the high global demand for Indonesian timber and oil palm has led to illegal logging and the enlargement of permanent agricultural areas (Tsujino et al. 2016). In 2013, the Ministry of Forestry (now The Environment and Forestry Ministry) also reported that 53% of the timber harvested from natural forests came from illegal logging—4000% more than the ministry predicted in 2007 (Purwanto 2014).

Despite government plans to eradicate illegal logging that has been occurring since 2011 (Murdiyarso et al. 2011), encroachment on the country's forests remains rampant. The significant growth of forest deprivation in the country has dire consequences for climate change mitigation and the government's efforts in biodiversity conservation (Margono et al. 2014). Palmer (2001) indicated that the institutional failures—high culture of corruption and weak law enforcement and monitoring capacity across government sectors—contributed to logging concession practices. Moreover, actors at the district, sub-district, and village levels in Indonesia have shown increasing political and economic interest in governing the forest resources, including timber operation, due to the illegal logging practices, particularly in forest on the Island of Sumatra (Ardiansyah et al. 2015; McCarthy 2002a, b).

Illegal logging leads to unsustainable outcomes for the forest's natural resources—with economic, social, and environmental consequences. This is further complicated by the government's 'double standard' of, on one hand, seeking to reduce deforestation, but on the other, placing forestry and agriculture as key economic pillars in government policy. The forestry and agriculture focus has produced inadequate and ineffective measures to address the increasing emissions that resulted from forests and peatland fires (Miranda 2018). From an economic perspective, deforestation has led to changes in livelihood and uptake of unsustainable economic practices, such as massive opening of forest areas for agricultural production which initially aimed to support economic growth. However, local people appear not to have been the ones receiving the primary benefits (Idris 2016). Indonesia experienced major economic loss annually due to illicit trade of timbers overseas, leading national timber companies to collapse (Hein 2018; Margono et al. 2014; Tsujino et al. 2016); furthermore, the palm oil plantations have contributed to the 300,000 ha of forest loss in the country (Buttler 2013).

Environmental impacts from the deforestation brought on by illegal logging have occurred particularly on Sumatra and in the Province of Aceh (Joy 2010; Panaligan 2010; Swainson 2016b), causing unique habitats or ecosystem types within the Sumatran rainforest to be at heightened risk of disappearing altogether (e.g. tigers, elephants and orangutans) (Guardian 2013; Nellemann 2007). In Aceh province (the north-most province on the island of Sumatra), massive illegal logging practices caused livelihood transformation particularly during the rehabilitation and reconstruction following the disastrous earthquake and tsunami in 2004.

Aceh Takes Actions to Mitigate Climate Change Pressures

The Indonesian environmental crisis is increasingly gaining political and media attention both nationally and globally. Of recent significance is former President Susilo Bambang Yudhoyono's (SBY) pledge to reduce greenhouse gas emissions by 26% independently by 2020, and by 41% with outside assistance. The country's recent level emission assessment showed the unconditional reduction target is 29% of the business-as-usual scenario by 2030. However, the Government of Indonesia is mindful of changes due to national and global economic circumstances (UNCC 2019). Driven by serious political, administrative, and economic demands for reforms, the central government has begun to decentralise, transferring new powers to the district and municipal levels.

Decentralisation in the forestry sector included transferring income from permits, logging and reforestation fees, as well as authorising lower levels of government to issue logging permits (Sahide et al. 2016). Despite the controversy of the local environmental policy, the Aceh Province, under the former governor Irwandi Yusuf, has shown progressive action to resolve environmental issues in the region, especially tackling illegal logging in the Sumatran forest within the Leuseur conservation area in Aceh. The governor announced a policy on environmental sustainability and protection for Aceh—the Aceh Green Vision (AGV)—in 2007, aiming to improve human wellbeing and social equity, while at the same time reducing environmental risks and ecological scarcities, as suggested by the UNEP in 2011 (BRR 2008; Swainson 2016b).

Aceh Green Vision: A Response to Environmental Destruction

Aceh Green Vision (hereafter AGV) was the first Indonesian provincial government attempt to transform economic and environmental management systems in an integrated, cross-sectoral manner, as envisaged by green economy engineers (Swainson 2016a, p. iii). It has become a critical battleground between proponents of the existing economic system and those actors who espouse a more sustainable approach to development which indicates challenges of green economy policy reform in Aceh (Swainson and Mahanty 2018; Swainson 2016a). AGV includes eight cross-sectoral components:

(a) **Land Use, Land Use Change, and Forest (LULUCF) Management**: (1) Primary forest protection and management, (2) Reforestation and forest restoration, (3) Community forestry and agro-forestry development,
(b) **Sustainable Economic Development**: (4) Smallholder estate crop development in partnership with private,
(c) **Sector and Parastatal Estate Crops and Associated Infrastructure**: (5) Spatial planning, management, and development of capture fisheries and aquaculture, (6) Public infrastructure development,
(d) **Renewable Green Energy**: (7) Geothermal energy, (8) Hydropower.

These mutually-dependent strategies are aimed to reduce environmental pressures and resource extraction, thereby also reducing climate pressures due to deforestation, by generating greater financial value and increased job opportunities from the resources already under use outside of protected areas (BRR 2008, pp. 4–5). Despite the strategies to generate revenue for the province and employment for rural communities, deforestation through illegal logging in this last major tract of intact Indonesian rainforest continues to concern the global environmental community and to threaten the vital ecosystem in the province (and especially the remaining populations of Sumatran tigers, rhinos, orangutans, and elephants).

The AGV has attracted attention from international media and non-governmental actors (Brummitt 2007; News 2007); yet, academic literature is lacking on how widespread the Aceh Green Initiative is, and the extent to which it has become prominent with the Acehnese people. People often interpret and make sense of issues like environment and climate change by relying on media representations as a point of departure (Boykoff 2011). Mass media play a central role in providing information about climate change, for example, with respect to scientific context and government programmes/policies in relation to protecting the environment (Anderson 2009; Boykoff 2011; Smith 2005). Members of the public expect adequate, accurate, and meaningful information to help them understand the complexity of issues such as climate change. So-called 'efficacy information'—information that explains what can be done, and how, to address a problem—may enable and encourage members of the public to take individual, group, or community action to support the government's goal and programme to preserve the environment, such as the AGV.

The objective of this chapter is *first* to examine and analyse coverage from 'Serambi Indonesia' about the AGV—this is the most read newspaper by the locals in Aceh, and therefore the most relevant source when thinking about dissemination of efficacy information on the AGV Initiative and the related environmental concerns that the AGV seeks to address. Below we first critically review: (1) the relevant studies, literature, and scholarship about media coverage and climate change policy, and (2) the implications of the Aceh Green policy for raising awareness, increasing knowledge, and influencing policymakers on the issue of deforestation, assessed through the efficacy information provided by the newspaper. Second, we present our methodology: the background of the local newspaper—Serambi Indonesia, the study objectives, the data collection process, and data analysis procedures. We then share our data and, discuss and synthesise the findings, and finally conclude by reflecting on the relevance of this research to policy and communication in Indonesia (and perhaps in other nations). We consider the role of mass media in sharing efficacy information, and offer recommendations for how mass media outlets can better attend to and use efficacy information in reporting on environmental and climate issues.

The 'Green' in the News

AGV was initiated as a response to the rise of climate change as one of the most imminent threats to the security and prosperity of the Asia Pacific region. According to the Forest, Nature and Environment of Aceh (HAkA), forest loss in Aceh especially in the Leuser Ecosystem between 2003 and 2012 about 173 km^2 or equivalent to an area the size of Hong Kong (Topsfield and Rompies, 2016). The valuable initiative to protect Aceh's forestry and environment faces challenges. For example, since the withdrawal of many aid agencies in April 2009, Aceh needs more sustainable public–private partnerships, non-governmental oversight, and private sector involvement to manage clean energy, biofuel feedstock, reforestation and avoided deforestation, all of which involve issues related to carbon and a commitment to green development (BRR 2008; Swainson 2016b).

These challenges led to former Aceh Governor Irwandi declaring a moratorium on all logging as a message to the international community that the province is willing to stop deforestation, but not without receiving something in return, that is, new revenues from trade, not only aid (BRR 2008). The AGV program is strongly supported under the Undang-Undang Pemerintahan Aceh No. 11 tahun 2006 (law of the Aceh Government) that governs the management of Aceh's forest. The Vision was shortlisted by The World Future Council for Future Policy Award, and the Whitley Fund for Nature (WFN) awarded its 'inspiring conservation champions' award to Farwiza Farhan—an Acehnese woman who fought to save endangered Sumatran species and the local livelihood in the Leuser Ecosystem (WFN 2016).

Besides global and national coverage, the AGV was also covered by local print newspaper—*Serambi Indonesia* since launched in 2007 (BRR 2008; Swainson 2016b). *Serambi Indonesia*—(hereafter SI) is the district printed newspaper in Aceh, and aims to provide an informative and educative message for the people. SI is a subsidiary of the prominent national newspaper—Kompas or the Tribune Network (Riezal 2016).

The offices and staff of SI were destroyed by the massive earthquake and Tsunami on 26 December 2004. The media industry was forced to stop publishing due to the loss of 55 employees—13 of whom were senior editors and missing journalists, and also due to the severely damaged printing machines. SI started to re-publish using its printing press in Lhokseumawe city in January 2005. The printed circulation increased significantly from 7000/day to 40,000/day in 1989–2014, and is now distributed throughout 18 districts and five cities in the Aceh province. Haryanto (2011) argued that as a subsidiary of the oligarchy Kompas Group, SI's agenda setting holds to the status quo political interest and has an impact on the quality of journalists and the journalism in Indonesia. Thus, this predictably influences how the reporters present and frame news on national issues, including environment and climate change.

Mass Media Coverage in 'Green' Risks

Communication scholars agree that for people to be better informed about important societal issues, efficacy information of appropriate quantity and quality is a significant aspect of useful mass media coverage. The concept of efficacy information has been widely used by scholars to address the issue health and environmental risk as an approach to help people to better respond to risks at both personal and societal levels (Covello et al. 2001). Individuals who are more likely to effectively respond to risks, for instance climate change, are those who perceive the dangers and feel self-assured that they are able to mitigate them (Turner et al. 2006). Efficacy information can: (1) inform people of actions to limit risk, (2) emphasize whether/how those actions are effective at achieving desired outcomes, and (3) facilitate confidence that one can perform those actions (Stryker et al. 2008).

We draw our approach to data collection and analysis from Feldman et al. (2017), who constructed an approach examining efficacy information in mass media coverage of climate change. They differentiate between negative and positive information within media coverage of climate change in the related but distinct realms of *self-efficacy*, *external efficacy*, and *response efficacy* (Feldman et al. 2017). In our content analysis, we expand upon Feldman et al.'s work by adding the concept of *group efficacy* (see definitions of the four efficacy types below).

We examine not only discussion of actions that individual readers can take to reduce exposure or mitigate the effects of local and global environmental degradation (self-efficacy) but also the actions that organizations can and should take to address environmental risks—a concept we call group efficacy, both in a negative and positive tones/influences. We make no claims, however, about how audiences receive efficacy information in print media; such understanding is essential for demonstrating the importance of efficacy information empirically, but is beyond the scope of this research. We assume, based on the literature reviewed in this paper, that increasing provision of efficacy information is an important communication goal. Although descriptive, our study extends existing research and serves as a foundation for future work on mediated communication of efficacy information.

Types of efficacy information in our study:

1. **Self-efficacy** reflects one's perceived ability to engage in a protective behaviour (Floyd et al. 2000). Positive levels of self-efficacy can help people set goals for attaining desirable outcomes, adopt and maintain recommended behaviour changes, and persevere when problems arise (Rimal and Real 2003). Negative levels of self-efficacy, in turn, can be a barrier against change, especially if people perceive a risk but feel unable to respond to it. One may ignore it, downplay its

potential effects, and/or avoid information on risk-reducing behaviours (Witte and Allen 2000).

2. **Response efficacy** reflects a person's belief that a behaviour will achieve a desired outcome, such as avoiding or reducing the magnitude of a threat to the environment, like reducing carbon emissions (positive), or helping to prevent illegal logging (negative).

Both response and self-efficacy are important components of many risk-related behaviour theories, including protection motivation theory, social cognitive theory, and the extended parallel processing model, and the risk information seeking and processing framework (Bandura 2004; Danna and Griffin, 1999; Mead et al. 1999; Stigler and Becker 1977; Witte and Allen 2000). They have proven to be robust predictors of behavioural intention and actual behaviour across theories and study types, although other factors also play an important role (i.e. perceived risk susceptibility and severity) (Milne et al. 2000).

3. **Group efficacy** explores efforts that groups of people in the society take to mitigate their fear of the risk. For instance, a community planting trees to address climate change (positive), or a community group or organisation eschewing need for action due to denying that deforestation will contribute to climate change (negative).
4. **External efficacy** or societal efficacy (Clarke and Durrant 2010; Evensen and Clarke 2012) relates to government or politicians' effort to take both positive and negative actions or responding to public opinion. These messages can also provide citizens with means to hold such actors and entities accountable for actions taken or not taken. Indeed, environmental issues like climate change or large-scale land conversion often involve contentious debates regarding (a) the role of government officials in ensuring environmental and human protection and (b) the trust the public places in these officials to effectively do their jobs (McComas 2004).

Methodology

This content analysis focuses on coverage on environmental issue in general, and in particular coverage on AGV. As the Aceh's government initiative to protect the conservation in Aceh, the goals of the AGV necessarily can be achieved through other government's environmental programmes and policies. The content analysis of SI seeks to understand the ways in which members of the public receive information about the extent to which the government is carrying out the initiative; and how the public can contribute to the achievement of the government's goals.

The AGV seeks to promote conservation in Aceh. Even though the AGV is a policy and initiative from the government, there are many ways (through government programmes and/or otherwise) that the goals of the AGV can be achieved. The media content analysis we undertook seeks to understand the ways in which the public is hearing about the extent to which those goals are being achieved, and how the public is receiving information on how they can contribute to the achievement of those goals.

Our study relies on a content analysis of Indonesia news items (articles and opinion pieces) available from SI online news (http://aceh.tribunnews.com/). We included articles in Bahasa Indonesia, with the keyword strings: *ekologi, emisi, iklim,* lingkungan *hidup* (In English: ecology, emission, climate change, and environment), and the following keywords in English—*green* and *illegal logging.* The time period for data collection was news/articles between 2007 and 2018, since the initiative launched in the year 2007. Duplicate articles and articles that did not include any discussion about Aceh Green Vision (AGV), environmental, or climate change were excluded from the final set following an initial reading.

We included 16 codes for distinct forms of efficacy information related to local and global issues. We based our codes on Feldman et al.'s (2017) six codes for efficacy information (negative and positive each for the following: self-efficacy, response efficacy, and external efficacy). We turned these six codes into 12 by coding each separately in relation to whether the topic being covered was a global environmental issue or a local environmental issue (e.g. we included 'self-efficacy—positive—local' and 'self-efficacy—positive—global' in the place of their single code for 'self-efficacy—positive'). Furthermore, we added a set of codes for 'group efficacy', which is a form of societal efficacy in addition to the 'external efficacy' measured by Feldman et al. and Hart and Feldman (Feldman et al. 2017; Hart and Feldman 2014). 'Group efficacy' refers to actions that groups of people (any collective more than one person) can take to protect the environment or reduce environmental risks. Thus, we added four additional codes to the 12 we already had: (1) group efficacy—positive—local, (2) group efficacy—positive—global, (3) group efficacy—negative—local, and (4) group efficacy—negative—global. For this content analysis, we coded and presented *only* for the type of efficacy information that was the most prominent in each article. The coding and definition are summarised in Table 14.1.

Results

This content analysis is the first to examine systematically how a local newspaper in Indonesia covers important and alarming environmental issues, such as environmental risk and climate change, which were exemplified in this study in the work of and responses to the AGV's initiative. Our results resonate with previous studies about efficacy information on contentious environmental issues (Feldman et al. 2017; Hart and Feldman 2014).

Overall, the sample was slightly more negative than positive in terms of efficacy information, with 68 positive efficacy and 84 negative efficacy articles (Table 14.2). The overall valence is particularly attributable to 2017–2018, as the coverage became far more negative in these final two years. Of the 88 articles published in 2011–2016, the ratio of positive to negative efficacy information was 47/41; of the 64 articles

Table 14.1 Conceptual definition of efficacy information

Concept	Operational definition	Examples[a] of the concepts
Self-efficacy for global issue—positive	Indicates that individual action to address global climate change is possible, easy, simple, etc.	'In addition, the green roof concept is already prevalent in developed countries such as Germany and the Netherlands, for example, may also be able be applied…'
Self-efficacy for global issue—negative	Indicates that individual action to address global climate change is impossible, difficult, hard, expensive, etc.	[No examples existed in the study sample]
Self-efficacy for local issue—positive	Indicates that individual action to address environmental causes and effects at local level is possible, easy, simple, etc.	'The development of organic farming systems is very important to support farmers' independence…'
Self-efficacy for local issue—negative	Indicates that individual action to address environmental causes and effects at local level is impossible, difficult, hard, expensive, etc.	'Customary institutions or traditional leaders who are patrons and believed by the people in preserving the local wealth, values, and 'spirit' has disappear…'
External efficacy for global issue—positive	Refers to the responsiveness of political or corporate elites to public opinion and/or calls for action to global environmental issues	'The Aceh Government's policy on carbon issues is to support national policies. We are given a target of 7 percent to reduce carbon emissions, from 26 percent of the national target, "Abubakar Karim…'
External efficacy for global issue—negative	Refers to the failure of political or corporate elites to respond to public opinion and/or calls for action global environmental issues	'Greater use of land for organic production leads indirectly to high carbon dioxide emissions, due to deforestation,' Stefan Wirsenius told the IFL Science
External efficacy for local issue—positive	Refers to the responsiveness of political or corporate elites to public opinion and/or calls for action to environmental causes and effects at local level	'…Therefore, the ecological theology approach is a way to stop the practice of environmental destruction by attacking the predatory nerve edge of the developers and miners of excavation C to stop the action of environmental destruction…'

(continued)

Table 14.1 (continued)

Concept	Operational definition	Examples[a] of the concepts
External efficacy for local issue—negative	Refers to the failure of political or corporate elites to respond to public opinion and/or calls for action environmental causes and effects at local level	'According to the survey, Indonesia is in the top 20 largest emission producing countries. That's all because illegal logging is everywhere," said Minister of Public Works, Djoko Kirmanto…'
Response efficacy for global issue—positive	Refers to the potential or actual success of a specific action or policy to address global climate change	'To protect water resources from being damaged and fish from extinction is a noble task for all parties. Not only water and fish must always be monitored and protected, but all environmental components must be protected…'
Response efficacy for global issue—negative	Refers to the lack of potential or actual success of a specific action or policy to address global climate change	'A report published in the Journal Nature Communications revealed researchers detailed on the effects of climate change in northern China, where major cities like Beijing and Tianjin are located'
Response efficacy for local issue—positive	Refers to the potential or actual success of a specific action or policy to address environmental causes and effects at local level	'Efforts to protect the forest through customary law/institutions showing positive result. This is due to the recognition from the state of the existence of the community of customary law in forest areas as stipulated in the Decision of the Constitutional Court (MK) Number 35/PUU-X/2012 concerning Customary Rights of the Customary Law Community'

(continued)

Table 14.1 (continued)

Concept	Operational definition	Examples[a] of the concepts
Response efficacy for local issue—negative	Refers to the lack of potential or actual success of a specific action or policy to address environmental causes and effects at local level	'The government must consider to be impossible for them to manage all damaged and degraded forests. Thus, there is a need to reach multi-stakeholders to help ease the burden. This must begin with the Aceh Government giving trust to local communities to manage forests and to multi stakeholders to improve damaged forests'
Group efficacy for global issue—positive	Indicates that group action to protect the global environment or reduce environment risk is possible, easy, simple, etc.	'Awareness of the current earth phenomenon makes the word green so familiar and shifts its function into a greening spirit, a spirit of maintaining natural purity, and a passion for saving energy'
Group efficacy for global issue—negative	Indicates that group action to protect the global environment or reduce environment risk is impossible, difficult, hard, expensive, etc.	'The struggle to get control to customary land has become a 'trend' in our society. The mountains, slopes, ravines, and swamps have all been "fenced off" with barbed and even electric wired…'
Group efficacy for local issue—positive	Indicates that group action to protect the local environment or reduce environment risk is possible, easy, simple, etc.	'Coffee farmers in the Gayo highlands, especially Central Aceh and Bener Meriah, need to be aware of global climate change that has an impact on the decline in coffee production in these two regions…'
Group efficacy for local issue—negative	Indicates that group action to protect the local environment or reduce environment risk is impossible, difficult, hard, expensive, etc.	'The infringement of the Forests Limited Production (HPT) and illegal logging in the Babahrot crossing, Aceh Barat Daya (Abdya) towards Terangun, Gayo Lues (Galus), is increasingly prevalent'

[a]The examples were translated from Bahasa Indonesia

Table 14.2 Type of article, valence, geographic focus, efficacy type, and year of publication

Media coverages	Number (and % of sample) of articles
Type of article • News • Opinion	124 (78%) 35 (22%)
Valence • Positive • Negative	68 (45%) 84 (55%)
Geographic focus of issues • Local • Global	111 (70%) 41 (26%)
Efficacy type • Self-efficacy • Response efficacy • External efficacy • Group efficacy	3 (2%) 43 (27%) 89 (56%) 17 (11%)
Year of publication • 2011 • 2012 • 2013 • 2014 • 2015 • 2016 • 2017 • 2018	13 (8%) 14 (9%) 5 (3%) 16 (10%) 23 (15%) 18 (11%) 32 (20%) 38 (24%)

**Note* some boxes only total to 152 articles and 96% because seven articles contained no efficacy information

from 2017 to 2018, the ratio was 21/43 (this difference in valence across these two groupings of publication dates had a chi-square of 6.4, $p < 0.01$, Cramer's V [effect size] = 0.21). The negative bent to efficacy information is also due to far more articles featuring response and external efficacy information, and these two types of efficacy information being negative far more often than positive (see Table 14.2; a chi-square test of difference between the categories of efficacy information and valence had a value of 30.3, $p < 0.001$, Cramer's V = 0.45). In contrast, every instance of self-efficacy and group efficacy information was positive (Fig. 14.1).

Further differences manifest in the sample include efficacy information associated with local issues being substantially more positive (57 of 111 articles) than efficacy information associated with global issues (11 of 41 articles) (this difference in valence across these two geographies had a chi-square of 7.3, $p < 0.01$, Cramer's V = 0.22). Additionally, articles written by editors were significantly more likely to have negative information than articles written by other journalists or members of the public (i.e. letters to the editor) (chi-square = 5.2, $p < 0.05$, Cramer's V = 0.19) (Fig. 14.2).

Local external efficacy was by far the most prevalent type of efficacy information (36% of all efficacy information) and was split almost evenly between positive and

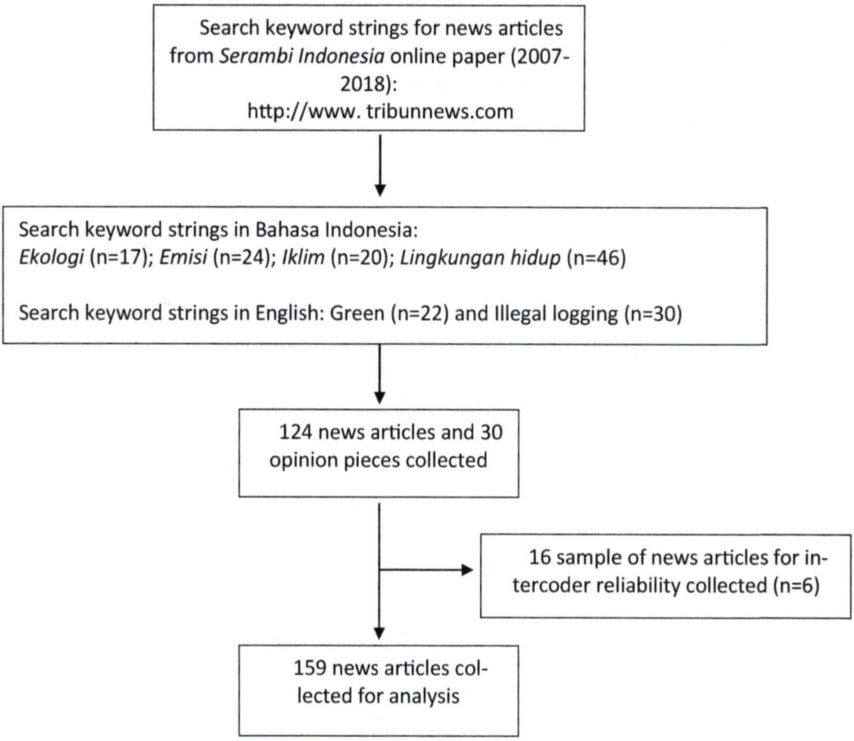

Fig. 14.1 Flow diagram of search keyword strings strategy

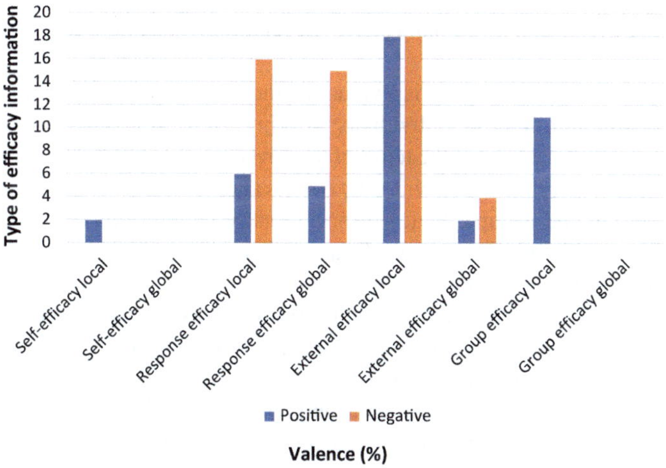

Fig. 14.2 Efficacy information by type in *Serambi Indonesia* coverage (% of total coverage). *Note* the bars do not total to 100%, because seven articles contained no efficacy information

negative (Fig. 14.1). Conversely, six of the 16 possible types of efficacy information for which we coded were not the focus of any article in our sample. The only theme that was present to any meaningful extent in the global efficacy information was negative external efficacy (24 articles—over half the global sample).

Discussion

Our content analysis examined how *Serambi Indonesia* (SI) reported on the issues of environment and climate change in general, and AGV in particular, focusing on use and valence of efficacy information for action to address the Aceh's government 'green' initiative. The results reveal that SI afforded most attention to external efficacy and response efficacy. Functionally, this means that SI was more likely to report on the capacity of authoritative actors (government, policymakers, NGOs, religious, and customary leaders) to act on environmental issues (local, national, and global level) than to focus on efficacy of individuals or civil society groups, who could take action to address the realities and risks of environmental challenges in Aceh.

The significant *differences in valence between types of efficacy information* has practical implications—for example, it may suggest that SI journalists (and Indonesian journalists generally) should consider whether more self-efficacy and group efficacy information can be given, and think about whether there are additional positive examples of response and external efficacy that can be provided. This empirical finding corresponds with the work of Feldman et al. (2017), but adds the important category of group efficacy—which could be particularly important for incentivising action on climate change, due to always being positive in the sample of articles we analysed here (although notably only in relation to local environment, and not global).

The *differences in authorship* is a valuable finding, because it shows that targeting a relatively small number of individuals working for a local newspaper such as SI might be valuable in changing the tenor of the coverage. Analysis of the differences in types of efficacy information provided by editors and non-editors shows similar levels of self, external, and group efficacy information by both groups of writers, but that editors offer notably more response efficacy information, which tends to be the most negative (i.e. highest percentage of negative information of any form of efficacy). Importantly, the more negative information from editors cannot be accounted for by the difference between news and opinion articles, as there is no significant difference in valence between those two article types.

The increased focus on environmental and 'green' issues in SI in 2017–2018 might be due, at least in part, to this topic featuring prominently in the election campaign In Aceh. Irwandi Yusuf was elected governor in the 2006 Aceh regional election; after taking office in 2007, he quickly implemented the AGV. He left office in 2012, but then ran again for governor in 2017. Political issues in relation to the election become a substantial topic that SI covered heavily. In Irwandi's campaign in 2017, he used AGV and his actions and stance on other environmental issues to win the election (Prang 2017; Zainuddin 2017).

Once Irwandi took office in July 2017, he and his staff prepared a follow-up program to the former AGV to address environmental issues in Aceh. Irwandi even allocated a specific budget for 'Green Aceh' from the Aceh Revenue and Expenditure Budget (APBA) for programs related to reducing climate change risks and programs to strengthen emissions mitigation system and to support disaster risk reduction (Mustakim 2018). SI gave much attention to these new programmes during his first year of his second term as governor. Irwandi was supposed to serve as governor from 2017 to 2022; however, in 2018 he was arrested on allegations of corruption. Irwandi's leadership likely influenced the sustainability of not only the environmental policies themselves, but also the attention that was given to them in coverage in SI. The current acting governor has not prioritised 'green' issues (Bakri 2018).

Another finding from our data suggests that, consistent with the far more negative efficacy information in 2017–2018, we also see a very notable increase in the amount of response efficacy information provided in those years (recall that most response efficacy information was negative). The negative efficacy information likely stems from circumstances surrounding the governor election in Aceh: the SI coverage highlighted social and political tension, focusing more on the political conflicts during campaign instead of addressing in detail the strategies and programmes envisioned by each candidate. The shifting political interest in these environmental issues may reinforce how the government has not played enough of a role in solving environment problems in the province; thus, most of the SI coverage in 2017–2018 addresses the failed approaches of policymakers who sought to address the issues.

Individuals and Groups Affirm the 'Green'

Our findings reveal a huge difference between positive and negative efficacy information based on the type of efficacy information used. There was more negative than positive efficacy information, which generally indicates that efforts and action to address and solve both local and global environmental risks remain a challenge for Indonesia, and the Aceh Province in particular. Nevertheless, whilst response efficacy and external efficacy were highly negative, every instance of group and self-efficacy for which we coded had positive valence.

The examples of positive self and group-efficacy information from our sample reveal more coverage of group efficacy than self-efficacy actions being taken to protect the environments [e.g. collecting trash/garbage, which was initiated by local non-government organizations (NGO), civil society organizations (CSO), or higher education institutions (universities)]. The findings indicate that delivering messages about the importance to reduce environmental risks or retain the Aceh Green that target individuals and groups is conceivable. Although most actions are carried out occasionally and within specific events, such as 'Car-Free Day' or commemorating the ‘Earth Day’, media coverage can target and engage individuals, groups, and sub-groups in the community to address the issue of ‘green’ and take action to preserve the environment across districts, sub-districts, and villages in the province.

Furthermore, the negative coverage linked to response and external efficacy intimates a social perception that the government and the policymakers failed in their efforts to address Aceh’s environmental issues that were targeted through policies such as the AGV. The AGV faltered in achieving its goals for three primary reasons: increase in actors’ self-interest, lack of private sector engagement, and problems in building capacity and coalitions (Swainson and Mahanty 2018).

Swainson and Mahanty (2018) argue that heightened self-interest among powerful actors in the government has sometimes led to ‘bribing, rent-seeking, dispossessing others, or avoiding action.’ This practice has been shown to strengthen the ‘culture of corruption’ (Dermawan et al. 2011; King 2000; McCarthy 2002a, b; Palmer 2001) and prevent the development of positive relationships with private sector actors, who would be less willing to invest in the province due to an unfavourable risk-reward balance. Similarly, international funding agencies are cautious to provide support for addressing environmental issues that are embedded in such a problematic political context.

The absence of developing better early-stage-capacity and coalitions resulted in the AGV’s shortage of funds (due to inability to attract external support); therefore, the AGV was recently closed as an independent institution. Nevertheless, the spirit (vision and mission) remain acknowledged and applied in the province’s medium- and long-term development plans. AGV arguably failed to develop effective networking between relevant institutions at local, national, or global levels, which has contributed the AGV’s shortcomings to establish further partnerships and collaborations to reduce environmental risks in the region.

This finding may also indicate that the AGV’s eight cross-sectoral approaches to addressing environmental issues are simply not working; thus, there is a need to apply other reporting strategies. However, there is also a shifting paradigm on environmental issues, a more community-based participatory approach instead of a top–down approach from the government by providing regulation about forest conservation, for instance. The bottom–up approach arguably supports community participation in protecting the devastating forest loss and serious environment issues in Aceh. This also signals increase potential for focus on information associated with group efficacy.

Global Environmental Not Noteworthy in Coverage

The findings indicate that SI has not prioritised sharing information with its readers about environmental issues from other countries, especially those that succeed in demonstrating best-practice to reduce environmental risks, such as Sweden. Likewise, the global effects of Indonesia’s environmental practices and challenges are sidelined. Less global efficacy news coverage might stem from local journalists’ lack of competence to interpret or translate external information, which is usually covered by English-language newspapers.

As explained in relation to the differences in authorship, the study findings have practical implications especially for journalists of SI—there is a need to consider whether more self-efficacy and group efficacy information can be provided in coverage of environmental issues. For example, journalist could consider including more positive information that endorsed self- and group-efficacy to act within the AGV's spirit. If members of the public became more active in taking action on environmental concerns, and pushing for societal action on such issues, this may help mitigate the negativity of the response and external efficacy in news coverage about environmental issues.

Conclusion

We selected SI not only due to its prominent position as the leading local newspaper circulated in Aceh and in nearby provinces such as Medan, South Sumatera but also because SI remains amidst the collapse of other print and online news in the province. Nevertheless, the findings derived from this single newspaper may not accurately represent coverage of environmental issues and the AGV across other newspapers. This study indicates the importance for climate change communication, and potentially for societal action on climate change, to understand the role of media in addressing climate change.

Future research could also analyse and compare the coverage of AGV and other environmental rhetoric from other local newspapers, including other media platforms such as blogs, to examine and understand how these platforms present efficacy information on the issues (e.g. types of efficacy information and valence). This descriptive content analysis of newspaper coverage of the AGV gives an overall evaluation of media advocacy efforts in the province. The study provides valuable information for policymakers and media personnel to report and (re)frame the environmental issues by considering how the coverage includes efficacy information that can foster (or close off) people's awareness, knowledge, and concrete action on the issues. To gain positive public engagement, there would be a need to extend the study through interviews with journalists and readers, to investigate their perceptions and interpretations on the significance of the 'green'.

References

Abood SA, Lee JSH, Burivalova Z, Garcia-Ulloa J, Koh LP (2015) Relative contributions of the logging, fiber, oil palm, and mining industries to forest loss in Indonesia. Conserv Lett 8(1):58–67. https://doi.org/10.1111/conl.12103

Anderson A (2009) Media, politics and climate change: towards a new research agenda. Sociol Compass 3(2):166–182

Ardiansyah F, Marthen AA, Amalia N (2015) Forest and land-use governance in a decentralized Indonesia: a legal and policy review 132, CIFOR

Atmadja S, Wollenberg E (2010) Chapter 5 Indonesia. In: Springate-Baginski O, Wollenberg E (eds), REDD, forest governance and rural livelihoods the emerging agenda. Center for international forestry research (CIFOR)

Bakri (2018) Program Aceh Green Masih Sebatas Wacana. Serambi Indonesia. Retrieved from https://aceh.tribunnews.com/2018/01/02/program-aceh-green-masih-sebatas-wacana

Bandura A (2004) Health promotion by social cognitive means. Health Educ Behav 31(2):143–164. https://doi.org/10.1177/1090198104263660

Boykoff MT (2011) Who speaks for the climate?: making sense of media reporting on climate change. Cambridge University Press

BRR (2008) Green economic development and investment strategy for Aceh, Indonesia "Aceh Green". Badan Rehabilitasi dan Rekonstruksi NAD-Nias

Brummitt C (2007) Carbon trade seen as lifeline for Indonesian forests. Retrieved from https://www.ap.org/en-gb/

Buttler R (2013) Palm oil now biggest cause of deforestation in Indonesia. Retrieved from http://mongabay.com/2013/09/palm-oil-now-biggest-cause-of-deforestation-in-indonesia

Clarke TM, Durrant JR (2010) Charge photogeneration in organic solar cells. Chem Rev 110(11):6736–6767. https://doi.org/10.1021/cr900271s

Covello VT, Peters RG, Wojtecki JG, Hyde RC (2001) Risk communication, the West Nile virus epidemic, and bioterrorism: responding to the commnication challenges posed by the intentional or unintentional release of a pathogen in an urban setting. J Urban Health 78(2):382–391

Danna K, Griffin RW (1999) Health and well-being in the workplace: a review and synthesis of the literature. J Manage 25(3):357–384. https://doi.org/10.1177/014920639902500305

Dermawan A, Petkova E, Sinaga A, Muhajir M, Indriatmoko Y (2011) Preventing the risks of corruption in REDD+in Indonesia, vol 80. CIFOR

Evensen DT, Clarke CE (2012) Efficacy information in media coverage of infectious disease risks: an ill predicament? Sci Commun 34(3):392–418

Feldman L, Hart PS, Milosevic T (2017) Polarizing news? Representations of threat and efficacy in leading US newspapers' coverage of climate change. Publ Underst Sci 26(4):481–497

Floyd DL, Prentice-Dunn S, Rogers RW (2000) A meta-analysis of research on protection motivation theory. J Appl Soc Psychol 30(2):407–429

Guardian T (2013) The Sumatran rainforest will mostly disappear within 20 years. The Guardian

Hart PS, Feldman L (2014) Threat without efficacy? climate change on US network news. Sci Commun 36(3):325–351

Haryanto I (2011) Media ownership and its implications for journalists and journalism in Indonesia. In: Politics and the media in twenty-first century Indonesia: decade of democracy, pp 104–118

Hein JI (2018) Political ecology of REDD+in Indonesia: agrarian conflicts and forest carbon, Routledge

Idris ND (2016) A win-win solution for economy, environment. The Jakarta Post. Retrieved from https://www.thejakartapost.com/academia/2016/06/09/a-win-win-solution-for-economy-environment.html

Joy A (2010) Illegal Logging in Indonesia: report from Aceh. Retrieved from

King DY (2000) Corruption in Indonesia: a curable cancer?. J Int Aff 603–624

Margono BA, Potapov PV, Turubanova S, Stolle F, Hansen MC (2014) Primary forest cover loss in Indonesia over 2000–2012. Nat Clim Change 4(8):730

McCarthy JF (2002a) Power and interest on Sumatra's rainforest frontier: clientelist coalitions, illegal logging and conservation in the Alas Valley. J SE Asian Stud 33(1):77–106
McCarthy JF (2002b) Turning in circles: district governance, illegal logging, and environmental decline in Sumatra, Indonesia. Soc Nat Res 15(10):867–886
McComas WF (2004) Keys to teaching the nature of science. Sci Teacher 71(9):24
Mead PS, Slutsker L, Dietz V, McCaig LF, Bresee JS, Shapiro C, Tauxe RV (1999) Food-related illness and death in the United States. Emerg Infect Dis 5(5):607
Milne S, Sheeran P, Orbell S (2000) Prediction and intervention in health-related behavior: a meta-analytic review of protection motivation theory. J Appl Soc Psychol 30(1):106–143
Miranda RJ (2018) Double standards in Indonesia's climate policy. The Jakarta Post. Retrieved from https://www.thejakartapost.com/academia/2018/01/04/double-standards-in-indonesias-climate-policy.html
Murdiyarso D, Dewi S, Lawrence D, Seymour F (2011) Indonesia's forest moratorium: a stepping stone to better forest governance?. Cifor
Mustakim (2018) Irwandi Anggarkan Program Aceh Hebat Dalam APBA 2018. Acehsatu.com. Retrieved from https://acehsatu.com/irwandi-anggarkan-program-aceh-hebat-dalam-apba-2018/
Nellemann C (2007) The last stand of the orangutan: state of emergency: illegal logging, fire and palm oil in Indonesia's national parks, UNEP/Earthprint
News R (2007) Indonesia's Aceh, papua pledge to protect forests. Retrieved from http://reuters.com
Palmer C (2001) The extent and causes of illegal logging: an analysis of a major cause of tropical deforestation in Indonesia
Panaligan R (2010) Another tragedy in Aceh: illegal logging
Prang AJ (2017) 'Welcome Back' Bang Wandi. Serambi Indonesia. Retrieved from https://aceh.tribunnews.com/2017/04/10/welcome-back-bang-wandi
Purwanto E (2014) Environmental values of forest resources. The Jakarta Post. Retrieved from https://www.thejakartapost.com/news/2014/11/04/environmental-values-forest-resources.html
Riezal C (2016) Perkembangan Surat Kabar Serambi Indonesia (1989–2015). ETD Unsyiah
Rimal RN, Real K (2003) Understanding the influence of perceived norms on behaviors. Commun Theory 13(2):184–203
Sahide MAK, Supratman S, Maryudi A, Kim Y-S, Giessen L (2016) Decentralisation policy as recentralisation strategy: forest management units and community forestry in Indonesia. Int For Rev 18(1):78–95
Smith J (2005) Dangerous news: media decision making about climate change risk. Risk Anal An Int J 25(6):1471–1482
Springate-Baginski O, Wollenberg E (2010) REDD, forest governance and rural livelihoods: the emerging agenda, CIFOR
Stigler GJ, Becker GS (1977) De gustibus non est disputandum. Am Econom Rev 67(2):76–90
Stryker JE, Moriarty CM, Jensen JD (2008) Effects of newspaper coverage on public knowledge about modifiable cancer risks. Health Commun 23(4):380–390
Swainson L, Mahanty S (2018) Green economy meets political economy: lessons from the "Aceh Green" initiative, Indonesia. Glob Environ Change 53:286–295
Swainson LJ (2016a) The complexities of green economy policy reform: a case study in Aceh, Indonesia
Swainson LJ (2016b) The complexities of green economy policy reform: a case study in Aceh, Indonesia (Ph.D), Australian National University
Topsfield J, Rompies K (2016) Haven in peril: Aceh activists lose class action over Indonesia's Leuser rainforest. The Sydney Morning Herald. Retrieved from https://www.smh.com.au/world/haven-in-peril-aceh-activists-lose-class-action-over-indonesias-leuser-rainforest-20161130-gt10u9.html
Tsujino R, Yumoto T, Kitamura S, Djamaluddin I, Darnaedi D (2016) History of forest loss and degradation in Indonesia. Land Use Policy 57:335–347

Turner MM, Rimal RN, Morrison D, Kim H (2006) The role of anxiety in seeking and retaining risk information: testing the risk perception attitude framework in two studies. Human Communication Research 32(2):130–156

UNCC (2019) intended nationally determined contribution Republic of Indonesia. United Nations Climate Change. Retrieved from https://unfccc.int/files/focus/indc_portal/application/pdf/indonesia_indc.pdf

WFN (2016) Saving Sumatra's iconic species in the leuser ecosystem. Retrieved from https://whitleyaward.org/winners/sumatras-iconic-species-in-the-leuser-ecosystem/

Witte K, Allen M (2000) A meta-analysis of fear appeals: Implications for effective public health campaigns. Health Education & Behavior 27(5):591–615. https://doi.org/10.1177/109019810002700506

Zainuddin Z (2017) A renewed hope for Aceh. The Jakarta Post. Retrieved from https://www.thejakartapost.com/academia/2017/04/07/a-renewed-hope-for-aceh.html

Chapter 15
Peatland Protection in Indonesia: Toward the Right Direction?

Andri G. Wibisana and Savitri Nur Setyorini

Abstract As a country with an extensive area of peatlands, Indonesia provides abundant sources of natural resources and a vast amount of carbon stocks. However, the uncontrolled logging and the expansion of timber and palm oil plantations have changed Indonesia's peatlands into a source of disasters, that is, wildfires. For years, Indonesia has been accused of violating state responsibility about transboundary haze resulting from wildfires in peatlands. This chapter discusses recent legal development in the protection and management of peatlands in Indonesia and analyzes how regulatory measures addresses the issue of peatland rehabilitation. The chapter argues that requiring the water-table level of a minimum of 0.4 m from the surface is justified according to the precautionary principle. The chapter is also of the opinion that the recovery of peatlands needs to be conducted under the polluter-pay principle, leading to the use of strict liability against peatland draining activities.

Keywords Logging · Timber and palm oil · Haze · Peatland · Legal development

Introduction

It has been argued that annual fires that occurred in Indonesia have a great deal to do with lack of law enforcement and indicate inadequate government responses toward peatlands degradation. One could, for example, refer to Tan (2015, pp. 5–6, 9) and Nurhidayah et al. (2015, pp. 195–197), who argue that in the context of international law haze caused by forest and land fires in Indonesia has indicated a violation of state responsibility principle, namely the responsibility to prevent transboundary pollution from forest fires. Following the 1997 fires, Tan (1999, pp. 847–848) concluded that Indonesia violated the principle of state responsibility, and hence:

> ...Indonesia is internationally responsible for the occurrence of large-scale fires and consequent transboundary injury to neighbouring States by failing to control the actions of its citizens within its territory. It first failed to prevent transboundary harm by not using its

A. G. Wibisana (✉) · S. N. Setyorini
Faculty of Law, Universitas Indonesia, Depok, Indonesia
e-mail: andri.gunawan@ui.ac.id

R. Djalante et al. (eds.), *Climate Change Research, Policy and Actions in Indonesia*,
Springer Climate, https://doi.org/10.1007/978-3-030-55536-8_15

> legislative and administrative powers to the fullest extent possible to prevent the fires from being started by the commercial enterprises. Once these fires had been detected and transboundary harm occasioned to the injured States, Indonesia further failed to control the actions of the commercial enterprises and to compel them to cease their harmful conduct. Further, based upon the lenient penalties imposed to date, it would appear that Indonesia has breached its responsibility to punish the offenders adequately in order to prevent future violations.

A similar opinion is also put forward by Tay, arguing that according to the Principle 21 of the Stockholm Declaration and Principle 2 of the Rio Declaration, "Indonesia should be held responsible and accountable to the countries which have been affected by the fires for the consequences of its failure to enforce its own laws and to control the fires. This would arguably be so even if the activities causing the haze may be traced to private companies in distant areas, over which the government in Jakarta may have little effective control" (Tay 1998, 206).

In the last few years, however, legal reforms have taken place related to Indonesia's peatland management. Having realized the importance of peatlands, the Indonesian government has initiated various efforts in improving the protection and management of peatlands. New regulations have been issued to strengthen the peatland protection and the prevention and control of fires. These include Peatland Regulation, that is, Government Regulation (GR) Number 71 of 2014 as amended by GR 56 of 2017 on peatland protection, and various regulations of Minister of the Environment and Forestry (MoEF) in 2017. As a response to the 2015 fires, the President signed the Presidential Regulation Number 1 of 2016 on the Peatland Restoration Agency, as the agency is responsible for accelerating peatland restoration in seven provinces worst affected by the fires.

The regulatory responses above take place along with the increase in the number of law enforcement against companies accused of triggering fires. In this regard, one could note that since 2015, the Ministry of the Environment (MOE), later becomes Ministry of Environment and Forestry (MoEF), has imposed various administrative sanctions to some companies. There are also increased criminal prosecutions against the companies and their CEOs since in recent years. In September 2015 alone, the MoEF imposed four administrative sanctions in terms of government coercion, four suspensions of permits, and two revocations of permits. During the same period, there were also 18 companies named as suspects for forest fires (MoEF 2016). This is a significant increase, because from 2010 to 2014, there were no administrative sanctions imposed, and there were only two corporations that were named suspects (MoEF 2016, 191).

This chapter aims to analyze the legal aspects of peatland protection and management in Indonesia. In this regard, the chapter discusses the importance of the various regulations on peatland, and the challenges faced in implementing these regulations. The chapter also analyzes how such regulatory measures addresses the issue of peatland recovery. The chapter argues that the recovery of peatlands needs to be conducted under the polluter-pay principle. Consequently, law enforcement, including lawsuits against concession holders, should play an essential role in financing the recovery of peatlands.

The term "recovery" (*pemulihan*) is used to refer to various activities, including rehabilitation, remediation, restoration, and other measures in accordance with scientific and technological development (Law Number 32 of 2009, art. 54). With respect to peatlands, the term restoration refers to very specific activities, including the implementation of restoration techniques, such as water management, the construction, operation, and maintenance of rewetting infrastructures of peatlands, and the implementation of cultivation methods according to local practices (Government Regulation Number 57 of 2016, art. 30A par. 1).

This chapter is structured as follows. After this introduction, Section "Peatland Degradation in Indonesia" will briefly explain the extent and impacts of peatland degradation in Indonesia. Section "Peatland Protection and Indonesia's GHG Emissions" shows the importance of peatland protection and the prevention of wildfires on peatlands to the fulfillment of Indonesia's emission reduction target. Section "Indonesia's Responses to Peatland Degradations" explains several responses undertaken by the government. They include provisions on the 1999 Forestry Law, the 2009 Environmental Law, and their respective regulations. The explanations indicate that progressive changes toward both fire prevention and control and peatland protection and recovery are made to implement the 2009 Environmental Law. Hence, from a legal perspective, the 2009 Environmental Law plays a more important role compared to the 1999 Forestry Law in providing the legal basis for peatland protection and recovery. Section "Legal Challenges for Peatland Protection: The Importance. Roles of the Principles of Environmental Law"analyzes the roles of legal principles in shaping Indonesia's responses to the protection and management of peatlands. Section "Conclusion" provides some concluding remarks.

Peatland Degradation in Indonesia

Peatlands are organic soils formed through decomposition of plant materials, accumulated for thousands of years under waterlogged and acidic conditions. Peatlands are typically characterized by their moisture content, hydraulic conductivity, water tension, and water yield (Adesiji et al. 2014, p. 483). According to Hooijer and colleagues, undisturbed peat consists of plant remains and water, respectively, of around 10 and 90% (Hooijer 2010, p. 1505).

Since the 1970s, the vast areas of lowland tropical peatland in Southeast Asia have been converted into agriculture, following forest clearance and drainage (Wösten et al. 2008, p. 212). Agriculture and industrial forestry have induced land-use changes in peatlands, which require drainage of the water-saturated peat. The process of subsidence will immediately follow the drainage of peatlands. Unfortunately, the subsidence of every cm leads to the release of 13 t CO_2 per hectare per year (Wösten et al. 2008, p. 213).

Indonesia's peat swamp forests have been deforested and converted continuously in recent decades. As a result, the proportion of peat swamp forests in Sumatra and Kalimantan declines dramatically from 76% in 1990, to 40% in 2007, and to 29% in

2015 (Hergoualc'h 2017, p. 204). The decline is due to unsustainable deforestation, uncontrolled forest fires, and unsupervised agricultural development, which lead to a drastic reduction in the peat swamp forest. Between 2007 and 2015, as much as 73% of forests were converted to smallholdings and industrial agriculture. The transition from forest to industrial agriculture is the primary type of conversion occurring in Kalimantan and Sumatra, at 64 and 44% (Hergoualc'h 2018, p. 1). Eventually, Indonesia's peatlands have declined to a critical condition, with an average decrease of 2.6% per year (Hergoualc'h 2017).

The moisture characteristics of peat do not allow natural fires to occur in tropical peatlands. However, the lands are usually actively drained, or even burned, for agricultural purposes (Chamorro et al. 2017; Risnandar and Fahmi 2018). The fire history, according to the Global Forest Watch Fires' data, confirms that fires tend to concentrate on agricultural and concessions in Indonesia. Between 2013 and 2014, about 75% of forest fires did not occur in forests (Dayne 2018). As much as 82% of fires occurred in peatlands which had not been planted or covered with scrub, fern, and burning deadwood (Gaveau et al. 2017, p. 259). The data confirms that agricultural expansion is the most significant cause of peatland fires (Chamorro et al. 2017; Mudiyarso et al. 2010, p. 19655; and Hergoualc'h 2017, p. 204).

Hence, the conversion of peatlands into an agricultural area or other purposes is the primary indication of degraded peatlands. For example, commenting on Suharto's Mega Rice Project, Aldhous states that the peatlands were drained by the construction of canals for the Project. "Even the dictator's decree," so the author continues, "can't make water run uphill, so the canals simply sucked the peat dry" (Aldhous 2004, p. 144).

Other causes of peatland degradation could include the clearing of peatlands allocated for plantations through the use of the slash-and-burn method. Also, the burning of land could be used as an instrument to claim occupation and "ownership" of land during tenurial conflicts (Harrison 2009, p. 157).

These forest and peat fires also affect several things, ranging from environmental losses such as damage to forests, soils, plants, to losses and threats to Indonesia's endemic species, such as orangutans. Another problem that could also arise from forest/land fires is related to health effect from the haze. This might be the case when fires occur for months, causing severe respiratory problems in society. Fires have disrupted Indonesia's economic activities, from which Indonesia suffered economic losses amounting to 6.1 billion dollars by 2015 (Chamarro et al. 2017) and experienced difficult diplomatic relations with other neighboring countries (Hergoualc'h 2018, p, 3).

Peatland Protection and Indonesia's GHG Emissions

Peatland contributes a significant portion of Indonesia's greenhouse gas (GHG) emissions. According to Indonesia's second National Communication of 2011, total GHG emissions in 2000 for the three main greenhouse gases (CO_2, CH_4, and N_2O, PFC) without LULUCF (LUCF and peat fires) reached 556,728.78 Gg CO_2e. When the emissions from the LULUCF sector is accounted for, total GHG emissions increase to about 1,377,982.95 Gg CO_2e. The GHG emissions were distributed unevenly between the three gases recorded, that is, net CO_2 was about 80.8%, methane (CH_4) was about 17.2%, and nitrous oxide (N_2O) was 2.0% of the National GHG emissions. The primary contributors of Indonesia's emissions are emissions from the LULUCF sector, energy, wildfires on peatlands, waste, agriculture, and industry (Indonesia's Second National Communication of 2011, p. II–3).

Due to its significant contribution to Indonesia's GHG emissions, it is logical for the government to put emphasis also on emission reduction from peatlands. One could see emission reduction target from various sectors in the President Regulation (PR) Number 61 of 2011 on the National Action Plan on GHG Emission Reduction (hereinafter referred to as the 2011 NAP-GHG). The NAP-GHG should be read in conjunction with Indonesia's voluntary emission reduction targets pledged in 2009, that is, emission reduction by 2020 of at least 26% from the business-as-usual (BAU) level without international support, and up to 41% from the BAU level with international supports (Moeliono et al. 2014, p. 9).

PR Number 61 of 2011 defines the 2011 NAP-GHG as a working plan to undertake various activities that will directly or indirectly reduce Indonesia's GHG emissions (Art. 1 point 1). The 2011 NAP-GHG consists of various core activities in agriculture, forestry and peatlands, energy and transportation, industry, waste management, and various supporting activities (Art. 2). The 2011 NAP-GHG is expected to function as guidance for relevant ministries and government agencies to plan, implement, monitor, and evaluate various measures to reduce GHG emissions. It also guides local governments in formulating regional action plans (PR No. 61 of 2011, Art. 3). In addition, the 2011 NAP-GHG also functions as guidance for the public and business entities in conducting their plan and implementation activities to reduce GHG emissions (PR No. 61 of 2011, Art. 4).

Appendix I of the PR provides several action plans in the agricultural, forestry and peatlands, energy and transportation, industry, and waste management sectors. The 2011 NAP-GHG in the agricultural sector is directed to reduce 0.008 and 0.011 $GtCO_2$e, respectively, for 26 and 41% emission reduction targets. One example of emission reduction activities in this sector is the transformation of degraded or deforested land into palm oil, rubber, or cacao plantations. The conversion of this land into plantation is expected to cover an area of 860 thousand hectares for oil palm plantation and 105 thousand hectares for rubber plantation. The conversion is to be conducted between 2011 and 2014, in 19 provinces, with an expected emission reduction of 74.53 $MtCO_2$e. The Ministry of Forestry is in charge of this conversion of degraded or deforested land.

The 2011 NAP-GHG in the forestry and peatland sector is expected to reduce 0.672 and 1.039 $GtCO_2$e, respectively, for the 26 and 41% emission reduction targets. One example of mitigation activities in this sector is a plan related to the designation of forest areas. In this regard, it is expected that from 2010 to 2014 forest boundaries of 25,000 km should have been established. The establishment of forest boundaries will be carried out in all provinces, and are expected to result in emission reduction as much as 123.41 $MtCO_2e$. The Ministry of Forestry will be responsible for this activity.

Another example of the mitigation activities within the forestry sector is the development of sustainable agriculture in peatland. It is expected that between 2011 and 2020, sustainable agriculture has been practiced in 325,000 ha of peatlands in 11 provinces. This activity is also expected to be able to reduce emissions by 103.98 $MtCO_2e$. The Ministry of Agriculture will be the responsible ministry for the implementation of such sustainable agriculture.

Based on these targets, one could see the importance of peatland protection and the prevention of wildfires on peatlands to the fulfillment of Indonesia's emission reduction target. The needs for peatlands protection and prevention of fires on peatland become even more important after the ratification of the 2015 Paris Agreement. As stated in the First Nationally Determined Contribution, submitted in November 2016, Indonesia's emission reduction targets are extended to the post-2020 targets, that is, the 29% against business-as-usual and 41% with International help in 2030 (Toccani 2016, p. 643).

Indonesia's Responses to Peatland Degradations

As explained earlier, peatlands degradation has resulted not only from land clearing and burning activities but also from land conversion from forests into plantations. Surprisingly, peatlands degradation is very likely to result from both legal and illegal activities. This section explains how the Indonesian government has responded to peatland degradation. It seems that the responses for peatland protection can be found not in forestry provisions, but in environmental management provisions. More importantly, the 2015 forest fires have also boosted the enactment of stricter regulations related to fire prevention and peatland protection. The provisions concerning peatland protection and management can be summarized in Table 15.1.

Forestry Law and Forest Protection Regulation in the 2000s: Business-as-Usual

From a legal point of view, provisions on peatland protection could fall into provisions within forestry management. In this regard, one could refer to Law Number 41 of 1999 on Forestry, hereinafter referred to as the 1999 Forestry Law. The Law defines forests as an ecosystem in the form of landscapes containing biological resources dominated by trees integrated with its natural environment (the 1999 Forestry Law, art. 1 point 2).

The 1999 Forestry Law provides several provisions prohibiting certain activities, including illegal logging; forest burning; carrying equipment for logging within

Table 15.1 Summary of peatland protection and management provisions

Year	Provision(s)
1997	Law Number 23 of 1997 on Environmental Management (The 1997 Environmental Law)
1999	Law Number 41 of 1999 on Forestry (The 1999 Forestry Law)
2001	Government Regulation Number 4 of 2001 on Environmental Damage and Pollution Control Related to Forest and Land Fires
2004	Government Regulation Number 45 of 2004 on Forest Protection
2009	Law Number 32 of 2009 on Environmental Protection and Management (The 2009 Environmental Law)
2011	Presidential Instruction Number 10 of 2011 on Moratorium of the Issuance of Licenses within Primary Forest and Peatlands (extended through Presidential Instruction Number 6 of 2013, Presidential Instruction Number 8 of 2015 and Presidential Instruction Number 6 of 2017)
2013	The Decree of Indonesia's Chief Justice Number 036/KMA/SK/II/2013on Guideline for Environmental Case Handling
2014	Government Regulation Number 71 of 2014 on Peatland Protection and Management (amended by Government Regulation Number 57 of 2016)
2016	Presidential Regulation Number 1 of 2016 on Peat Restoration Agency
2017	The Regulation of the Minister of Environment and Forestry Number 14 of 2017 on Procedures for the Inventory and Determination of Peat Ecosystem Functions
	The Regulation of the Minister of Environment and Forestry Number 15 of 2017 on Procedures for Water-table Measurement at the Compliance Point within Peat Ecosystem
	The Regulation of the Minister of Environment and Forestry Number 16 of 2017 on Technical Guidelines for Recovery of Peat Ecosystem Functions
	The Regulation of the Minister of Environment and Forestry Number 17 of 2017 on Amendment of The Regulation of the Minister of Environment and Forestry Number P.12/MENLHK-II/2015 on Development of Industrial Forest

Source authors

forest that is not covered by a license; cutting the trees within a certain radius from a river, valley, lake, or coast; and owning, selling, or buying timbers that have no valid document (the 1999 Forestry Law, art. 50 part. 3). The violation against these obligations, responsibilities, and prohibitions can result in administrative sanctions and civil and/or criminal liability (the 1999 Forestry Law, arts. 71–76, 77–79 and 80).

Concerning forest fires, the 1999 Forestry Law states that the protection of forest is an effort to, among others, prevent and limit forest damage caused by human actions, livestock, fires, nature powers, pests as well as diseases (the 1999 Forestry Law, art. 47). In this context, the 1999 Forestry Law further obliges the license holders, as well as parties who have the authority to manage forests, to protect forests under their control or authority from damage due to human actions, livestock, and fires (the 1999 Forestry Law, art. 48 part. 3). Accordingly, it could be concluded that the responsibility to prevent and control forest fires lies with the license holders. They are the parties who have the obligation to ensure that areas within their control will not experience fires. The 1999 Forestry Law also strengthens this responsibility by stating that the rights of license holders are responsible for fires taking place in their area of work (the 1999 Forestry Law, art. 49).

In addition to these responsibilities, the 1999 Forestry Law also contains provisions related to forest fires. The Law prohibits every person from burning any forest (the 1999 Forestry Law, art. 50 part. 3 letter d) and from throwing any object that has the potentials of causing fires (the 1999 Forestry Law, art. 50 part. 3 letter l). The burning of forest falls within the so-called *delik formal* (*formal delict*), meaning that an activity is considered a crime regardless of the consequences of the activity. Burning is punishable by the imprisonment of a maximum of 15 years and a fine of maximum IDR 5 billion for violations of article 50 par. 3 letter d (the 1999 Forestry Law, art. 78 part. 3) and imprisonment of maximum three years and fine of maximum IDR 1 billion for violations of article 50 sub-article 3 letter l (the 1999 Forestry Law, art. 78 sub-art. 11).

Provisions on forest protection in the 1999 Forestry Law are further stipulated in Government Regulation Number 45 of 2004 on Forest Protection. The Regulation holds the forest license holders responsible for taking forest protection measures within their respective working area. Specifically, the Regulation requires the license holders to take measures to prevent and control fires (GR Number 45 of 2004, art. 6 letters a and art. 8 par.4), regardless of whether the fires were triggered by human actions or natural factors (GR Number 45 of 2004, art. 18). Despite the provisions on the prevention of forest fires, law and regulations on the forestry sector remain silent on the protection and management of peatlands. No specific provision on peatlands can be found in these laws and regulations.

The Importance of Environmental Law in the 2010s

In 2009, the government enacted Law Number 32 of 2009 on Environmental Protection and Management (the 2009 Environmental Law). The Law replaces the previous law on environmental management, that is, Law Number 23 of 1997. New provisions are introduced in the 2009 Environmental Law, including those related to forest or land fires and peatlands.

(a) The 2009 Environmental Law and Peatland Protection

The 2009 Environmental Law provides provisions related to fires and peatlands. In the first place, the provisions on fires and peatlands can be found in those related to environmental damage. The Law stipulates that these criteria are used to determine whether a particular environment has been damaged. These criteria are further divided into criteria for determining ecosystem damage on the one hand and climate change damage on the other hand (the 2009 Environmental Law, art. 21, paras 1–2). In this regard, the Law introduces damage due to forest fires on the one hand and peatland damage on the other, as two types of standards to determine ecosystem damage (the 2009 Environmental Law, art. 21 part. 3 letter c and f).

Introducing peatland damage in the 2009 Environmental Law is justifiable since peatland could be found not only inside but also outside the forest areas. Hence, once the criteria to define peatland damage have been exceeded, it will be considered that environmental damage has taken place. This damage will stimulate necessary measures for peatland protection and recovery. More importantly, provisions on peatland stipulated in the 2009 Environmental Law prove to play a very important role in providing the legal basis for further regulation on peatland protection.

Furthermore, wildfires are also addressed in the 2009 Environmental Law. In addition to provisions related to criteria of environmental damage, the 2009 Environmental Law also sets forth prohibitions to cause pollution/environmental damage (the 2009 Environmental Law, art. 69 par. 1 letter a). Read in conjunction with the standard of environmental damage, the provision might mean a prohibition of conducting any action that might lead to wildfires. To further strengthen this point, the 2009 Environmental Law also prohibits land clearing by burning the forested land (the 2009 Environmental Law, art. 69 par. 1 letter h). If a person conducted an activity that causes forest/land fires and these fires lead to pollution/environmental damage, then this person has committed a *delik materil* (material delict), namely an activity that is considered a crime because of its consequences. This *delik materil* is punishable with the imprisonment of 3–15 years and the fines of IDR 3–15 billion (the 2009 Environmental Law, arts. 98 and 99). In addition, the 2009 Environmental Law also provides another provision which regards the burning of forested land a *delik formil*, that is, a criminal act regardless of the effect of the act, which is punishable with imprisonment of 3–10 years and the fines of IDR 3–10 billion (the 2009 Environmental Law, art. 108).

With respect to civil liability, it could be argued that some plantation activities, especially the drainage of tropical peatlands, might be considered as activities that could be held liable according to strict liability. Since it is likely that the utilization of forested land or peatlands for timber and plantations does not use hazardous substances, nor produce hazardous wastes, one might wonder how strict liability could be applicable for losses from wildfires allegedly resulted from the utilization. To answer this question, one needs to look at whether the operation of timber and palm oil plantations, particularly related to land or peatland clearing, is indeed posing serious threats to the environment. One could first refer to the definition of "serious threats" set forth in the 2009 Environmental Law. The law defines "serious threats" as threats that have the potentials to give rise to a wide range of environmental impacts and public anxiety (the 2009 Environmental Law, art. 1 point 34). Furthermore, the term "serious threats" is also explained in the Chief Justice Decree Number 36 of 2013 as pollution or environmental damage likely to be irreversible or with widespread consequences to human health, water surface, underground water, soil, air, plants, or animals (The Decree of Indonesia's Chief Justice Number 036/KMA/SK/II/2013, p. 39).

It could further be argued that strict liability is applicable if the clearing of the land or peatland falls into an unnatural use of land, or if the activity is exposing the area into threats of forest fires so significant that an exercise of utmost care cannot eliminate the threats (US Restatement (Second) of Torts § 520). The case for this type of argument is stronger when the activities of land or peatland clearing for timber and palm oil plantations cover a very large area. Eventually, it is the court who will judge whether the activities have the potential to give rise to serious threats to the environment.

Another way to argue that strict liability is applicable to wildfires is by referring to regulations related to wildfires. This type of argument indicates that it is the decision-makers, that is, the regulators or lawmakers, who decide whether strict liability is applicable for forest fires. Surprisingly, there are few provisions indicating that the license holders in forest-related activities are strictly liable for forest fires. For this purpose, one might refer to Article 49 of the 1999 Forestry Law, stating that the license holders are responsible for forest fires occurring within an area under their control. A similar provision can also be found in Article 21 par. 1 of the GR Number 4 of 2001, which determines that the license holders are not only responsible for the fires in an area under their control, but also are also under an obligation to control the fires and conduct remediation programs for environmental impacts of the fires.

Another important policy related to peatland protection is Presidential Instructions on the moratorium of permits for activities within primary forest and peatlands. To indicate that Indonesia is serious enough in carrying out REDD activities, especially those related to the implementation of the LoI with the Norwegian Government, the President has recently issued a Presidential Instruction Number 10 of 2011 on The Moratorium of the Issuance of Licenses Within Primary Forest and Peatlands. This instruction states that all ministries and regional governments should take necessary measures to support the moratorium of the issuance of timber license on the "primary" forest and peatlands. In this case, the President instructs to postpone the granting of new licenses for primary natural forests and peatlands in conservation forests, protection forests, production forests (limited production forests, regular/fixed production forests, convertible production forests), and other areas of use as indicated in the *Peta Indikatif Penundaan Izin Baru* (PIPIB, the map indicating areas where the moratorium for the issuance of new licenses is applicable). The instruction states that the moratorium program will be implemented for two years, except for the implementation of nationally vital development, including the development of geothermal, oil and gas, electricity, and land for paddy, rice, and sugarcane. The moratorium has been then extended three times, each for two years, respectively through Presidential Instruction Number 6 of 2013, Presidential Instruction Number 8 of 2015, and Presidential Instruction Number 6 of 2017.

These instructions have triggered mixed opinions. On the one hand, some commentators argue that the moratorium is less optimal since it renders no additionality to the existing level of forest and peatland protection. In this regard, it has been argued that the moratorium is directed on the primary forest and peatlands, which most of them are already protected under Indonesian laws (Simamora 2011, p.3). In addition, although initially the proposed moratorium also covered the so-called "secondary forest", in addition to the primary forest and peatland, finally the President excluded the secondary forest from the Presidential instruction. It should be noted that terms primary and secondary forests are problematic since there is no reference to these terms in the 1999 Forestry. Hence, a clear definition of what constitutes the primary and secondary forests remains to be seen. On the other hand, the government states that the moratorium is a necessary first move in the implementation of REDD. In the future, there will be more regulations on forest preservation (The Jakarta Post, May 23rd 2011, p. 2).

(b) Fire Prevention and Control

The prevention of forest/land fires is also addressed in Government Regulation Number 4 of 2001 on Environmental Damage and Pollution Control Related to Forest and Land Fires. In addition to prohibiting forest/land fires (art. 11), this GR also requires every person to prevent environmental damage in relation to forest/land fires (GR Number 4 of 2001, art. 12). The responsibility and obligation to prevent forest fires are vested in those who carry out business or activities (GR Number 4 of 2001, art. 13). In this case, they have to ensure that adequate facilities and infrastructures are in place, in order to prevent fires in their business location. The facilities or infrastructures include: (a) An early detection system which informs the event of forest/land fires; (b) Equipment to prevent forest/land fires; (c) Standard operation procedure to prevent and control forest/land fires; (d) An organization that is responsible for preventing and controlling forest/land fires; and (e) Regular training for controlling forest/land fires (GR Number 4 of 2001, art. 14). Further, the GR rules that those who carry out business/activities also bear the obligation to conduct supervision and to provide a regular report to the government concerning the result of the supervision (GR Number 4 of 2001, art. 15). They are also responsible for the forest/land fires occurring in an area covered by their business/activities and hence are required to immediately conduct necessary actions to control forest/land fires and to conduct recovery of environmental impacts arising out of forest fires (GR Number 4 of 2001, art. 18 par.1 and art. 22 par.1).

With respect to civil liability, the GR Number 4 of 2001 provides provision on strict liability (art. 51). This article is actually a provision on strict liability that is similar to the formulation of strict liability under article 35 of the 1997 EMA. However, it appears that the applicability of strict liability to forest fires is not straightforward. It requires more explanation regarding whether or not the defendant's activity belongs to the category of abnormally dangerous activity, to which the application of strict liability is justified.

The regulation is of high importance for two reasons. For one thing, it is the first government regulation specifically enacted to address forest and land fires. Hence, the regulation not only provides stipulations on responsibility for fire prevention and control but also on emergency responses to be taken by the government and business entities. For another, the regulation provides standards (criteria) relevant for determining environmental damage due to forest fires. The standards are for damage in mineral land, peatlands, and flora and fauna, as established, respectively, in Annexes A, B, and C of GR Number 4 of 2001. These standards, in turn, play a determining role in some rulings on forest and land fires.

(c) Peatland Regulations

In 2014, the President signed GR Number 71 of 2014 on Protection and Management of Peatland, later amended by GR Number 57 of 2016. The regulation considers the protection and management of peat ecosystem as systematic and integrated efforts to maintain the function of peat ecosystems and prevent damage to the peat ecosystem (GR Number 71 of 2014, art. 1 point 1). The GR defines peat as organic materials formed naturally from the remains of decomposing plants accumulating in the swamp (GR Number 71 of 2014, art. 1 point 2). The definition is revised later into organic materials naturally formed from the imperfect decomposition of plant remains accumulated in the swamp with a depth of 50 cm or more (GR Number 57 of 2016, art. 1 point 2). Hence, according to the definition above, peatlands should have the depth of minimum 50 cm, and should also consist of organic materials from plant decomposition.

The planning of peatland protection and management consists of three stages, namely peat ecosystem inventory, determination of peat ecosystem function, and the preparation and establishment of peat ecosystem protection and management plan. The peat ecosystem inventory is undertaken by the Minister of the Environment, currently the MoEF, and carried out with satellite imagery and aerial photographs, taking into account the indicative map of the distribution of the national peat ecosystem. Meanwhile, the determination of the function of peat ecosystem is carried out by the MoEF after coordination with the minister responsible for water management (i.e. Minister of Public Works) and minister responsible for spatial arrangement (i.e. the Minister of Agraria and Spatial Planning) (GR Number 71 of 2014, arts. 4, 5 and 9 part. 1).

The GR Number 71 of 2014 gives the Minister of the Environment the authority to determine the functions of peatland ecosystem, which comprise the function of protection (*fungsi lindung ekosistem gambut*) and the function of utilization (*fungsi budidaya ekosistem gambut*) (art. 9 paras 1–2). More importantly, the Minister is obliged to designate no less than 30% of hydrological peat unit as the protection function of the peatland ecosystem. In addition, the Minister is also obliged to designate as protection function of peatland ecosystem areas located beyond the hydrologically unified peatland areas which contain peat of more than three-m depth, specific and endemic germplasm, protected species, or considered as a protected area under spatial planning, or as conservation or protected forests (GR Number 71 of Government Regulation Number 71 2014, art. 9 paras 3–4). Areas allocated for utilization function can be changed into those for protection function, but not the other way around (GR Number 71 of 2014, art. 11).

The GR Number 71 of 2014 urges the formulation of peatland management plan at national, provincial, and district/city levels. The plan is hierarchically formulated, in the sense that the plan at a district/city level should be based on national and provincial plans, while the provincial plan should be based on the national plan. The national plan will, in turn, be based on the national map of the peatland ecosystem (*peta fungsi Ekosistem Gambut nasional*) (art. 9 paras 14–15).

The GR Number 71 of 2014 also sets forth some criteria for determining damage on peatland. Peatland designated for protection function is damaged if there is man-made drainage, if pyritic sediment and/or quartz underneath the peat are exposed, or if the area or land cover is reduced (art. 23 part. 2). Meanwhile, peatland designated for utilization function is damaged if the water-table level is deeper than 0.4 m from the peat surface, or if the pyritic sediment and/or quartz underneath the peat are exposed (art. 23 part. 3). However, the criteria are not applicable for peatland designated for utilization function whose depth is less than 1 m (art. 24).

The GR Number 71 of 2014 prohibits land clearing within the peatland ecosystem designated for protection function, building canals likely to drain the peatlands, burning the peatland, or conducting activities that cause damage to peatland ecosystem (art. 26). The license holders are obliged to control damage to the peatland ecosystem occurring within the area under their control (art. 27). If they fail to control the damage within the specified time, the Minister of the Environment, Governor, or Mayor/Head of district can appoint a third party to control the damage, at the expense of the license holders (art. 28). Similarly, the license holders are also obliged to undertake environmental recovery for the damage occurring within the area under their control (arts 30–31).

Interestingly, the GR Number 71 of 2014 stipulates that the government may impose an administrative sanction, that is, the government's coercive order, to those who violate the prohibitions and obligations above (arts. 41–44). The threat of administrative sanction under this GR is, however, questionable since it has the potentials to be in contradiction with the sanction under the Law Number 32 of 2009, which imposes criminal sanctions for those whose activities have given rise to environmental pollution or damage (the 2009 Environmental Law, arts. 98–99).

The Peatland Regulation divides the peatland protection and management plan into three hierarchical plans, namely the national plan, the provincial plan, and the district plan, which at least includes plans (GR Number 71 of 2014, art. 14 part. 1 and art. 17 part. 1):

4.1 the use or preservation of peatland ecosystems;
4.2 the maintenance and protection of the quality or function of the peatland ecosystem;
4.3 the control, monitoring, utilization, and conservation of peatland ecosystems; and
4.4 adaptation and mitigation to climate change.

Furthermore, the Regulation determines that the utilization of peatlands with protection function is limited to research activities, science, education, and environmental services. Meanwhile, peatlands with utilization function can be used for any activity according to the protection and management plant of peatlands (GR Number 71 of 2014, arts. 20 and 21). The GR Number 71 of 2014 also stipulates that the control of peatlands shall be undertaken according to the national, provincial, and district/municipal protection and management plans, which comprises the prevention of peatland damage, the mitigation of damage, and the recovery of degraded peatlands (art. 22).

GR Number 71 of 2014 stipulates that the utilization of peatlands shall have an environmental permit (art. 25). Since the permit requires an environmental impact assessment, the Article implies that every utilization of peatlands requires the proponent to have its environmental impact analysis approved. Meanwhile, GR Number 57 of 2016 sets forth the prohibition of the land clearing until the establishment of zoning of peatlands, that is, either as protected functions or utilization function. The GR also prohibits the construction of drainage canals likely to drain the peatlands, the burning of peatlands, and the execution of other activities which might result in peatland degradation (GR Number 57 of 2016, art. 26).

Furthermore, the regulation specifies that the maintenance of peatlands is carried out through the preservation and conservation of peatlands. In this regard, the MoEF holds the authority to determine the peatland ecosystem that cannot be utilized within a certain period. The area of peatlands excluded from utilization includes (GR Number 71 of 2014, art. 33 part. 3 and art. 34):

(a) peatlands with a protected function that is less than 30% of the total hydrological area of peat in the province or district/city;
(b) peatlands with a utilization function, half of which has been granted a business license or activity beyond the damage criteria;
(c) peatlands which have been designated for to utilization moratorium according to other relevant regulations; or
(d) peatlands with a utilization function which have been changed into protection function.

The supervision for compliance of the permit holders shall be undertaken by the MoEF, Governor, or Regent/Mayor according to their respective authority. The authorities can further delegate their power to officers or institution responsible for environmental management (GR Number 71 of 2014, art. 36 paras 1–2). Administrative sanctions could be imposed by the MoEF, Governor, or Regent/Mayor. The sanctions consist of written warnings, coercive government order, suspension of environmental permits, and revocation of environmental permits (GR Number 71 of 2014, art. 40 paras 1–2).

(d) Ministerial Regulations Related to Peatland Protection

To implement the Peatland Regulation, the MoEF issued four inter-related regulations in 2017. These are the MoEF Regulation Number P.14/MENLHK/SETJEN/KUM.1/2/2017 on Procedures for the Inventory and Determination of Peat Ecosystem Functions (hereinafter referred to as MoEF Regulation Number 14 of 2017), MoEF Regulation Number P.15/MENLHK/SETJEN/KUM.1/2/2017 on Procedures for Water-table Measurement at the Compliance Point within Peat Ecosystem (hereinafter referred to as MoEF Number 15 of 2017), MoEF Regulation Number

P.16/MENLHK/SETJEN/KUM.1/2/2017 on Technical Guidelines for Recovery of Peat Ecosystem Functions (hereinafter referred to as MoEF Number 16 of 2017), and MoEF Regulation Number P.17/MENLHK/SETJEN/KUM.1/2/2017 on Amendment to the Regulation of the Minister of Environment and Forestry Number P.12/MENLHK-II/2015 on the Development of Industrial Forest (hereinafter referred to as MoEF Number 17 of 2017). These regulations are explained as follows.

1. The MoEF Regulation Number 14 of 2017

The regulation is issued to regulate procedures for the inventory and determination of peatland functions according to peatland protection and management plans both at the national and local levels (art. 2). The procedures include the inventory and final map of peatland hydrological units, the determination of peatland functions, and the changes in peatland functions (art. 3).

The inventory is carried out using various maps, namely: the indicative map of the distribution of peatlands, the national map of peatlands, the map of river hydrology networks and altitude, the map of land cover obtained through the medium resolution of remote sensing image, the map of elevation model of 30 m height, the landscape map, the watershed maps, and other valid sources in geospatial mapping methods (art. 4).

After the inventory, the subsequent step is the determination of peatland functions. To determine peatlands with the protection function, several criteria need to be observed, namely: (a) the minimum depth of 3 m, (b) located within protected areas, protected forests, and conservation forests, and (c) peatlands already designated to undergo the moratorium of utilization. Peatlands with the protection function have to be at least 30% of the entire peatland hydrological unit, of which location starts from one or more peaks of peat domes. Meanwhile, peatlands with utilization function are those not included in the peatlands with protection function (arts. 9–10).

The MoEF Regulation Number 14 of 2017 also states that peatlands with utilization function can be converted into those with protection function with the following criteria: (i) Peatlands with a minimum depth of 3 m, peatlands in protected areas, protected forests, and conservation forests, and peatlands already designated for the moratorium of utilization; (ii) There is an ecological need to undertake measures to prevent or recover environmental damage in peatlands, (iii) There is a need to preserve peatlands in provinces or districts/municipalities, and (iv) The existence of protected species or area in an area beyond 30% of the total area of peat hydrological unit (art. 16).

The MoEF Regulation Number 14 of 2017 requires permit holders to revise their *Rencana Kerja Umum* (RKU, literally means *General Operational plan*), Operational plan Document, Management Plan Document to conform with the MoEF Regulation. They must also submit the proposal to change their environmental permit due to changes in peatland functions (art. 21). The permit holders whose more than 40% of their working area have been designated as peatland with protection function can apply for a land-swap, that is, the new area from mineral land allocated to compensate the "lost" area resulting from changes in peatland functions (art. 22).

2. MoEF Regulation Number 15 of 2017

The MoEF Regulation Number 15 of 2017 addresses procedures for measuring the water-table level at the point of observation or compliance. The point of compliance is one or more locations used as the basis for carrying out the measurements of the water-table level in peatlands. This measurement constitutes a control point for supervision, set at least 15% of the total number of trees or production blocks and located in the middle of the crops or production block (art. 2).

The measurement of the water-table level in peatlands is conducted to determine whether damage has occurred in peatlands (art. 4 par. 1). The measurement should take into account the location, coordinate, and elevation of the points of compliance, the water-table level, the precipitation data, the time and date of measurement, and the rate of peat subsidence (art. 5 par. 1). Furthermore, the analysis of the water-table level will be used as the basis for the government to conduct supervision, to issue orders to permit holders to undertake recovery measures, and to evaluate the functions of peatlands (art. 10).

The MoEF Regulation Number 15 of 2017 also sets forth the responsibility of permit holders to revise their general operational plan, operational plan document, and management plan document, and to apply for a revision of environmental permit accordingly (art. 13).

3. MoEF Regulation Number 16 of 2017

The MoEF Regulation Number 16 of 2017 is perhaps the most important one, with respect to the recovery of peatlands. The Regulation aims to provide technical guidance on the recovery of peatlands (art. 2). The recovery measures are carried out for peatlands, in both protection and utilization functions, that have been degraded (art. 3). In case of a peat-dome is located within a concession area that has not been utilized, the permit holders shall maintain the peat-dome as protected peatland. However, if the area has been utilized, the permit holders are still allowed to cultivate the crops in one cultivation cycle. Afterward, the permit holders are prohibited from replanting the area on the one hand, and are obliged to undertake peatland recovery through the construction of canal blocking and natural succession on the other hand (art. 4).

It is important to note that the MoEF Regulation Number 16 of 2017 divides the responsibilities to conduct peatland recovery according to whether the peatland has been allocated for permits. In this regard, government agencies, that is, the MoEF and local governments, are responsible for the recovery of peatland located in an area that has not been allocated for permits. Specifically, the MoEF is responsible for the recovering of peatlands located within conservation forests area. The local governments are responsible for the recovery of peatlands located in a protected forest, production forest, forest parks, and other areas as long as they have not been allocated for permits. These areas include areas controlled by the forest-dependent community or indigenous people. Meanwhile, the permit holders are responsible for the recovery within their working area and are required to conduct recovery within 30 days since the damage is known or since the fire occurred (arts. 5–6). The permit holders could undertake the recovery based on the recovery plan previously consulted with and reported to the director-general of environmental pollution and damage control at the MoEF (arts. 9 and 12 par. 5).

The recovery conducted by permit holders in the forestry sector shall be made on the basis of spatial change and revision of the general operational plan, and shall be executed within 30 days after the damage was found or fires occurred. If the permit holders fail to do so, the MoEF and local governments shall appoint a third party to conduct recovery of the function of the peat ecosystem with the expenses of the permit holders (arts. 8 and 10 par. 1).

The recovery of the peatlands is carried out in three stages, namely planning, implementation, and evaluation (art. 11). Planning for the recovery of peatlands is conducted in accordance with the field surveys or analysis of spatial data, spatial changes, or changes in the general operational plan. It has to be arranged within a maximum of 14 working days (art. 12). Afterward, the recovery of peatlands is conducted within 14 working days through:

(a) Rehabilitation, which will be conducted through revegetation with primarily endemic plants. The rehabilitation takes place in areas that have been burned, areas experiencing clear-cutting, an open area with limited vegetation, or burned areas that have experienced natural succession (art. 14 paras. 1 and 2);
(b) Natural succession, which shall take place in peatlands with blocked canals and have disturbance from human activities (art. 15);
(c) Restoration, directed toward the complete or partial re-functioning of peatlands. It is conducted through the construction of peatland rewetting infrastructures, including canal blocking, water catchment, canal stockpiling, or water pumping (arts. 16 and 17 par. 1); or
(d) Other means consistent with the development of science and technology, within 14 working days (art. 13).

The MoEF Regulation Number 16 of 2017 also determines that the execution of recovery plan should be reported to the director-general of environmental pollution and damage control every six months (art. 19 par. 2). The director-general has the authority to evaluate the execution of the recovery plan (art. 19 par. 3). The recovery of peatlands is declared successful if: (a) there is no exposure to pyrite or quartz sediments under the peatlands layer measured at the point of compliance; (b) the water-table level on peatlands is less than 0.4 m under the peatland surface measured at the point of compliance; (c) the recovered peatland quality is better than the criteria for peatland damage specified in the environmental permit; (d) the recovered peatland quality is better than criteria for peatland damage according to spatial analysis of data from field survey, or analysis from data with 1:250.000 scale, or data collected from compliance points; or (e) the number of healthy growing plants is at least 500 trees per hectare in the third year (art. 18 par. 1).

The MoEF Regulation Number 16 of 2017 requires that water and water management systems for the recovery of peat ecosystems be established in the first six months, whereas the improvement of the water-table level to reach less than 0.4 m from the peatland surface be visible within three months after the construction of the water infrastructures (art. 18 paras. 2 and 3). Information concerning the achievement of these targets should be reported to the director-general, who will accordingly assess the success of the recovery.

If the recovery is considered wholly or partially unsuccessful, the director-general shall provide warnings, or take measures according to relevant regulations, or appoint other parties to recover the functions of the peat ecosystem at the expense of the permit holders (art. 20).

4. MoEF Regulation Number 17 of 2017

According to MoEF Regulation Number 17 of 2017 the permit holders in the forestry sector, namely the holders of Timber Forest Product Utilization License in Industrial Timber Estate (IUPHHK-HTI), whose working areas become part of peatland with protection function shall make adjustment of the spatial planning of the permit and shall propose the revision of general operational plan. The revisions should be conducted within 30 working days after the permit holders receive the map of peatland functions. The revisions should also be submitted to the director-general for consultation and approval (art. 8A).

The map of the peatland function is the basis for determining the spatial planning of the timber plantation permit. The overlay of the spatial planning within the permit and the map of peatlands functions will indicate whether peat-domes exist. If there is peat-dome within the working area, and the area has not been planted, the MoEF Regulation requires the area be maintained as protected peatlands (arts. 8B and 8C par. 1). Furthermore, the regulation states that overlay might indicate changes of the working areas, including changes of areas or plants into peatland with protection function or peatland with utilization function, as well as changes of protected areas or protected forest into protected peatlands (art. 8D).

If a change needs to be made in an area already planted, in which the area falls into the protected peatlands, the permits holders are allowed to cultivate the land for one cultivation cycle. However, they are prohibited from replanting and are required to undertake peatland recovery measures (art. 8E par. 1).

The MoEF Regulation Number 17 of 2017 is very important in that it sets forth detailed provisions on land-swap. In this regard, the Regulation states that the permit holders of which at least 40% of their working areas are designated as protected peatlands have the rights to apply for land-swap, that is, replacing land (art. 8G par. 1). It further states that 40% of the replacing land should be allocated for community management, and 60% for the permit holders (art. 8G par. 2).

Interestingly, the Regulation instructs that timber permits that were issued and had been implemented prior to the enactment of GR Number 71 of 2014 and GR Number 57 of 2016 be declared valid until the termination date of the permits, subject to the adjustment of the general operational plan. Meanwhile, the timber permits that had been issued before the entry into force of the Peatland Regulations, but had not been in operation, are required to make adjustments for their general operational plan and to maintain the preservation of hydrological functions of peatlands (art. 23A).

However, in 2017 there was a petition for a judicial review against the MoEF Regulation Number 17 of 2017. The petitioners challenged the legality of MoEF Number 17 of 2017, especially Article 1 point 15d, Article 7 letter d, Article 8A, Article 8B, Article 8C par. 1, Article 8D letter a, Article 8E par. 1, Article 8G, and Article 23A.

The petitioners seem to worry that the Regulation might force them to change the operational plan, which will eventually allow for the change of working area into protected peatlands. The Supreme Court granted the petition and declare the Regulation to have no legal basis and should be revoked accordingly. However, the Supreme Court's ruling seems to miss the target since the obligations to make a revision in the spatial planning of timber permits and to make an adjustment in the operational plan are also instructed in other regulations, especially the MoEF Number 16 of 2017 and the Peatland Regulation. In this regard, the ruling does not eliminate the obligations, since the obligations are also mandated in other regulations. To make it worse for the timber permit holders, the annulment of the MoEF Number 17 of 2017 means that they no longer have the right to apply for land-swap.

(e) Authorities in Peatland Protection: the MoEF, the Peatland Restoration Agency (Badan Restorasi Gambut), and Local Government

Following the 2015 fires, the President signed the Presidential Regulation on *Badan Restorasi Gambut* (BRG, Peat Restoration Agency) in 2016 (hereinafter referred to as PR Number 1 of 2016). The BRG is a non-structural agency under and directly responsible to the President (PR Number 1 of 2016, art. 1 par. 1). The Agency has to coordinate and facilitate peat restoration in Riau Province, Jambi Province, South Sumatra Province, West Kalimantan Province, Central Kalimantan Province, South Kalimantan Province, and Papua Province (Ibid. arts. 2 and 4). BRG is obliged to formulate and implement peatland restoration for two million hectares, within five years. The restoration target should be met as much as 30% by the year 2016, 20% by the year 2017, 20% by the year 2018, 20% by the year 2019, and 10% by the year 2020. The restoration targets kicked off at Pulang Pisau Regency in Central Kalimantan Province, Musi Banyuasin Regency, and Ogan Komering Ilir in South Sumatera Province, as well as Meranti Islands Regency in Riau Province. The target area of restoration efforts are peatlands in the concession area (58%), in state-owned forests, or in the area controlled by individuals or community (Hergoualc'h 2018, op. cit. p. 5). BRG aims to systematize peatland restoration through rewetting by blocking canals to wet peatlands in priority areas such as burning land and peat-dome, and replanting using endemic plants or other plants which can stand against floods (PR Number 1 of 2016, op. cit. art. 3 letters e).

With such a new institution, one might conclude that the institutions for peatland protection and recovery are the MoEF, the BRG, and local government. Without clear authority and without well-functioning coordination and cooperation system among the institutions, different institutions might lead to ineffectiveness of peatland protection and recovery, where every institution throws the responsibility to others.

Legal Challenges for Peatland Protection: The Importance Roles of the Principles of Environmental Law

The peatland regulations have been faced with various challenges and critiques; two of them are selected for the discussions in this section. First, the use of strict liability for forest fires has been criticized on the grounds that forestry activities and plantation do not fall into abnormally dangerous activities, and should hence be excluded from the implementation of strict liability (Sudarsono 2017). Second, critiques have also been directed toward the requirement of water-table level of 0.4 m in the Peatland Regulation (Gor 2017; Sabiham et al. 2016, p. 39.). This section replies these critiques and defends the improved government responses against peatland degradation.

1. Strict Liability for Plantation?

Recalling some provisions regarding forest or land fires, one could find the following conclusions. First, forest/land fire is prohibited and considered a crime according to Indonesian law. Second, permit holders have the legal obligation and responsibility to prevent the forest/land fires within the area under their control. Third, permit holders have the obligation and responsibility to control forest fires and conduct recovery for the resulting environmental damage arising out of forest fires within the area under their control. Fourth, the legal construction of various provisions on forest fires in Indonesia does not allow excuses, because the obligations to prevent forest/land fires and to control as well as to conduct environmental recovery are linked to permits or authority to manage a certain area of forest.

Unfortunately, such powerful provisions were not fully utilized until 2013. Indeed, 2013 marks an important moment in law enforcement in forest fires. For the first time ever, the government (the MoE/MoEF) brought a lawsuit against a concession holder. Following the success of the lawsuit, the MoEF has initiated several other suits against the concession holders and won the majority of them.

Interestingly, there are attempts to apply strict liability to these cases. In *MoE v. PT. Kalista Alam* (District Court of Meulaboh Decision Number 12/PDT.G/2012/PN.MBO, 2013), for example, the Minister of Environment (MoE), used strict liability to hold the defendant liable (*MoE v. PT. Kalista Alam*, District Court of Meulaboh, p. 25) in addition to the conventional liability rule for unlawful conducts (Ibid. pp. 20–25 and 26–33).

The court was, unfortunately, silent on the question of whether strict liability was applicable to the case. The *MoE v. PT. Kalista Alam* (2013) ruling is, nevertheless, a notable case since it is the first case of wildfires ruled in favor of the government. In this case, the court found the defendant liable according to the unlawful act rule. The court then ordered the defendant to pay a compensation of IDR 115 billion and to conduct restoration worth at least IDR 250 billion (Ibid. p. 231). The ruling went through the Supreme Court, which eventually upheld the lower court rulings (Supreme Court Decision Number 651 K/Pdt/2015).

Similarly, a call for strict liability was also briefly discussed by the plaintiff in *MoEF v. PT. Jatim Jaya Perkasa* (District Court of North Jakarta Decision Number

108/Pdt.G/2015/PN.Jkt.Utr, pp. 22–23). Similar to the plaintiff's logic in MoE v. PT. Kalista Alam, here the plaintiff also relied on the unlawful act rule, by arguing on the one hand that the defendant has intentionally conducted an unlawful act by setting the fire on its land in order to make preparation for land clearing and the planting of oil palm, and on the other hand that the defendant has negligently conducted an unlawful act by failing to comply with various obligations to prevent and control fires (*MoEF v. PT. Jatim Jaya Perkasa,* pp. 18–24, and 24–27). Although the court held the defendant liable and consequently ordered the defendant to pay a compensation of IDR 7.2 billion and to conduct rehabilitation measures worth of IDR 22.2 billion (Ibid. pp. 166 and 170), the court remained silent on the issue of liability.

Furthermore, strict liability was briefly discussed by the court in *MoEF v. PT. Bumi Mekar Hijau* (High Court of Palembang Decision Number 51/PDT/2016/PT.PLG, pp. 168–169). However, the High Court was eventually of the opinion that the defendant had negligently failed to comply with obligations related to the prevention and control of forest fires (*MoEF v. PT. Bumi Mekar Hijau,* pp. 168–169 and 171–172). Accordingly, the court held the defendant liable for conducting an unlawful act and ordered the defendant to pay a compensation of IDR 78.5 million (Ibid. p. 190).

Strict liability was finally seriously taken into account in the *MoEF v. PT. Waringin Agro Jaya* (District Court of South Jakarta Decision Number 456/Pdt.G-LH/2016/PN.Jkt.Sel). Like in previous cases, the plaintiff employed both the unlawful act rule and strict liability for the defendant's liability (*MoEF v. PT. Waringin Agro Jaya,* pp. 11–23 and 23–30). Concerning strict liability, the plaintiff argued that the defendant's activities pose serious threats to the environment, and hence, the application of strict liability is justified (Ibid. pp. 38–40). In its ruling, the court found that the defendant's activities related to oil palm plantation posed serious risks to the environment, and hence strict liability could be applied to this case (Ibid. pp. 293–294). The court held the defendant strictly liable for the losses arising out of fires in the defendant's controlled area, and awarded the plaintiff with compensation of IDR 173.4 billion as well as ordered the defendant to conduct restoration measures of at least IDR 293 billion (Ibid. p. 304).

Critiques have been directed toward the use of strict liability in fire litigation. Wijoyo and Hartiwiningsih, for example, argued that strict liability is not applicable to oil palm plantation because the plantation does not constitute dangerous activity (Sudarsono 2017).

Interestingly, on 19 May 2017, the *Asosiasi Pengusaha Hutan Indonesia* (APHI), that is, the Indonesian association of forest concession holders, and the *Gabungan Pengusaha Kelapa Sawit Indonesia* (GAPKI), that is, the Indonesian Palm Oil Organization, filed a petition to Constitutional Court against several provisions on the 2009 Environmental Law and the 1999 Forestry Law, including article 88 of the 2009 Environmental Law regarding strict liability. According to the petitioners, strict liability (i.e. article 88 of the 2009 Environmental Law) is in contravention with article 28D of the Constitution regarding the guarantee that everyone has the rights to fairness, legal certainty, and equality before the law (Petition to Constitutional Court Number 25/PUU.XV/2017, p. 16). In the petitioners' point of view, strict liability is unconstitutional because it departs from the conventional liability based on fault. Furthermore, the departure from liability based on fault not only renders strict liability unconstitutional but also makes it result in legal uncertainty for the petitioners (Ibid. pp. 16–17). The petitioners argue that article 88 should be void unless it is applied according to liability based on fault (Ibid. p. 17). In conclusion, the petitioners ask the Constitutional Court to declare article 88 of the 2009 EPMA on strict liability conditionally unconstitutional, in the sense that it is unconstitutional unless interpreted to place liability for losses proportionally to those whose faults have contributed to the losses in question (Ibid. p. 18).

Following Wibisana (2019), this chapter finds that charges against the use of strict liability in Indonesia's peatland fire are unfounded for several reasons. First, strict liability has already been in Indonesian environmental law at least since 1982, that is, through Law Number 4 of 1982 on General Provisions of Environmental Management. In the 2009 Environmental Law, strict liability is stated in article 88, by which those who carried out activities using hazardous substances, producing hazardous wastes, or creating serious threats to the environment are strictly liable for the resulting damage (Wibisana 2019, p. 187).

Second, although it is likely that the utilization of forested land or peatlands for timber and oil palm plantations does not use hazardous substances, nor produce hazardous wastes, one could still apply strict liability for the activity as long as it poses serious threats to the environment. The 2009 Environmental Law defines "serious threats" as the types of threats that have the potentials to give rise to a wide range of environmental impacts and public anxiety (art. 1 point 34). Furthermore, the term "serious threats" is also explained in the Chief Justice Decree Number 36 of 2013 as pollution or environmental damage likely to be irreversible or with widespread consequences to human health, water surface, underground water, soil, air, plants, or animals (The Decree of Indonesia's Chief Justice Number 036/KMA/SK/II/2013, p. 39).

In 2010, Hooijer et al. argue that land-use activities might have impacts on the net greenhouse gas balance of peatlands, dominated by net CO_2 uptake by vegetation, CO_2 emissions from drainage-related peat decomposition, CO_2 and other emissions from fires, exports of dissolved and particulate organic carbon, and likely emissions of methane and nitrous oxides (Hooijer et al. 2010, p. 1506). If fires occurred, the impacts are devastating. The 1997 fires, for example, released as much as 40% of annual global emissions from fossil fuels (Aldhous 2004, p. 145). Therefore, it is plausible to argue that timber and oil palm plantations indeed pose serious threats to the environment.

Third, it could further be argued that strict liability is applicable if the clearing of the land or peatland falls into an unnatural use of land, or if the activity is exposing the area into threats of forest fires so significant that an exercise of utmost care cannot eliminate the threats. The case for this type of argument is even stronger when the activities of land or peatland clearing for timber and oil palm plantations cover a very large area.

In this regard, one could refer to Harrison et al. who argue that the reduction in the water-table level through the construction of drainage is a major trigger for fire. In contrast, undisturbed peatlands are typically waterlogged and flooded for much of the year, making it "naturally fire-resistant". Thus, peatland drainage is the principal cause of peatland fires. In addition, fire does not normally spread in the undisturbed peatlands, which are typically moist with the closed-canopy environment. Once fires occurred, positive feedbacks could take place, by which the burned peatlands become more susceptible to fires (Harrison et al. 2009, p. 158).

Hence, timber and plantations might indeed constitute abnormally dangerous activities, especially if the activities involve the drainage of peatlands. Categorizing peatland draining activity as an abnormally dangerous activity is scientifically sound. It is a clear statement that from a legal perspective, the draining activity is a dangerous human interference to the otherwise naturally wet and moist environment. Consequently, those who gain benefits from such a dangerous activity should bear the costs incurred by others resulting from the activity.

More importantly, the use of strict liability in fire litigation is defendable under the polluter pay principle. Principle 16 of Rio Declaration urges that states "endeavour to promote the internalization of environmental costs and the use of economic instruments, taking into account the approach that the polluter should, in principle, bear the cost of pollution…" From Principle 16, one could see the relation between the polluter-pays principle and the idea of internalization of environmental costs. From an economic perspective, an externality is a form of market failure, indicating that the market fails to consider the total cost, that is, pollution, caused by a production process. Thus, an externality is evidenced by prices that do not reflect environmental costs. Externality gives the wrong direction to the individuals when making a decision because with this externality the market is unable to reflect the actual price of a product or activity (Pindyck and Rubinfield 2001, p. 592. Solberg 1982, p. 540. Faure and Skogh 2003, p. 95). Therefore, from an economic perspective, environmental law is primarily aimed at internalizing the externalities. This process is designed to induce parties to include considerations on environmental costs in their decision-making process. In this way, consumers are expected to face real product prices (Faure 2001, p. 10; Hunter et al. 1998, p. 108).

Theoretically, Polluter-Pays Principle is an application of economic theory to allocate costs for pollution and environmental damage, but then has implications for the development of international and national environmental law, especially when damage occurs or when environmental costs are borne by the public. This principle was first noted in some of the recommendations of the OECD in the 1970s, which states that the principle requires polluters to bear the necessary costs in the framework of efforts taken by public officials to keep the environmental conditions at acceptable conditions. Put it differently, the prices of goods and services that cause pollution should reflect the costs incurred to carry out environmental measures (Boyle 1994, pp. 179–182; Birnie and Boyle 2002, op. cit. pp. 92–95).

The chapter argues that one way to implement the Polluter-Pays Principle is through the implementation of strict liability. Certainly, strict liability is applicable only for activities that belong to abnormally dangerous activities; and as discussed earlier, peatlands draining activities can be seen as abnormally dangerous.

2. The Precautionary Principle and Water-Table Level

As mentioned earlier in section C, Peatland Regulation requires that the water-table level for peatland utilization function be set no deeper than 0.4 m from the peat surface. If the level is found to exceed 0.4 m, the peatland is considered to have been damaged (GR Number 71 of 2014, art. 23 par. 3). Once the peatland is damaged, the permit holders are obliged to take recovery measures, which will be declared successful only if, among others, the water-table level is less than 0.4 m from the surface (MoEF Regulation Number 16 of 2017, art. 18 par. 1).

The requirement of a minimum water-table level has also been fiercely criticized. For example, scholars argue that the requirement has no scientific basis. It is also impossible to be implemented, as the requirement could dramatically reduce the productivity of oil palm plantations (Gor 2017, loc. cit. Sabiham 2016, loc. cit.). Similarly, Sumawinata, in an interview, argues that the requirement of 0.4 m is an excessive requirement unlikely to be met since the water-table level could fall below 0.4 m even in health peatlands (Faperta IPB 2017). Rather differently, however, Rahutomo and Winarna, argue that oil palm productivity is optimal when the water-table level is maintained between 0.4 and 0.6 m from the peat surface (Rahutomo and Winarna 2017).

However, there is also an opinion arguing the importance of maintaining the maximum water level at 0.4 m from the surface. In this regard, Page observes a clear relationship between the water-table level and the drainage, and eventually with undrained peatlands are indicated by constantly high water-table level, with levels following an annual cycle that involves flooding or near-surface water levels during the wet season and drawdown during the dry season usually to between 20 and 40 cm below the peat surface (Cochrane (ed) 2009, p. 269). Drainage lowers the water-table level, which not only promotes oxidation of the organic peats, the release of CO_2 emissions, and the subsidence of peat surface but more frighteningly also increases the risk of fires since drainage provides an abundant fuel of dry peat (Ibid. pp. 269–270).

More importantly, Wösten, et al. argue that when the water-table level drops below 40 cm from the surface, the moisture content of peat top layer decreases, and hence making peat vulnerable to fire. Furthermore, the initial fire could increase temperature and humidity on the one hand, and decrease soil moisture on the other, which eventually lead to the increased susceptibility of peatlands to subsequent fires (Wösten et al. 2008, op. cit. p. 216–217). Similarly, Susilo et al. argue that the water-table level cannot be lower than 40 cm below the surface. They observed that if the water-table level falls below 40 cm, the moisture content decreases drastically. The drop in the moisture of peatlands is likely to lead to wildfires on peatlands as the dray peat soil will facilitate the fire to spread rapidly (Susilo et al. 2013, p. 124). Shortly, as Nazir Foead, the head of BRG, puts it, data from previous fires indicates that the peak of the number of fires is reached when the water-table level was below 0.4 m (Afrianto 2017).

This chapter suggests that from a legal perspective, the required water-table level of 0.4 m is justifiable based on the precautionary principle. According to Principle 15 of the Rio Declaration, "[i]n order to protect the environment, the precautionary approach shall be widely applied by States according to their capabilities. Where there are threats of serious or irreversible damage, lack of full scientific certainty shall not be used as a reason for postponing cost-effective measures to prevent environmental degradation." (For general discussion on the implementation of the precautionary principle in Indonesia, see: Wibisana 2011, pp. 169–202).

Assuming that differences in expert opinions on water-table level constitute scientific uncertainty, one could argue that the required water-table level of 0.4 m is a "safe level" of the water-table level, necessary to prevent serious irreversible threats to the environment, either in the form of peatland drainage and subsistence or peatland fires. The case of such a safe-level is even stronger given the fact that peatland fires take place almost annually.

One might, however, accuse the water-table level requirement violates sustainable development by sacrificing economic interests for the sake of environmental protection. This opinion could be misleading. Following Voigt, it could be argued that sustainable development requires both fairness to the current generation, and also to the future generations. However, so Voigt further argues, within the framework of sustainable development, the balance of justice could only be carried out through protection of the integrity of the Earth's ecosystem. As a result, development must be the development that also protects the life-supporting system, especially if the system is under threats (Voigt 2009, p. 51). In this sense, the concept of justice within the framework of sustainable development does not only focus on the issue of fair allocation of rights and obligations but also recognizes that the unity and integrity of the environment are an integral part of the concept of justice (Ibid. p. 52). From Voigt's view, one could see that if environmental conditions are already in a critical stage, then the decisions taken can no longer be neutral. In this situation, the 40 cm requirement could be seen as a pro-environment decision intended to recover the damage in Indonesia's peatlands and to prevent further damage of them.

Conclusion

Peatland degradation and peatland fires in Indonesia have triggered various responses from the Indonesian government in the last five years. During these years, the government has enacted various regulations related to peatland protection, ratified the Paris Agreement, formed an institution specifically aimed at accelerating peatland restoration in seven provinces seriously affected by fires called Peat Restoration Agency, and launched severe law enforcement against companies contributing to forest fires since 2013. However, there are two critiques related to the response on the peatland protection, namely the use of strict liability for plantation along with the "abnormally dangerous activities" criteria and the requirement of water-table level of a minimum of 0.4 m from the surface which constitutes a scientific uncertainty. Apart from those critiques, this chapter shows at least two arguments supporting the government responses. First, the polluter pays principle justifies the implementation of strict liability for peatland draining activities in fire litigation since the draining activities itself can be seen as abnormally dangerous activities. Secondly, the precautionary principle supports the establishment of the water-table level at a maximum of 0.4 m from the surface.

References

Adesiji RA et al (2014) Effects of water table fluctuations on peatland: a review. Scholars J Eng Technol 2(3C):482–487

Afrianto D (2017) Aturan Diprotes Pengusaha, Ini Reaksi Badan Restorasi Gambut. Okezone-finance. Available at: https://economy.okezone.com/read/2017/04/26/320/1677218/aturan-diprotes-pengusaha-ini-reaksi-badan-restorasi-gambut. Accessed in April 2018

Aldhous P (2004) Borneo is burning. Nature 432:144–146

Andres C et al (2017) Exploring Indonesia's long and complicated history of forest fires. Glob Forest Watch. Available at: http://blog.globalforestwatch.org/fires/indonesias-fire-history-provides-insights-to-prevent-future-fires.html. Accessed on March 31, 2018

Boyle A (1994) economic growth and protection of the environment: impact of international law and policy. In: Boyle A (ed) Environmental regulation and economic growth. Clarendon Press, London

Boyle A, Birnie P (2002) International law and the environment, 2nd edn. Oxford University Press, Oxford

Dayne S (2016) Re-examining Indonesia's land and forest fires. CIFOR. Available at: https://forestsnews.cifor.org/46097/lead-up-to-cop22-re-examining-indonesias-forest-fires?fnl=en. Accessed on March 30, 2018

Faure M (2001) Environmental law and economics. METRO—Maastricht University

Faure Michael, Skogh Goran (2003) The economic analysis of environmental policy and law: an introduction. Edward Elgar, Cheltenham, UK

David LA et al (2017) Overlapping land claims limit the use of satellites to monitor no-deforestation commitments and no-burning compliance. Conserv Lett 10(2):257–264

Gor (2017) Akademisi Nilai PP Gambut Tak Punya Dasar Ilmiah. Investor Daily Indonesia. Available at: http://id.beritasatu.com/agribusiness/akademisi-nilai-pp-gambut-tak-punya-dasar-ilmiah/157785. Accessed in April 2018

Harrison ME, Page SE, Limin Suwido H (2009) The global impact of Indonesian forest fires. Biologist 56(3):156–163

Hergoualc'h K et al (2017) Total and heterotropic soil respiration in a swamp forest and oil palm plantations on peat in central kalimantan, Indonesia. Biogeochemistry 135:203–220

Hergoualc'h K et al (2018) Managing peatlands in Indonesia: challenges and opportunities for local and global communities. CIFOR Info Brief No. 205

Hooijer A et al (2010) Current and future CO2 emissions from drained peatlands in Southeast Asia. Biogepsciences 7:1505–1514

Hunter D et al (1998) International environmental law and policy. Foundation Press, New York

Indonesia's constitutional court (2017 May 19) Petition to Constitutional Court No. 25/PUU.XV/2017

Indonesia's Supreme Court. The decree of Indonesia's chief justice on guideline for environmental case handling. The decree of Indonesia's chief justice No. 036/KMA/SK/II/2013

Institut Pertanian Bogor (2017) Wawancara Dr. Ir. Basuki Sumawinata, M.Agr: Lahan Sawit Terdampak Aturan Gambut. Buletin Faperta IPB. Available at: <http://faperta.ipb.ac.id/buletin/2017/09/06/lahan-sawit-terdampak-aturan-gambut/>. Accessed in April 2018

Managing Peatlands in Indonesia: challenges and opportunities for local and global communities. Cifor Info Brief (2018) 205:1–8

Ministry of Environment. Indonesia's second national communication of 2011. Available at: http://unfccc.int/files/national_reports/non-annex_i_natcom/submitted_natcom/application/pdf/indonesia_snc.pdf Accessed in August 2019

Ministry of Environmental and Forestry (ed) (2014) Statistik Kementerian Lingkungan Hidup dan Kehutanan. Available at: www.menlhk.go.id/downlot.php?file=STATISTIK_2014.pdf. Accessed in March 2016

Ministry of Environmental And Forestry. Perkembangan Penanganan Penegakan Hukum Kebakaran Lahan dan Hutan. Available at: http://www.menlhk.go.id/berita-16-perkembangan-penanganan-penegakan-hukum-kebakaran-lahan-hutan.html. Accessed in March 2016

Moeliono M et al. (2014) Information networks and power: confronting the "wicked problem" of REDD+in Indonesia. Ecol Soc 19(2):9

Mudiyarso D et al (2010) Opportunities for reducing greenhouse gas emissions in tropical Peatlands. PNAS 107(46):19655–19660

Nurhidayah L et al (2015) The influence of international law upon ASEAN approaches in addressing transboundary haze pollution in Southeast Asia. Contemporary SE Asia 37(2):183–210

Page S et al (2009) Tropical peatland fires in Southeast Asia. In: Mark C (ed) Tropical fire ecology: climate change, land use and ecosystem dynamics. Springer, Berlin

Pindyck RS, Rubinfield DL (2001) Microeconomics, 5th edn. Prentice Hall, New York

Rahutomo S, Winarna (2017) Restorasi gambut dan tinjauan terhadap PP No.57 Tahun 2016. PPKS Note. Available at: http://www.iopri.org/wp-content/uploads/2017/04/PPKS-NOTE-APRIL-2017.pdf

Rakhmat Muhammad Zulfikar and Dikanaya Tarahita Indonesia's jokowi seeks to slow forest fire devastation. https://www.asiasentinel.com/econ-business/indonesia-jokowi-seek-slow-forest-fire-devastation/. Accessed on March 31, 2018

Risnandar C, Ali F (2018) Peatlands. https://jurnalbumi.com/knol/lahan-gambut/ Accessed on March 30

Sabiham S et al (2016) What is the way forward on Indonesian Peatland 15th international peat congress pp 39–42

Simamora AP (2011) Moratorium issued to protect primary forests, Peatland. The Jakarta Post

Solberg Eric J (1982) Intermediate microeconomics. Business Publication, Plano, Tex

Sudarsono (2017) Strict Liability Tidak Berlaku Untuk Industri Sawit. Available at: https://ekbis.sindonews.com/read/1251727/34/strict-liability-tidak-berlaku-untuk-industri-sawit-1508956539. Accessed in April 2018

Supreme Court. Decision No. 651 K/Pdt/2015. *MoE v. PT. Kalista Alam* (2015)

Susilo GE et al (2013) Modeling groundwater level fluctuation in the tropical peatland areas under the effect of el nino. The 3rd international conference on sustainable future for human security SUSTAIN 2012, Procedia Environmental Sciences 17:119–128

Tan AK (2015) The 'Haze' crisis in Southeast Asia: assessing Singapore's transboundary haze pollution act 2014. NUS law working paper 2015/002

Tan AKJ (1999) Forest fires of Indonesia: state responsibility and international liability. Int Compar Law Q 48(4):826–855

Tay SS (1998) South East Asian forest fires: haze over Asean and international environmental law. RECIEL 7(2):202–208

The Jakarta Post (2011) Govt defends decree as first step in saving forest. The Jakarta Post

Toccani L (2016) Preventing fires and haze in Southeast Asia. Nat Clim Change 6(7):640–643

Voigt C (2009) Sustainable development as a principle of international law: resolving conflicts between climate measures and wto law. Martinus Nijhoff Publishers, Dordrecht

Wibisana Andri G (2011) The development of the precautionary principle in international environmental law and indonesian environmental law. Asia Pac J Environ Law 14(1 & 2):169–202

Wibisana AG (2019) The many faces of strict liability in Indonesia's wildfire litigation. Rev Eur Compar Int Environ Law 28:1–12

Wösten JHM et al (2008) Peat–water interrelationships in a tropical peatland ecosystem in Southeast Asia. CATENA 73:212–224

List of Laws and Regulations

Government Regulation Number 4 of 2001 on Environmental Damage and Pollution Control Related to Forest and Land Fires
Government Regulation Number 45 of 2004 on Forest Protection
Government Regulation Number 71 of 2014 on Peatland Protection
Government Regulation Number 57 of 2016 on the Amendment of Government Regulation on Protection and Management of Peatlands
Law Number 23 of 1997 on Environmental Management
Law Number 41 of 1999 on Forestry
Law Number 32 of 2009 on Environmental Protection and Management
Presidential Instruction Number 10 of 2011 on Moratorium of the Issuance of Licenses within Primary Forest and Peatlands
Presidential Instruction Number 6 of 2013 on Moratorium of the Issuance of Licenses within Primary Forest and Peatlands
Presidential Instruction Number 8 of 2015 on Moratorium of the Issuance of Licenses within Primary Forest and Peatlands
Presidential Instruction Numbewr 6 of 2017 on Moratorium of the Issuance of Licenses within Primary Forest and Peatlands
Presidential Regulation Number 1 of 2016 on Peat Restoration Agency
The Regulation of the Minister of Environment and Forestry Number 14 of 2017 on Procedures for the Inventory and Determination of Peat Ecosystem Functions. MoEF Regulation
The Regulation of the Minister of Environment and Forestry Number 15 of 2017 on Procedures for Water-table Measurement at the Compliance Point within Peat Ecosystem
The Regulation of the Minister of Environment and Forestry Number 16 of 2017 on Technical Guidelines for Recovery of Peat Ecosystem Functions
The Regulation of the Minister of Environment and Forestry Number 17 of 2017 on Amendment of The Regulation of the Minister of Environment and Forestry No. P.12/MENLHK-II/2015 on Development of Industrial Forest

List of Indonesian Court Decisions

District Court of Meulaboh. Decision No. 12/ PDT.G/ 2012/ PN.MBO. *MoE v. PT. Kalista Alam* (District Court of Meulaboh, 2013)
District Court of North Jakarta. Decision No. 108/Pdt.G/2015/PN.Jkt.Utr. *MoEF v. PT. Jatim Jaya Perkasa* (District Court of North Jakarta, 2016)
District Court of South Jakarta. Decision No. 456/Pdt.G-LH/2016/PN Jkt. Sel.. *MoEF v. PT. Waringin Agro Jaya* (2017)
High Court of Palembang. Decision No. 51/PDT/2016/PT.PLG. *MoEF v. PT. Bumi Mekar Hijau* (High Court of Palembang, 2016)

Correction to: Climate Change Research, Policy and Actions in Indonesia

Riyanti Djalante, Joni Jupesta, and Edvin Aldrian

Correction to:
R. Djalante et al. (eds.), *Climate Change Research, Policy and Actions in Indonesia*, Springer Climate, https://doi.org/10.1007/978-3-030-55536-8

The original version of this book has been revised.

In the original version of chapter 9, the reference section was omitted. This has now been rectified and the full reference section has been added below.

The original version of chapter 10 was previously published as non open access chapter. This has now been changed to open access under a CC BY 4.0 license, and the copyright holder for this chapter has been updated to "The Author(s)". The book has also been updated with these changes.

References

Abubakar I (dkk) (2009) Modul Pelatihan Agama dan Hak Asasi Manusia, CSRC UIN-Jakarta, Jakarta, hal. 23

Adolf H (1990) Aspek-aspek Negara Dalam Hukum Internasional, Rajawali Pers, Jakarta

Agoes ER (Penyunting) (2000) Dimanahkah Batas-Batas Wilayah Kita Di Laut ?, Departemen Kelautan dan Perikanan, Jakarta, 13 Desember 2000.

Aldrian et al (2012) Perubahan Iklim Upaya Peningkatan Pengetahuan Dan Adaptasi Petani Dan Nelayan Melalui Radio, Indonesia Climate Change Trust Fund Badan Meteorologi, Klimatologi Dan Geofisika Lembaga Ilmu Pengetahuan Indonesia, Sarana Komunikasi Utama, Bogor

---

The updated version of these chapters can be found at
https://doi.org/10.1007/978-3-030-55536-8_9
https://doi.org/10.1007/978-3-030-55536-8_10

R. Djalante et al. (eds.), *Climate Change Research, Policy and Actions in Indonesia*, Springer Climate, https://doi.org/10.1007/978-3-030-55536-8_16

Alwiah, Utina R (2013) Bapongka: Studi Nilai Pendidikan Pelestarian Ekosistem laut dan Pesisir pada Masyarakat Bajo

Artanto YK (2017) BAPONGKA, Sistem Budaya Suku Bajo Dalam Menjaga Kelestarian Sumber Daya Pesisir, Sabda 12(1), ISSN 1410–7910 E-ISSN 2549-1628, Juni 2017, hal. 52

Athena D, Anwar M (2007) Penelitian/Pengembangan Model/Sistem Surveilans Dampak Kesehatan Perubahan Iklim, Buletin Penelitian Kesehatan, vol 42, no 1, Jakarta, 2014, hal. 47–47. Lihat juga, IPCC, Climate Change Impacts, Adaptation and Vulnerability, Report of IPCC, Geneva, 2007, hal. 85–107

Athena D, Anwar M (2014) Penelitian/Pengembangan Model/Sistem Surveilans Dampak Kesehatan Perubahan Iklim, Buletin Penelitian Kesehatan, vol 42, no1, Jakarta

Aulia T, Dharmawan AH (2018) Kearifan Lokal Dalam Pengelolaan Sumber Daya Air Di Kampung Kuta, 2(2):345–355, https://download.portalgaruda.org/2010, (Diakses 25 April 2018). Lihat juga dalam Loc.cit., Mirnawati Firdaus, (diakses tanggal 25 April 2018). Lihat juga dalam Loc.cit., Yohanes Kristiawan Artanto, 2017, hal. 56

Ayatohaedi (ed) (1986) Kepribadian Budaya Bangsa (Local Genius), Pustaka Jaya, Jakarta hal 46s

Bachtiar I, Husni L, dan Romdiana (2018) Penyusunan Awig-Awig Pengelolaan Sumber Daya Perikanan di Pesisir Selatan kabupaten Lombok Timur, https://www.researchgate.net (diakses tanggal 26 Apr 2018)

Bahtiar (2012) Kearifan Lokal Orang Bajo Dalam Pengelolaan Sumber Daya Laut. Jurnal Mudra 27(2):178–185, ISSN 0854-3461. https://repo.isi-dps.ac.id/Bahtiar.pdf

Baskara B, Astuti O (2011) The Pamali of Wakatobi Bajo and its Role for Marine Conservation. J Indonesia Coral Reefs 1(2). https://coastalunhas.com/pdf. 2011, (Diakses 12 Januari 2018). Lihat juga dalam Op.Cit., Yohanes Kristiawan Artanto, 2017, hal. 54

Basuni S (2012) Mengelola Konservasi Sumberdaya Alam Hayati Berbasis Pengetahuan Tradisional dan Kearifan Lokal, Makalah Seminar Hasil-hasil penelitian Balai Penelitian Tekonologi Konservasi Sumberdaya Alam, 2012. (Diakses 12 Januari 2016). Lihat juga dalam Loc.cit., Yohanes Kristiawan Artanto, 2017, hal. 56

Dahuri (2001) Pengelolaan Ruang Wilayah Pesisir Dan Lautan Seiring Dengan Pelaksanaan Otonomi Daerah, Makalah ini disampaikan pada acara Semiloka dan Pelatihan Penataan Ruang Wilayah Propinsi, Kabupaten dan Kota Dalam Rangka Otonomi Daerah, kerjasama Bappeda Propinsi dan LPPM Unisba, Bandung, 2–3 Mei 2001, hal. 5

Diposaptono S, Budiman, Firdaus A (2009) Menyiasati Perubahan Iklim di Wilayah Pesisir dan Pulau-Pulau Kecil, Sarana Komunikasi Utama, Bogor (ID), 2009. Lihat juga, Loc.cit., Ari Wibowo, 2015, hal. 1

Finger M (2006) Which Governance for Sustainable Development? An Organizational and Institutional Perspective, dalam. In: Park J, Conca K, dan Finger M (eds) The crisis of global environmental governance: towards a new political economy of sustainability, Routledge Taylor & Francis Group, New York, hal 125

Firdaus M (2017) Kearifan Lokal: Dari Tano Batak ke Maluku, hal. 1. Lihat juga Minarwati Firdaus, Kearifan Lokal Masyarakat Suku Bajo Di Sulawesi Tenggara Mendukung Kelestarian Sumber Daya kelautan dan Perikanan, https://www.penyuluhperikanan.id, 09 Januari 2017 (diakses tanggal 25 Apr 2018)

Gerungan LKFR (2016) Penegakan Hukum Di Wilayah Perairan Indonesia. Jurnal Lex et Societatis IV/No. 5/Mei, Fakultas Hukum Unsrat, Manado

Gusti Ngurah Jayanti I, Made Sumertha I, Ritual Dan Kepercayaan Masyarakat Pulau Bungin Kabupaten Sumbawa, Propinsi Nusa Tenggara Barat, https://varianwisatabudayasundakecil.blogspot.co.id, Rabu, 10 Desember 2014 (diakses tanggal 23 Apr 2018)

Hasanah A, Gustini N, Rohaniawati D (2016a) Nilai-nilai Karakter Sunda (Internalisasi Nilai-nilai Karakter Sunda di Sekolah), Deepublish, Jokyakarta, Juni 2016, hal. 42

Hasman FR (2009) Kekerabatan Masyarakat Bajou, Fak. Sastra Universitas Sumatera Utara, Medan, https://repository.usu.ac.id, 2009, (Diakses 25 April 2018). Lihat juga dalam Op.cit., Yohanes Kristiawan Artanto, 2017, hal. 55

Hidayati D (2015) Adaptasi & Mitigasi Masyarakat Pesisir Terhadap Perubahan Iklim dan Degradasi SDL, Adaptasi & Mitigasi Masyarakat Pesisir, Leuser Cita Pustaka, Jakarta (ID), 2011, hal. 120–155. Lihat juga, Ibid., Ari Wibowo

Hidayati D, Aldrian E (dkk) (2012) Perubahan Iklim Upaya Peningkatan Pengetahuan Dan Adaptasi Petani Dan Nelayan Melalui Radio, Indonesia Climate Change Trust Fund Badan Meteorologi, Klimatologi Dan Geofisika Lembaga Ilmu Pengetahuan Indonesia, Sarana Komunikasi Utama, Bogor, 2012, hal. iii

Hidayati D, Deny (2011) Adaptasi & Mitigasi Masyarakat Pesisir Terhadap Perubahan Iklim dan Degradasi SDL, Adaptasi & Mitigasi Masyarakat Pesisir, Leuser Cita Pustaka, Jakarta (ID)

Indra-Fardhan. Kearifan Lingkungan Suku Bajo Dalam Pemanfaatan Pesisir Dan Laut Secara Berkelanjutan, https://dokumen.tips.html, 28 Sept 2015 (diakses tanggal 25 Apr 2018)

IPCC (2007) Climate change impacts, adaptation and vulnerability. Report of IPCC, Geneva

Kaligis OC (2003) & Associates. Sengketa Sipadan-Ligitan, Mengapa Kita Kalah, Jakarta

Kementerian Agraria dan Tata Ruang/Badan Pertanahan Nasional, Rencana Strategis Direktorat Jenderal Tata Ruang 2015–2019, Jakarta, 2015, hal. 2.1

Keraf S (1997) Hukum Kodrat dan Teori Hak Milik Pribadi, Kanisius, Yogyakarta, hal. 16

Konvensi Perserikatan Bangsa-Bangsa Tentang Keanekaragaman Hayati, Terjemahan Resmi Salinan Naskah Asli

Lampe M (2011) Dinamika Kelembagaan Sosial Ekonomi Orang Bajo, Jagad Bahari Nusantara. https://centerformunawareducation.files.wordpress.com (Diakses 28 Apr 2018), Lihat juga dalam Ibid., Yohanes Kristiawan Artanto, 2017, hal. 56

Mahmassani S (1993) Konsep Dasar Hak-hak Asasi Manusia, Suatu Perbandingan Dalam Syariat Islam dan Perundang-undangan Modern, Tinta Mas, Jakarta, hal. 206

Makmur A (ed) (2011) Kearifan Lokal di Tengah Modernisasi, Pusat Penelitian dan Pengembangan Kebudayaan, Jakarta, 2011,. Lihat juga dalam Loc.cit., Yohanes Kristiawan Artanto, 2017, hal. 57

Marzuki PM (2005) Penelitian Hukum, Kencana Prenada Media Group, Jakarta, hal 93

Numberi F (2009) Perubahan Iklim: Implikasi terhadap Kehidupan di Laut, Pesisir dan Pulau-pulau Kecil, Citrakreasi Indonesia, Jakarta (ID), 2009. Ibid., Ari Wibowo, Lihat juga

Nurlaili (2012) Strategi Adaptasi Nelayan Bajo Menghadapi Perubahan Iklim: Studi Nelayan Bajo Di Kabupaten Sikka, Flores, Nusa Tenggara Timur. Jurnal Masyarakat & Budaya 14(3) Tahun

Nurlaili (2012) Strategi Adaptasi Nelayan Bajo Menghadapi Perubahan Iklim: Studi Nelayan Bajo Di Kabupaten Sikka, Flores, Nusa Tenggara Timur. Jurnal Masyarakat & Budaya 14(3):609–611 (data tersebut diolah kembali)

NZ Human Rights, The rights of indigenous peoples: what you need to know. NZ Human Rights, New Zealand

Obie M (2016a) Perubahan Sosial Pada Komunitas Suku Bajo Di Pesisir Teluk Tomini, IAIN Sultan Amai Gorontalo. Jurnal Al-Tahrir 16(1):163

Paino C (2013) Lihat juga dalam Alwiah & Ramli Utina, Bapongka: Studi Nilai Pendidikan Pelestarian Ekosistem laut dan Pesisir pada Masyarakat Bajo, 2013. https://repository.ung.ac.id (diakses tanggal 23 Apr 2018). Baca juga dalam Op.cit., Indra-Fardhan, (diakses tanggal 25 Apr 2018)

Paino C (2014) Kearifan Suku Bajo Menjaga Kelestarian Pesisir dan Laut, 26 Jan 2014, https://www.mongabay.co.id (diakses tanggal 23 April 2018)

Park J, Conca K, Finger M (eds) (2006) The crisis of global environmental governance: towards a new political economy of sustainability. Routledge Taylor & Francis Group, New York

Poedjowibowo D, Waani JO, Warouw F. Teritorialitas Pada Permukiman Suku Bajo Di Desa Tumbak (Studi Kasus Permukiman Diatas Air), jurnal yang tidak terpublikasikan, hal. 41. Lihat juga dalam www.indonesia.travel/suku-bajo-kisah-manusia-perahu-di-sulawesi

Pratikto AS, Hariadi MH, Aldrian E (dkk) Konsep Implementasi Adaptasi Sektoral Perubahan Iklim, Pusat Penelitian dan Pengembangan Badan Meteorologi, Klimatologi, dan Geofisika, Jakarta

Pratikto AS, Hariadi MH, Aldrian dkk (Penulis) (2014) Konsep Implementasi Adaptasi Sektoral Perubahan Iklim, Pusat Penelitian dan Pengembangan Badan Meteorologi, Klimatologi, dan Geofisika, Jakarta, 2014, hal. 3

Primyastanto M (2015) Pengelolaan Sumber Daya Perikanan Melalui Kelembagaan Lokal Dalam Rangka Pemberdayaan Masyarakat Berbasis Kearifan Lokal di Pesisir Selat Madura, Gunung Samudra Cet. I, Malang, hal 14

Pusat (2005) Pengembangan Bahasa DEPDIKBUD, Kamus Besar Bahasa Indonesia, Balai Pustaka, Jakarta, hal 530

Raka Dalem GAA et al (2007) Kearifan Lokal dalam Pengelolaan Lingkungan Hidup, Cetakan Pertama, Pusat Penelitian Lingkungan Hidup (PPLH) UNUD, Denpasar, hal. 72

Rani M, Stalker P (2009b) Sisi Lain Perubahan Iklim: Mengapa Indonesia Harus Beradaptasi Untuk Melindungi Rakyat Miskinnya, UNDP-Indonesia, 2009, hal. 3. Lihat juga dalam Ibid., Nurlaili, 2012, hal. 614 (data tersebut diolah kembali)

Saad S (2009) Bajo Berumah di Laut Nusantara, CoremapII, Jakarta Selatan (ID), 2009, tanpa halaman. Lihat juga dalam Ari Wibowo, Strategi Adaptasi Dan Mitigasi Nelayan Di Pulau-Pulau Kecil Terhadap Dampak Perubahan Iklim, Laporan Studi Pustaka (KPM 403) Departemen Sains Komunikasi Dan Pengembangan Masyarakat Fakultas Ekologi Manusia Institut Pertanian Bogor, 2015, hal. 1

Sabarno H (2016) Pelaksanaan Administrasi Pemerintahan dan Pengelolaan Pulau-Pulau Indonesia di Wilayah Perbatasan. Dalam O.C. Kaligis & Associates, Sengketa Sipadan-Ligitan, Mengapa Kita Kalah, Jakarta, 2003, hal. 60. Etty R. Agoes, (Penyunting), Dimanahkah Batas-Batas Wilayah Kita Di Laut ?, Departemen Kelautan dan Perikanan, Jakarta, 13 Desember 2000, hal. 1. Lihat juga, Lusy K. F. R. Gerungan, Penegakan Hukum Di Wilayah Perairan Indonesia, Jurnal Lex et Societatis, Vol. IV/No. 5/Mei, Fakultas Hukum Unsrat, Manado, 2016, hal. 7

Salim E (2009) 3.000 Pulau Bakal Tenggelam, seminar Kelautan dan Perubahan Iklim, yang diselenggarakan Program Studi Ilmu Lingkungan Pascasarjana Universitas Indonesia bekerja sama dengan Direktorat Jenderal Kelautan Pesisir dan Pulau-pulau Kecil Departemen Kelautan dan Perikanan, Kamis (30 April 2009), Copyright—Badan Informasi Geospasial, https://www.big.go.id, Jakarta, 2009, hal. 1

Soemitro RH (1990) Metode Penelitian Hukum dan Jurimetri, Ghalia Indonesia, Jakarta, hal 36

Sudarja D (2007) Reformasi Pengelolaan Lingkungan Hidup, artikel, Pada Diskusi Membangun Gerakan Lingkungan Skepo, Walhi

Supriadi H (2008) Lingkungan di Indonesia: Sebuah Pengantar, Sinar Grafika, Jakarta, hal. 104–107

Suryanegara E, Nahib I (2015) Perubahan Sosial Pada Kehidupan Suku Bajo: Studi Kasus Di Kepulauan Wakatobi, Sulawesi Tenggara, Majalah Globe vol 17, no 1

Suyuti N (1996) Kehidupan Sosial Budaya Komunitas Nelayan di Boenaga, Kabupaten Konawe, Sulawesi Tenggara, (Hasil Penelitian), Kerjasama FISIP UNHALU Depsos Sulawesi Tenggara di Desa Sulaho Kecamatan Lasusua Kabupaten Kolaka Utara, Kendari, tanpa halaman

Suyuti N (2011) Interaksi Orang Bajo dan Orang Bugis: dalam Konteks Kearifan Lokal-Global di Desa Sulaho, Kabupaten Kolaka Utara, Sulawesi Tenggara, Jagad Bahari Nusantara

Tahara T (2011) Politik Identitas Orang Bajo. Dalam: Yuga S (ed) Jagad Bahari Nusantara: Telaah Dinamika Pranata Sosial Terhadap Kearifan Lokal Masyarakat Pantai, Melestarikan Budaya Bahari dalam Pembangunan Kebudayaan dan Pariwisata, Kementerian Kebudayaan dan Pariwisata, Jakarta, 2011. Lihat juga dalam Ellen Suryanegara, Suprajaka, dan Irmadi Nahib, Perubahan Sosial Pada Kehidupan Suku Bajo: Studi Kasus Di Kepulauan Wakatobi, Sulawesi Tenggara, Majalah Globe Volume 17 No. 1 Juni 2015, hal. 74

Triadi BL, Gifariyono M (2014) Dampak Kenaikan Muka Air Laut Terhadap Kesesuaian Lahan Rawa Pasang Surut Tabunganen Kalimantan Selatan. Jurnal Sumber Daya Air, 10(2) Mei 2014, hal. 102

Wardana A (2007) Perusakan Lingkungan Sebagai Pelanggaran HAM, artikel, 2007, hal. 2

Wibowo A (2015) Strategi Adaptasi Dan Mitigasi Nelayan Di Pulau-Pulau Kecil Terhadap Dampak Perubahan Iklim, Laporan Studi Pustaka (KPM 403) Departemen Sains Komunikasi Dan Pengembangan Masyarakat Fakultas Ekologi Manusia Institut Pertanian Bogor

Wibowo A, Arif S (2015) Strategi Adaptasi Nelayan di Pulau-Pulau Kecil terhadap Dampak Perubahan Iklim (Kasus: Desa Pulau Panjang, Kecamatan Subi, Kabupaten Natuna, Kepulauan Riau), Sodality: Jurnal Sosiologi Pedesaan, Agustus 2015

Wibowo A dan Satria A (2015) Strategi Adaptasi Nelayan di Pulau-Pulau Kecil terhadap Dampak Perubahan Iklim (Kasus: Desa Pulau Panjang, Kecamatan Subi, Kabupaten Natuna, Kepulauan Riau), Sodality: Jurnal Sosiologi Pedesaan, Agustus 2015, hal. 107–124

Widayatun et al (2011) Adaptasi dan Mitigasi Masyarakat Pesisir Terhadap Perubahan Iklim dan Degradasi, PPK-LIPI, Jakarta

Zacot (1979) dalam Nasruddin Suyuti, Interaksi Orang Bajo dan Orang Bugis: dalam Konteks Kearifan Lokal-Global di Desa Sulaho, Kabupaten Kolaka Utara, Sulawesi Tenggara, Jagad Bahari Nusantara, 2011, https://centerformunawareducation.files.wordpress.com (diakses tanggal 23 April 2018). Dalam Loc.cit., Indra-Fardhan, (diakses tanggal 25 April 2018)

Zuliansyah (2010) Kelestarian Lingkungan Hidup Sebagai Hak Asasi Manusia, artikel, Kementerian Hukum dan Hak Asasi Manusia, Direktorat Jenderal Hak Asasi Manusia 20 Sept 2010

9783030555351